CYBERWAR

Law and Ethics for Virtual Confli

Cyberwar

Law and Ethics for Virtual Conflicts

Edited by
JENS DAVID OHLIN
KEVIN GOVERN
CLAIRE FINKELSTEIN

UNIVERSITY PRESS

OXFORD
UNIVERSITY PRESS

Great Clarendon Street, Oxford, OX2 6DP,
United Kingdom

Oxford University Press is a department of the University of Oxford.
It furthers the University's objective of excellence in research, scholarship,
and education by publishing worldwide. Oxford is a registered trade mark of
Oxford University Press in the UK and in certain other countries

First Edition published in 2015

Published in the United States of America by Oxford University Press
198 Madison Avenue, New York, NY 10016, United States of America

British Library Cataloguing in Publication Data
Data available

Library of Congress Cataloging in Publication Data
Data available

ISBN 978–0–19–871750–8

Foreword

In 2013, the Group of Governmental Experts (GGE), a collection of cyber experts from fifteen states, concluded that "International law, and in particular the Charter of the United Nations, is applicable and is essential to maintaining peace and stability and promoting an open, secure, peaceful and accessible ICT [information and communications technology] environment."[1] Although drawing world-wide attention, the statement hardly represented a jurisprudential epiphany. Earlier the same year, the International Group of Experts (IGE) that produced the *Tallinn Manual on the International Law Applicable to Cyber Warfare* agreed unanimously "both the *jus ad bellum* and the *jus in bello* apply to cyber operations."[2] Indeed, it is unfortunate that the GGE failed to explicitly pronounce on the applicability of the jus in bello (international humanitarian law (IHL)) to cyber operations occurring during an armed conflict.

Claims that cyberspace is a new domain to which international law is inapplicable (or inapplicable in part) persist but are steadily diminishing. The logic underlying the premise of international law's applicability to cyberspace is simply too compelling for such assertions to gain meaningful traction. For instance, in its *Nuclear Weapons Advisory Opinion*, the International Court of Justice confirmed that the UN Charter's Article 2(4) prohibition on the use of force and Article 51 acknowledgment of the "inherent" right of self-defense apply "regardless of the weapon used."[3] Today experts in the field universally accept this pronouncement as accurate. It is, therefore, difficult to sustain an argument that cyberweapons do not fall within its ambit. This is so despite occasional arguments that cyber operations do not involve the use of weapons. Such arguments, which tend to be advanced by those with little expertise in the jus ad bellum, were rejected by both the GGE and IGE.

Similarly, it is spurious to assert that IHL does not govern cyber operations during an armed conflict. Consider Article 36 of the 1977 Additional Protocol to the 1949 Geneva Conventions: "In the study, development, acquisition or adoption of a new weapon, means or method of warfare, a High Contracting Party is under an obligation to determine whether its employment would, in some or all circumstances, be prohibited by this Protocol or by any other rule of international law applicable to the High Contracting Party."[4] This provision generally reflects customary law, and thus binds

[1] Group of Governmental Experts on Developments in the Field of Information and Telecommunications in the Context of International Security, para 19, UN Doc A/68/98, June 24, 2013, at <http://undocs.org/A/68/98>. The experts came from Argentina, Australia, Belarus, Canada, China, Egypt, Estonia, France, Germany, India, Indonesia, Japan, the Russian Federation, the United Kingdom of Great Britain and Northern Ireland, and the United States of America.

[2] Michael N Schmitt (ed), *Tallinn Manual on the International Law Applicable to Cyber Warfare* (New York: Cambridge University Press, 2013) 5.

[3] Legality of the Threat or Use of Nuclear Weapons, Advisory Opinion, 1996 ICJ 226, para 39 (July 8).

[4] Protocol Additional to the Geneva Conventions of August 12, 1949, and Relating to the Protection of Victims of International Armed Conflicts, art 36, June 8, 1977, 1125 UNTS 3.

all states irrespective of party status.[5] Since cyber operations involve "new weapon[s], means or method[s] of warfare," they, therefore, require review for compliance with the extant IHL. The Article unambiguously demonstrates that IHL was intended to continue to apply as the nature and instruments of warfare evolved. This is the position that has been taken by the International Committee of the Red Cross;[6] it is one that is, quite frankly, indisputable.

Although less studied, the applicability of international law to "below the threshold" cyber operations, that is, those that neither constitute a "use of force" nor an "armed conflict," would likewise appear certain. For instance, although it may sometimes be difficult to attribute cyber operations to a particular state, non-state actor, or individual as a matter of *fact*, there is no reason to exclude application of the *law* of state responsibility's attribution principles to them.[7] Similarly, on what basis would cyber operations conducted from land, sea, air, and space escape the reach, for instance, of the law of sovereignty, the law of the sea, air law, or space law? On the contrary—the risks associated with cyber operations to states, economies, societal functions, and individuals, make the argument for application of existing law especially compelling. As examples, the principle of due diligence can act to impede malicious cyber operations mounted from other states' territories by third parties,[8] while the plea of necessity affords a meaningful basis for responding to cyber operations against critical infrastructure in situations where even the originator of the cyber operation cannot be determined.[9]

Acknowledging that international law is applicable to cyber operations is only, however, the initial step in the process of articulating and implementing the normative architecture. Two more are necessary.

First, it is obviously essential to identify *how* that law applies. For instance, while it is clear that pursuant to the UN Charter and customary law, cyber uses of force are prohibited and forceful responses are only available once a cyber operation rises to the level of an armed attack, it remains unclear when a cyber operation qualifies as either a use of force or armed attack if it causes no physical damage or injury. The most oft-cited case is a massive cyber operation directed against a state economic infrastructure. Would the operation be unlawful as a prohibited use of force and would it qualify as an armed attack that allowed the target state to respond with its own forceful kinetic or cyber operations? Opinions vary.[10] Similarly, although the due diligence principle requires all states to take feasible measures to stop ongoing malicious cyber operations emanating from their territory that harm other states, what obligations does

[5] Tallinn Manual, Rule 48 and accompanying commentary. Although some states do not acknowledge the customary nature of the norm vis-à-vis methods of warfare, this minor deviation from the text of Article 36 has little bearing on the general applicability of IHL to cyber operations.

[6] ICRC, International Humanitarian Law and the Challenges of Contemporary Armed Conflicts, 31st International Conference of the Red Cross and Red Crescent, November 28–December 1, 2011, Doc 31IC/11/5.1.2, 36–7.

[7] UN International Law Commission, Report of the International Law Commission, Draft Articles of State Responsibility, Articles 4–11, U.N. GAOR, 53rd Session, Supp. No 10, U.N. Doc. A/56/10 (2001).

[8] Tallinn Manual, Rules 6–8 and accompanying commentary.

[9] Draft Articles of State Responsibility, Article 25.

[10] See discussion in commentary accompanying *Tallinn Manual* Rules 11 and 13.

that principle impose on a transit state in light of the difficulty of identifying malicious packets of data transiting their cyber infrastructure and the fact that a blocked transmission will often simply traverse a different route to the intended target?[11] And in the field of IHL, it is clear that attacks, including cyber attacks, against civilians and civilian infrastructure, are prohibited.[12] But when does a cyber operation qualify as an "attack" in the meaning of Article 49 of Additional Protocol I such that it is unlawful? Must it cause physical damage or injury? Or does interference with functionality qualify as damage? If so, what degree of interference?[13]

Second, the exercise of applying international law in the cyber context will reveal lacunae in the law that may need to be addressed directly through treaty action or that will inevitably become the subject of state practice that will in turn contribute to the crystallization of either responsive interpretations of existing law or new customary norms. To illustrate, IHL protects civilian objects against direct attack. The majority of the IGE concluded that data did not constitute a civilian object since they were intangible.[14] While this conclusion may be sound as a matter of legal interpretation, the consequences of the interpretation were seen as problematic even by some of the experts who took the position. Some data are plainly of great significance both to the orderly functioning of societies during an armed conflict and the general well-being of individuals. It would accordingly appear likely that over time a broader interpretation of the notion of objects in IHL will, and should, gain traction. The process of identifying such lacunae is essential if law is to adapt itself to the new realities of cyberspace.

Moreover, legal norms are but one facet of the normative universe. They merely articulate the outer limits of permissible cyber operations. Once these boundaries are defined, policy-makers craft ethical, political, and operational norms that further refine the permissible scope of cyber activities. The norms will find expression in domestic law or policy. They may also evolve into regional or global prescriptive norms. Thus, the work in these fields is no less important than that which is ongoing in the legal field. On the contrary, ethical, political, and operational norms may prove to have greater influence on restricting the conduct of cyber operations since legal norms sometimes allow states and other actors in cyberspace great leeway.

Unfortunately, non-legal cyber norms are too often conflated with legal ones. For example, ethicists speaking at the last two global CyCon conferences convened by the NATO Cooperative Cyber Defence Centre of Excellence, a leader in the field of cyber law and policy, repeatedly proffered ethical standards as binding international law. In doing so, they badly mangled the law. As the normative tapestry of cyber operations develops, it is essential that the various bodies of normative strictures be defined with precision. The process begins with international law boundaries for cyber operations and then those boundaries contract based on other concerns.

Cyberwar: Law and Ethics for Virtual Conflict measurably contributes to the process. It contains highly sophisticated legal analyses that not only apply extant international

[11] *Tallinn Manual* Rule 8 and accompanying commentary.
[12] Additional Protocol I, Article 52(1); *Tallinn Manual*, Rule 37.
[13] See discussion in commentary accompanying *Tallinn Manual*, Rule 30.
[14] *Tallinn Manual*, 127.

law in such areas as cyber deception and criminal law, but also tease loose such interpretive dilemmas as the classification of cyber armed conflict, the meaning of the term "attack," and the application of legal causality principles to cyber operations. The book also focuses on cyber activities that do not lend themselves well to the simple application by analogy of legal principles and rules developed in the non-cyber context. These topics include how the law responds to cyber operations mounted by individuals or non-state groups, including cyber terrorists, and whether traditional understandings of borders in international law are suited to application in cyberspace.

Recognizing that normative constraints are not exclusively juridical in nature, the book also explores the nature of cyberwar and its unique ethical status. Additionally, it usefully places cyber activities into a technical context, for norms, whether legal or not, must take cognizance of the distinctive technical environment to which they have to respond.

In this book, Jens Ohlin, Kevin Govern, and Claire Finkelstein have gatherd a distinguished and diverse group of contributors. They include accomplished scholars from different disciplines, as well as experienced practitioners. All have offered an especially perceptive perspective on their respective topics. Together their contributions take the discourse, which is too often counter-normative and usually stovepiped, to a new level. I congratulate the distinguished editors and contributors on their role in producing this fascinating and useful work.

Michael N Schmitt
Director, Stockton Center, United States Naval War College
Chair of International Law, Exeter University
Senior Fellow, NATO Cooperative Cyber Defence Centre of Excellence

Introduction

Cyber and the Changing Face of War

Claire Finkelstein and Kevin Govern

I. War and Technological Change

In 2012, journalist David Sanger reported that the United States, in conjunction with Israel, had unleashed a massive virus into the computer system of the Iranian nuclear reactor at Natanz, where the Iranians were engaged in enriching uranium for use in nuclear weaponry.[1] Operation "Olympic Games" was conceived as an alternative to a kinetic attack on Iran's nuclear facilities. It was the first major offensive use of America's cyberwar capacity, but it was seen as justified because of the importance of preempting Iran's development of nuclear weapons. The so-called "Stuxnet" virus successfully wreaked havoc with Iran's nuclear capabilities, damaging critical infrastructure and spreading massive confusion among Iranian scientists and engineers. The damage was comparable to a direct physical attack on Natanz, though perhaps even more debilitating, given the difficulties of attribution and the extremely covert nature of the attack.

Operation Olympic Games issued in a new era in national defense. As former CIA Chief Michael Hayden reportedly remarked, "This is the first attack of a major nature in which a cyber attack was used to effect physical destruction."[2] He likened the transformation in warfare to that which occurred in 1945 with the release of the atomic bomb over Hiroshima. The computer infrastructure of North Korea sustained serious damage, just two days after President Obama warned that the United States would not accept North Korea's threats to attack the infrastructure of Sony pictures unless they cancelled plans to make the movie *The Interview*, intended to portray a CIA plot to kill North Korean President Kim Jong-Un. Sony capitulated and cancelled the movie premiere, much to the consternation of the U.S. government.[3]

Cyberwar, also known as cyberspace operations, is defined by a Department of Defense Memorandum as "[t]he employment of cyber capabilities where the primary purpose is to achieve objectives in or through cyberspace."[4] The Memorandum goes on to say that such operations "include computer network operations and activities to operate and defend the Global Information Grid." As this definition makes clear, the

[1] David E Sanger, "Obama Order Sped Up Wave of Cyberattacks Against Iran," *The New York Times*, June 1, 2012, at <http://www.nytimes.com/2012/06/01/world/middleeast/obama-ordered-wave-of-cyberattacks-against-iran.html>.

[2] See David E Sanger, *Confront and Conceal* (London: Crown Publishers, 2012), 200.

[3] Soraya N McDonald, "Sony tells theaters they can pass on showing 'The Interview.' Premiere canceled," *Washington Post*, December 17, 2014, at <http://www.washingtonpost.com/news/morning-mix/wp/2014/12/17/sony-tells-theaters-they-can-pass-on-showing-the-interview/>.

[4] Vice Chairman of the Joint Chiefs of Staff, "Memorandum for Chiefs of the Military Services: Joint Terminology for Cyberspace Operations," Washington, November 2010, at James E. Cartwright http://www.defense.gov/bios/biographydetail.aspx?biographyid=138; "Joint Terminology for Cyberspace Operations," in *Cyberwar Resources Guide*, Item #51, at <http://www.projectcyw-d.org/resources/items/show/51> (accessed November 26, 2014).

concept of cyberwar contains an implicit recognition that the US has a security interest in the operation of its electronic network that surpasses the immediate impact of military operations on the protection of human life. Protecting the Grid is comparable to protecting our physical borders: informational security and autonomy have thus become key attributes of national sovereignty.

The importance of defending our electronic infrastructure grows consistently as our dependence on information technology grows. Offensive cyber capacities are of increasing military importance, due to the converse dependence on information technology on the part of our adversaries. At the same time that cyber attacks are providing an increasingly attractive alternative to direct kinetic operations, US and other forces have independently been shifting from kinetic targeting strategies towards more multifaceted approaches, such as those involving diplomacy, economic assistance, education and communications. Cyber operations fit somewhat better with this approach than do traditional kinetic operations. The changing nature of warfare, as well as the changed circumstances in which war takes place, have enhanced the attractions of inflicting the damage of war by non-kinetic means. The methods of cyberwar have thus arrived at a propitious moment.

II. Placing Cyberwar in Historical Context

In 2006, the Department of Defense (DoD) Joint Staff developed a formerly-classified "National Military Strategy for Cyberspace Operations," reflecting a substantially developed and operationally integrated defense cyber capacity. [5] That strategy defined cyberspace as "a domain characterized by the use of electronics and the electromagnetic spectrum to store, modify and exchange information via networked information systems and physical infrastructures."[6] By 2011, the concept of cyber operations was well enough established that (retired) General Michael Hayden, former Director of the National Security Agency and Central Intelligence Agency, could comment: "[l]ike everyone else who is or has been in a US military uniform, I think of cyber as a domain. It is now enshrined in doctrine: land, sea, air, space, cyber."[7] However, the seeds of the cyber revolution were sown long before 2011, even prior to the development of computer technology. The techniques of cyberwar are a subset of a broader approach to national defense technology, one that involves the use of the electromagnetic spectrum. The more general category might aptly be called "Electromagnetic Warfare" (EW), of which both cyber and electromagnetic activities are a part.[8]

[5] DoD, "The National Military Strategy for Cyberspace Operations" (2006) 11.

[6] DoD, "The National Military Strategy for Cyberspace Operations" (2006) 11.

[7] Michael V Hayden, "The Future of Things 'Cyber'" (2011) 5 *Strategic Studies Quarterly* 3. By contrast, Libicki's conclusion is very much the opposite:

> [U]nderstanding cyberspace as a warfighting domain is not helpful when it comes to understanding what can and should be done to defend and attack networked systems. To the extent that such a characterization leads strategists and operators to presumptions or conclusions that are not derived from observation and experience, this characterization may well mislead.

Libicki, (n 4) 336.

[8] See Department of the Army, Field Manual (FM) 3-38, Cyber Electromagnetic Activities, February 2014 at <http://fas.org/irp/doddir/army/fm3-38.pdf>.

One of the first uses of the harnessed electromagnetic spectrum for communications in warfare was the telegraph.[9] The telegraph also became the first physical target of EW more than one hundred years prior to the cyber age. By the time of the civil war, 50,000 miles of telegraph cable had been laid for purely military purposes.[10] The telegraph was as much of a revolution in military affairs in the 19th Century as cyberwarfare is in the 21st. Mobile military telegraph wagons sent and received messages behind the front lines all the way to the first President Lincoln's War Department Telegraph Office.[11] Prior to this innovation, the ability to have rapid exchanges between a national leader at the seat of government and his forces in the field had been difficult to impossible. With the telegraph, however, there could be almost instantaneous communication between Washington and armies in distant fields.[12]

The demands on this valuable means of communication led to the first governmental seizure of electronic communications systems. Congressional Act of January 31, 1862 authorized the President to take possession of railroad and telegraph lines if in his judgment public safety so required.[13] Pursuant to this Act, on February 26, 1862, the President seized control of all telegraphic lines, thus laying the ground for executive control of electronic communications and technology as part and parcel of national defense efforts.[14]

By World War I, there was widespread use of wireless radios for civilian communications as well as military transmission of combat information. This was a great advantage, as wireless radios were less susceptible to damage from enemy artillery barrages than were wired telephone lines, and they were not subject to enemy listening by induction.[15] British intelligence was able to crack the code used for messages to and from the German station, and in this way intercepted the infamous German "Zimmerman telegrams" to Mexico, which invited Mexico to attack US territory. Technological advances in espionage had thus uncovered one of the crucial pieces of information that would contribute to bringing America into the war.[16]

By 1916, the British were experimenting with jamming enemy wireless intercept operations, and jamming began along the entire British front in October 1916.[17] Both sides experimented with early efforts at electronic deception, such as false transmissions,

[9] Since the telegraph operates using electrical signals transmitted across wire lines, telegraph operations are electromagnetic in nature, as are radio, telephone, radar, infrared, ultraviolet and other less used sections of the electromagnetic (EM) spectrum. See J B Calvert, *The Electromagnetic Telegraph* (2000), and Tom Wheeler, *Mr. Lincoln's T-Mails: The Untold Story Of How Abraham Lincoln Used The Telegraph To Win The Civil War* (2007)

[10] Daniel W Crofts, Communication Breakdown, New York Times May 21, 2011.

[11] David H. Bates, Lincoln in the Telegraph Office: Recollections of the United States Military Telegraph Corps during the Civil War (1995) ix.

[12] Bates (n 36) x. [13] 12 Stat. 334 (1862).

[14] A concession to private commercial demand for and access to the telegraph was made by the War Department, which articulated that the possession of the telegraph lines was "not intended to interfere in any respect with the ordinary affairs of the companies or with private business." Joshua R. Clark, Emergency Legislation Passed Prior to December, 1917, Collected, Annotated and Indexed Under The Direction of The Attorney General, Current Emergency Legislation 10.

[15] Sterling (n 43) 445 [16] Sterling (n 43) 445.

[17] Comint and Comsec: The Tactics of 1914-1918 – Part II Summer 1972 Vol 2, No. 3 11. The report also notes that "[t]he British soon found that jamming was costly and ineffective and it was discontinued." Comint and Comsec (n 47) 11.

dummy traffic and other similar ruses for misleading the enemy.[18] During World War II, the British began to equip their aircraft with noise jammers and passive electronic countermeasures (ECM) as an effort to foil the sophisticated Wurzburg gun-laying German radars.[19] The Japanese were meanwhile working on their own types of radars, though their efforts were hampered by a dearth of scientists and engineers, as well as by a shortage of materials.[20] Throughout the war, there was a fight between rudimentary EW capabilities and simple ECM,[21] such that each side would temporarily gain the upper hand in EW, only to lose it in a new countermeasure.[22]

It was not until well after the advent of the internet and the attacks of 9/11, however, that the development of cyberwar techniques began in earnest. Although the initial foray in this direction came from the Bush Administration, the biggest support for technological advance has come from the Obama Administration. In 2012, the Administration articulated the National Security Presidential Directive/NSPD-54, which remains the US policy definition for cyberspace.[23] There is now an agency—the Pentagon's Defense Advanced Research Projects Agency (DARPA)—that has the mission of protecting computer systems and developing the capability to disrupt or destroy enemy systems. We are thus transformed, not only in offensive uses of cyber, but also in our attempts to use cyber as a tool for defending against the heavily computer-dependent kinetic attacks of others.

Of course the United States and its allies are not the only world powers that have been developing a cyberwar capacity. In 2007, security firm McAfee estimated that 120 countries had already developed ways to use the internet to target financial markets, government computer systems, and utilities. In 2008, the Russian government allegedly integrated cyber operations into its conflict with Georgia. According to these accusations, Russian cyber intelligence units conducted reconnaissance and infiltrated Georgian military and government networks. When the conventional fighting broke out, Russia used cyberweapons to attack Georgian government and military sites as well as communication installations. Foreign militaries, such as China's, have conducted exercises in offensive cyber operations, both stealing information from other governments and simulating attacks on other countries command and control systems. In 2011, Iran boasted that it had the world's second-largest cyber army. With states around the globe improving their cyberwarfare capabilities, the world may experience a cyber arms race reminiscent of the nuclear arms race of the Cold War.

Arms races revolving around technological advances in war are nothing new, and from experience we know such moments often produce significant changes in the fundamental structure of war. Technological developments have consistently transformed the way wars are conducted as well as the nature of the risks to both combatant and civilian populations. Most notably, increasingly sophisticated and deadlier weapons have enabled combatants to keep a greater distance

[18] See, for example, Andrew Eddowes, The Haversack Ruse, and British Deception Operations in Palestine During World War I (1994)

[19] Peter J. Hugill, Global Communications Since 1844: Geopolitics and Technology (1999) 194

[20] Hugill (n 50) 162. [21] Hugill (n 50) 194. [22] Government of India DRDO (n 33) 14

[23] National Security Presidential Directive/NSPD-54 Homeland Security Presidential Directive/HSPD-23, Subject: Cybersecurity Policy (U), January 8, 2008 at 3.

from one another, thus diffusing the risks they face. At the same time, such technologies have often broadened the scope of war, further increasing risks to civilian populations. Hayden's comparison between cyber and the transformative power of the new technology reflected in the nuclear attack on Hiroshima seems apt. Less dramatic examples, such as the development of drones, demonstrate the same process.

Precision technologies have increased the distance between combatants at the cost of subjecting civilian populations to new risks in other ways, ironically risks that combatants no longer face. Despite their capacity for precision, mistakes in the use of such weapons have been common, due to inaccurate information, unintended effects on third parties, ranging from death and bodily injury to more diffuse effects, such as the repeated stress from exposure to drones in the vicinity of their targets. Where the use of drones is a persistent feature of everyday life, civilians report symptoms of trauma and anxiety from living in their midst. No technological change to date, however, appears to rival this transformative potential to the same degree as the development of cyber offensive capacities. Indeed, this is captured by the coining of a new label for the notion of war involving cyber attacks, namely "cyberwar," as though it were not only a new kind of weapon, but an entirely new genre of war. The possibility that we might be able to destroy a target like the Iranian nuclear reactor from the "inside out," avoiding detection for significant periods of time while an electronic virus works its way through the system's infrastructure, opens up the possibility of just such a dramatic change in our offensive capabilities. In addition, cyber technology creates the opportunity for a new kind of defense strategy, one designed both to counter cyber offensives and to pre-empt kinetic attacks, under scenarios that do not fit neatly within the traditional paradigm of war. When technological evolution is combined with geopolitical change, such as the demise of state sovereignty and the entrance of civilians or non-governmental actors into the arena of war, the transformative nature of cyber technology is enhanced.

III. Transformations in the Nature of War

The revisionary effect of technological change has conspired with dramatic changes in the basic structure of war, particularly since the United States engaged al-Qaeda in the wake of 9/11. The most significant shift in the demographics of war is the influx of civilians into battle. The US is increasingly drawn into conflict with ideologically driven populations, organized into powerful civilian militias, in lieu of governmental forces carrying out a concerted state policy of old. With this crucial shift in the landscape of war, the formerly bright-line distinction between state and non-state actors has been eclipsed, and with it the boundary distinction between combatants and civilians. However, we cannot satisfy the requirements of the Law of Armed Conflict (LOAC), in particular the crucial principle of distinction, without being able reliably to identify who is a legitimate target. In this way, changes in modern warfare have been attended by a breakdown of the traditional foundation on which adherence to the rule of law in war depends. There is a ripple effect: the widespread entry of civilians into the theater of war results in a corresponding disintegration of the boundary between military jurisdiction, on the one hand, and the jurisdiction of law enforcement, on the other.

Historically, the distinction between the civilian and combatant populations was a sharp one. The uniform was the most visible means of marking that distinction, but even without uniforms there would have been little doubt about who was military and who civilian. In addition, the civilian population was kept physically separate from war by the fact that the fighting took place on a battlefield, the boundaries of which were fairly clear. In modern conflicts, the historical distinction of roles is no longer applicable, as the enemy consists in non-state actors who blend nearly seamlessly into the civilian population. This is facilitated by the fact that there is no longer a distinct battlefield in war. Military operations now take place anywhere and everywhere. We might indeed say that modern war is characterized by a loss of location and the abolition of the traditional locus of battle, and with the advent of cyberwar we have that process brought to an exteme: cyber represents the complete loss of the physical battlefield. The advent of war in cyberspace is the culmination of that ebbing of historical boundaries around the concept of war.

It is crucial to understand the link between the availability of cyberwar technology and the role of civilians in war. Where the threat to national security comes primarily from non-state actors, it is reasonable to anticipate expanded use of technologies of war where the barriers to entry are low. Such is the case with cyberwar: members of al-Qaeda or ISIS may, in the long run, be particularly likely to turn to cyberweapons to compensate for the kinetic forces they lack. Despite the elaborate effort, planning, and expertise that went into Operation Olympic Games, destructive cyber attacks can be launched with little preparation or expense. Such attacks, potentially carried out by a small number of individuals with sharply limited resources, have the power to impose destruction on a level that only kinetic attacks have hitherto made possible. What enables such destructive capabilities for apparently slight intervention and remote causal impact is the highly technological infrastructure of modern life. We are, in effect, leveraged on technology. The same can be said for civilian life: we are dependent on computers, and breaches of our technological infrastructure can produce devastating results.

A second crucial change in the circumstances of war is the increasing importance of both military and personal data. In an age when individuals voluntarily transmit, store, and receive vast amounts of personal data through the internet, planting software in electronic devices to obliterate, alter, or appropriate data has become a crucial new tactic of warfare: a "fifth dimension battlefield," as it is often said, after air, sea, land and outer space. This shift in frameworks has resulted in the merging of military and corporate espionage functions, and for this reason, the militarization of cyberspace has created a legal and moral ambiguity regarding privacy rights, as well as a personal liability to be targeted in cyberspace by virtue of the mere position one occupies relative to a network of information. These tendencies have contributed to the shift in the structure of warfare, with the result that the line distinction between the military and the civil domains has faded. Cyberwar operations thus occupy a crucial position in the altered landscape of military conflict.

IV. Is Cyberwar an Act of War?

It is often debated whether cyber attacks constitute true acts of war. Those who offer a negative answer to that question maintain that since cyber attacks can cause only limited damage, mostly of an economic nature, such acts do not belong to the domain of war. Those who answer affirmatively maintain the irrelevance of the fact that cyber attacks do not cause physical damage on the grounds that this argument fails to consider the secondary effects of infrastructure failure, particularly where the quality of civilian life is concerned. They point out that one need only recall the loss of life routinely caused by systems failures from electrical surges during fairly routine temperature spikes in the summer months to recognize the destructive potential of cyber attacks.[24] In addition, they argue that the massive damage that cyber attacks can cause, and the serious use of such attacks as an alternative to kinetic attacks in war, belies such claims. Cyberweapons and cyberwarfare are now considered by the FBI to be the number one threat to national security.[25] On the side of the latter position, the US government has taken a functional view of the notion of war and declared cyber attacks as acts of war,[26] given that cyber attacks increasingly serve the function that kinetic attacks have historically served. It therefore becomes harder and harder to see such acts as limited to economic and financial destruction. In response, however, those who reject cyber attacks as acts of war may argue that the point is not that cyber attacks fail to cause destruction, but that the nature of the destruction is unclear. Brown outs are a case in point: while damage from such events may be significant, no one would label them acts of war as a result. Once again, cyber events appear to challenge the traditional categories in war, leaving our theoretical accounts of war in search of an object.

A further difficulty with the functional characterization of cyber attacks as acts of war is that it does not specify the theory of war against which this judgment is being made, and arguably such a theory is necessary in order to know whether a certain characterization of acts of destruction should qualify them as acts of war. Traditional models of warfare are problematic in this regard, since they have all been implicitly called into question by the dramatic changes in the nature of warfare itself. In addition to the advent of cyber attacks and the introduction of other new technologies, there is a fundamental shift in features that formerly characterized acts of war in the first place, and it may seem that the theory of war must evolve as quickly as the emergence of challenging marginal cases whose identity we are seeking to understand by that theory. Because so-called "cyberwar" puts a particular strain on our traditional conception of war, it forces us to return to the basic building blocks of just war theory and to re-examine the theory of war in light of striking new examples.

[24] See the Centers for Disease Control and Prevention website at <http://www.cdc.gov/mmwr/preview/mmwrhtml/mm6231a1.htm>.

[25] See, for example, "FBI: Cyber Attacks—America's Top Terror Threat," *RT.com*, March 2, 2012, at <http://www.rt.com/news/cyber-fbi-security-mueller-691/>; J Nicholas Hoover, "Cyber Attacks Becoming Top Terror Threat, FBI Says," *Informationweek.com*, February 1, 2012, at <http://www.informationweek.com/news/government/security/232600046.

[26] David Sanger and Elisabeth Bulmiller, "Pentagon to Consider Cyberattacks Acts of War," *The New York Times*, May 31, 2012, at <http://www.nytimes.com/2011/06/01/us/politics/01cyber.html>.

Cyber attacks, for example, put immense pressure on conventional notions of sovereignty and the moral and legal doctrines that were developed to regulate them. The problem stems from the fact that the traditional notion of sovereignty and the boundaries of states seem to disintegrate in the face of a type of conflict where boundaries are irrelevant. Could an electronic virus designed to destroy technological infrastructure without ever requiring the kinetic infiltration of the territory or another nation possibly violate the sovereign authority of that other nation? Article 2, Section 4 of the UN Charter promises that members will not use the "threat or use of force against the territorial integrity or political independence of any state." This provision, however, is likely inadequate in view of the increasing use of cyber attacks. It leads to questions of whether problems of cyberwarfare require new treaties and legal definitions. For example, does the cyberweapons race require treaties similar to the Treaty on the Non-Proliferation of Nuclear Weapons? As the country that controls the internet infrastructure, as well as the country with the highest percentage of internet business as a share of its economy, the US is in a uniquely difficult negotiating position in developing any treaties. In a world of attack and destruction without conventional military assets, do traditional notions of sovereignty based on geography and territorial integrity retain their relevance? Unlike past forms of warfare circumscribed by centuries of just war tradition and Law of Armed Conflict (LOAC) prescriptions, cyberwarfare occupies a particularly ambiguous status in the traditions and conventions of the laws of war.

Then there are the difficulties associated with maintaining the principle of distinction. If the threat we are facing stems from the engagement of non-state actors in hostilities, there is little choice but to fight the enemy on civilian territory and in and around the civilian population.[27] This has been one of the most serious developments in warfare since World War II. In cyberwar, however, the difficulty we have demarcating the military and civilian populations in modern warfare is exacerbated. Combatants and civilians are arguably more intertwined than in any other form of war, and, as discussed above, the physical identification of the battlefield, which has helped to mark a conflict as military in nature, has been eliminated. A possible ramification is that efforts to prevent and defend against cyber attacks will result in the complete effacement of the domains of civil and military authority—and, to an even greater degree than exists in modern kinetic warfare, national defense will invade the domain traditionally reserved for law enforcement. If this is true, might military action designed to protect against cyber attacks, for example, pose a serious threat to due process rights, or the moral equivalent of such rights in the international arena? These legal ambiguities, devoid of moral perspective, make adherence to the rule of law in cyberwarfare more challenging than in any other domain of warfare.[28]

V. A Look Ahead

The chapters in this volume grew out of a conference held at the University of Pennsylvania by the Center for Ethics and the Rule of Law (CERL). CERL was founded

[27] Stewart A Baker and Charles J Dunlap Jr, "What is the Role of Lawyers in Cyberwarfare?," *ABA Journal.com*, May 1, 2012, at <http://www.abajournal.com/magazine/article/what_is_the_role_of_lawyers_in_cyberwarfare>.

[28] Baker and Dunlap Jr, "What is the Role of Lawyers in Cyberwarfare?" (n 8).

in 2012 to address foundational legal and moral issues that arise in national security and modern warfare, particularly those that impact the rule of law. The conference was organized to explore questions about the degree to which engaging in war using the techniques of cyber technology is compatible with rule of law values. The central question was whether cyberwar is consistent with the idea that there are deep moral and legal principles, adherence to which successfully limits the permissibility of war to cases where those principles are observed. Can we both accept the legitimacy of cyberwar and maintain that war is fundamentally a constrained activity, one that can be justified according to a set of moral principles? Or does an acceptance of cyberwar, insofar as it requires us to relinquish our attachment to so many of the doctrines of just war theory, mean that we have given up on the idea that warfare can be limited, in favor of a more Clausewitzian vision that anything goes?

If one does allow that the use of techniques of cyberwar are compatible with the traditional laws of war, and hence with rule of law values, there are further and more fine-grained decisions to be made. One might ask whether the laws of war, such as those typically applied to kinetic war, must be understood as structured in parallel fashion when applied to cyberwar in lieu of conventional techniques of war. Do the laws of armed conflict apply to cyberspace in the same way they apply to traditional warfare?

Proportionality, for example, is a crucial question in military ethics, as well as in domestic criminal law. It requires that no more force be used than is necessary to repel an attack or meet other legitimate military objectives. But how does one determine what constitutes a necessary response to a cyber attack? Worse, how should we determine whether it is *ever* proportionate to launch an offensive cyberwar attack? Was the attack on the Iranian nuclear reactor at Natanz a permissible act of prevention or an illegitimate first strike between sovereign nations? More complicated still, would a cyber attack on the part of the US against North Korea be proportionate to North Korea's threat against Sony pictures?

The current volume brings together leading authorities in law, technology, and moral philosophy, as well as from multiple academic disciplines and representing many types of expertise in practice, to consider the law and morality of cyberwar. We have organized the volume into four parts. Part I contains chapters that attempt to expose foundational and conceptual issues in cyberwar. The chapters in this part primarily seek to answer the question whether acts of cyberwar should count as war according to the criteria of Just War Theory. Larry May's and James Cook's chapters directly contradict one another on this topic: May argues that so-called cyberwar is not in fact a part of the law at all, while Cook maintains that it can be so seamlessly considered a part of the law of war that no adjustment of the Just War Theory paradigm is even required to fit cyberwar in. May's argument draws attention to the aim and outcome of cyber attacks. He argues that insofar as such attacks do not cause, and do not aim to cause, massive loss of life and injury, they are too distant from the types of acts the laws of war have sought to regulate, and so cannot be considered acts of war. May's argument depends on a distinct characterization of war, one we have already identified as necessary if one is to consider whether the laws of war have proper application to cyber attacks. For example, for May, law must be a *public* phenomenon. But since cyber attacks are clandestine, May argues, they do not fit within the characterization of the norms of war that come down to us through the ages. Thus, May suggests that cyber attacks should be assessed according to the ordinary rules that govern the ethics of conduct in ordinary life, rather than according to the more permissive standard of the rules of war.

James Cook disagrees, and sees traditional Just War Theory as applying as readily to cyber technology as to any other type of attack or initiative in a conflict with another sovereign state. All that is required, Cook maintains, is the ability to identify the agents involved, the intentions with which they act, and the effects of their actions. Hence no revision or updating of Just War Theory is necessary in order to accommodate the central dilemmas of the cyber realm. Rather than argue for this thesis on the grounds that Just War Theory is capacious enough to accommodate the evolving nature of warfare, with cyberwar taking its place at the outer limits of those items to which Just War Theory can rightly apply, Cook reaches his conclusion by asserting the more ordinary nature of cyber attacks and cyberwar activities. Just War Theory applies to cyberwar, then, because there is nothing particularly special to accommodate that conventional war did not already require.

Cook's thesis makes for an interesting contrast with May's, particularly in the characterization of cyberwar itself. While May says that cyberwar is a form of embargo or economic constraint, thus characterizing it as an ordinary form of economic pressure, Cook takes precisely the opposite view. Not only do cyber attacks automatically count as acts of war, given their proximity to kinetic attacks in structure, they are more like war than the standard acts of war that form the paradigm of our treatment of war. Cyber attacks are more potent than other attacks, primarily because once unleashed they require no human intervention to release their potential. Although Cook does not put it this way, a virus like Stuxnet can be fruitfully thought of as a kind of autonomous weapons system, since it functions according to its programming and effectively "takes the human out of the loop."

Jens Ohlin's chapter examining the concept of causation as it applies to cyberwar identifies particular difficulties for the law of war in the context of cyberwar. Although causation is not in general an important concept for understanding the legal limits imposed by IHL, it becomes essential to understand the role of causation where one attempts to understand the limits IHL imposes in the cyber arena. Because cyber attacks are particularly causally complex, it is essential for identifying what Ohlin, following George Fletcher, calls a "pattern of manifest criminality," namely that the rules governing attribution are clear and are able to trace judgments of responsibility along causal lines. The greatest source of complexity lies in the causal role played by third parties, whose involvement may help produce acts and effects that violate the law of war. Until IHL has a more adequate account of causation, particularly as applied to intervening voluntary acts of other agents, it will not be able to clarify the permissibility of cyber interventions under the law of war.

The chapters in the second part of this volume focus on the civil–military divide and the difficulty disentangling these frameworks in the context of cyber security and cyberwar. Stuart Macdonald's paper, which deals with cyberwar and the criminal law, addresses a fundamental aspect of the concept of responsibility as it relates to cyber terrorism. Macdonald's distinction between domestic and enemy criminal law recapitulates the dual framework theories of earlier chapters in its distinction between ordinary rules of criminal prosecution and rules governing the treatment of suspected terrorists. Macdonald points out that by treating cyber attacks under the heading of "enemy criminal law," the British have enabled intervention earlier to prevent, rather than merely punish, dangerous acts of cyber terrorism.

Laurie Blank's chapter addresses the split between law-enforcement principles and the theory of war. Blank identifies the position of the notion of cyberwar relative to these two different jurisdictions as hovering in the middle. She allows that cyberwar could constitute an "armed conflict" within the prevailing use of the term and against the background of the LOAC. However, such instances are unlikely, as under the principles and authority of international law, the relevant conflict must be fairly intense before cyber conflict would be rightly seen as part of an armed conflict. As Blank implicitly recognizes, cyberwar is an in-between concept that hovers between law-enforcement and military paradigms. As such it provides a theoretically useful way to test out the boundaries of our conception of war.

Nicolò Bussolati's chapter echoes the thought that the possibility of significant destruction of infrastructure through cyberwar is particularly relevant given the increased role of non-state actors in current armed conflict. He goes yet further, however, and argues that the availability of cyber techniques in war has actually helped establish a place for non-state actors in international relations and, which in turn creates a challenge for how we think of cyberwar in relation to the traditional paradigm of war. Bussolati sees cyber attacks as constituting a use of force, and as posing a danger of the first order. Rather than view hacking as involving a lower risk of force than kinetic attacks, he might see such attacks as having comparably graver dangers and constituting a more, rather than less, invasive means of accomplishing military ends.

The third part of this volume deals with a somewhat more applied topic, namely the ethics of hacking and spying. Duncan Hollis begins by weighing in strongly on James Cook's side of the debate about the nature of cyberwar, and whether it is part of the LOAC. The concepts of the jus ad bellum—the principles of justice accounting for when it is permissible to go to war—and the jus in bello—the principles that identify the manner of one's conducting war—apply by way of analogy with the non-cyber law of war and find an adequate foothold in the world of cyberwar to make it reasonable to think of it as part of the law of armed conflict. International law does regulate cyberspace at the very least by analogy. However, Hollis laments the absence of a clear theory to explain why. The usual explanation—that IHL imposes boundaries on the permissibility of acts of war that should be thought of as extending to the cyber arena—strikes Hollis as a highly imperfect argument. A more flexible way of thinking of IHL, oriented towards principles rather than boundaries, might impose a duty to adopt the "least harmful means available" for achieving military objectives. And given that cyberwar can be less invasive and harmful than kinetic actions, IHL might impose an affirmative duty to attempt to achieve a military objective by hacking rather than kinetic action.

Christopher Yoo weighs into the debate about the choice of frameworks for assessing cyber activity in the context of war by discussing espionage. The argument falls on Larry May's side of the fence, in the debate between May and Cook. The chapter's claim is that cyber operations should mostly be thought more in the domain of espionage and law enforcement than acts of war. Yoo's argument for this position—that cyber acts do not cause the same level of damage as kinetic attacks—might send us debating again how we characterize the damage caused by the Stuxnet virus. However, the point can be reasserted in another guise: the essential role of personal, internet-based information in the fight against terrorism cannot be over-estimated. As such, espionage becomes a central tool in national security operations and many cyber operations are properly identified under this heading.

The third chapter in this section, William Boothby's, treats cyberwar as a critically important domain for future warfare. He writes:

> The rifle, the bayonet, mortars, bombs, missiles, and mines will remain critically important tools in the conduct of hostilities in many future, conventional armed conflicts...But cyberspace will...become the environment in which adversaries employing some degree of operational sophistication will seek to gain and to maintain military advantage by leveraging their own hostile activities while impeding the enemy's capacity to organize and operate

However, Boothby raises the concern that insofar as cyber operations typically involve the employment of deceptive tactics, they may be illegitimate and illegal under the laws of war.

A fourth section deals with the crucial and difficult question of responsibility for cyber attacks. Marco Roscini identifies particular challenges to assessing responsibility for cyber attacks due to evidentiary hurdles. The problems he identifies stem on the one hand from the famously difficult problem of attribution in the domain of cyber attacks, combined with the evidentiary rules that govern international law, particularly the rules on the use of force.

Sean Watts's chapter represents something of a digression from our themes thus far, though an interesting and relevant one. It addresses the conflict between the principle of non-intervention in international law and the existence of low-level cyber signals among states. The question he poses is whether the existence of these cyber connections violates the principle of non-intervention. He offers interpretations of the latter principle that would allow low-level cyber connections to persist.

Taken together, the chapters in this volume constitute a comprehensive examination of the challenges that have arisen in the law of armed conflict since the advent of cyber intervention as a means for achieving military objectives. While authors are not in perfect agreement with one another, they have broad consensus on the identity of the challenges the world faces as a result of the advent of cyber technology as a means of war. The major problems addressed in the chapters in this volume—whether cyber operations constitute acts of war, the civil–military divide and the choice between law enforcement and military frameworks, the ethics of hacking and espionage, and finally the attribution of legal and moral responsibility for cyber activities—cover the essential challenges in understanding the fifth dimension battlefield and in identifying the nascent legal and ethical boundaries within which national defense in cyber must operate. With cyber becoming an increasingly important technique for national self defense, clarity around the boundaries of war in the Fifth Dimension seems essential. We hope this volume contributes towards this end.

Table of Contents

PART IV: RESPONSIBILITY AND ATTRIBUTION IN CYBER ATTACKS

List of Contributors

Laurie R Blank is Clinical Professor of Law and Director of the International Humanitarian Law Clinic at Emory University School of Law. Professor Blank is the co-author of *International Law and Armed Conflict: Fundamental Principles and Contemporary Challenges in the Law of War*, a casebook on the law of war. She is also the co-director of a multi-year project on military training programs in the law of war and the co-author of *Law of War Training: Resources for Military and Civilian Leaders*. Before joining Emory, Professor Blank was a Program Officer in the Rule of Law Program at the United States Institute of Peace. At USIP, she directed the Experts' Working Group on International Humanitarian Law, in particular a multi-year project focusing on New Actors in the Implementation and Enforcement of International Humanitarian Law. Professor Blank received an AB in Politics from Princeton University, an MA in International Relations from The Paul H Nitze School of Advanced International Studies (SAIS) at The Johns Hopkins University, and a JD from New York University School of Law.

Air Commodore William H Boothby retired as Deputy Director of Legal Services (RAF) in 2011. In 2009 he took a Doctorate in International Law at Europa Universität Viadrina, Frankfurt (Oder) in Germany and published *Weapons and the Law of Armed Conflict* (Oxford University Press). His second book, *The Law of Targeting*, appeared with Oxford University Press in 2012. He was a member of the Group of Experts convened by the ICRC to discuss Direct Participation in Hostilities, was a member of the Group of Experts who produced the HPCR *Manual of the Law of Air and Missile Warfare*, and was a member of the Group of Experts and of the drafting committee of the CCD/COE project that produced the *Tallinn Manual on the Law of Cyber Warfare*. His third book, *Conflict Law*, was published in March 2014. He has an associate fellowship at the Geneva Centre for Security Policy and teaches at Royal Holloway College, University of London, Australian National University, Canberra, and University of Adelaide.

Nicolò Bussolati is a PhD candidate in International Criminal Law at the University of Amsterdam, Amsterdam Center for International Law. He received LLB and LLM degrees summa cum laude from the University of Turin and an LLM degree summa cum laude from Columbia Law School/Amsterdam Law School. His main research interests are in the field of cybercrime, cyberwar, and terrorism. In particular, his dissertation research addresses political hacking and implications for domestic and international criminal law.

Colonel James Cook was confirmed by the US Senate as permanent professor and head, department of philosophy, US Air Academy in 2002. A cyber officer and regional affairs specialist, he has served in NATO, at the Pentagon, and as the senior US advisor to the National Military Academy of Afghanistan. He received his PhD from the Universität-Heidelberg.

Claire Finkelstein is Algernon Biddle Professor of Law, Professor of Philosophy, and Director of the Center for Ethics and the Rule of Law at the University of Pennsylvania. She writes at the intersection of moral and political philosophy and the law. She has published extensively in the areas of criminal law theory, moral and political philosophy as applied to legal questions, jurisprudence, and rational choice theory, and has recently begun writing on the law and ethics of war. One of her distinctive contributions is bringing philosophical rational choice theory to bear on legal theory. She has focused in recent years on the implications of Hobbes' political theory for substantive legal questions. She is currently finishing a book entitled *Contractarian Legal Theory* and is the editor (with Jens Ohlin and Andrew Altman) of a volume entitled *Targeted Killings: Law & Mortality in an Asymmetrical World* (Oxford University Press, 2012), and of *Hobbes on Law* (Ashgate, 2005).

Kevin H Govern is Associate Professor of Law at the Ave Maria School of Law, and an instructor for John Jay College and California University of Pennsylvania. He earned a JD from Marquette University Law School, and LLM degrees from The Judge Advocate Generals' School US Army, and from Notre Dame Law School.

Duncan B Hollis is James E Beasley Professor of Law and the Associate Dean for Academic Affairs at Temple University's Beasley School of Law. Professor Hollis's scholarship focuses on treaties and international regulation of cyber threats. He is editor of the award-winning *Oxford Guide to Treaties* (Oxford University Press, 2012) as well as *National Treaty Law and Practice* (Martinus Nijhoff, 2005). He is an elected member of the American Law Institute and a Senior Team Member of *Metanorm: A Multidisciplinary Approach to the Analysis and Evaluation of Norms and Models of Governance for Cyberspace*, an MIT-led project that supported his work in this volume through Office of Naval Research Grant N000141310878 and the Department of Defense Minerva Research Initiative.

Stuart Macdonald is Associate Professor of Law at Swansea University, UK, where he is Deputy Director of the University's Centre for Criminal Justice and Criminology. His research interests lie in criminal law and criminal justice policy, with a particular interest in counterterrorism and cyber. Dr Macdonald is co-director of the multidisciplinary, multi-institutional cyberterrorism research project (www.cyberterrorism-project.org) and co-editor of *Cyberterrorism: Understanding, Assessment, and Response* (Springer, 2014) and *Terrorism Online: Politics, Law and Technology* (Routledge, 2015).

Larry May is the W Alton Jones Professor of Philosophy, Professor of Law, and Professor of Political Science, at Vanderbilt University. He has written widely in political philosophy and international law. His most recent books include: *Genocide* (Cambridge University Press, 2010), *Global Justice and Due Process* (Cambridge University Press, 2011), *After War Ends* (Cambridge University Press, 2012), *Limiting Leviathan* (Oxford University Press, 2013), and *Proportionality in International Law*, with Michael Newton (Oxford University Press, 2014).

Jens David Ohlin is Professor of Law at Cornell Law School. His published books include *The Assault on International Law* (Oxford University Press, 2015); *Defending*

Humanity: When Force is Justified and Why, with GP Fletcher (Oxford University Press, 2008); and *Targeted Killings: Law and Morality in an Asymmetrical World*, edited with C Finkelstein and A Altman (Oxford University Press, 2012. His articles and essays have appeared in the *Columbia Law Review, Cornell Law Review, American Journal of International Law, Harvard International Law Journal, Leiden Journal of International Law, Yale Journal of International Law, Chicago Journal of International Law*, and many other journals. Ohlin earned MA, MPhil, PhD, and JD degrees from Columbia University.

Marco Roscini has a PhD from the University of Rome "La Sapienza" and is currently Reader in International Law at the University of Westminster School of Law, as well as Visiting Fellow at King's College London. Dr Roscini has written extensively on international security law, including disarmament, the law of armed conflict, and cyber warfare. His most recent book, *Cyber Operations and the Use of Force in International Law*, was published by Oxford University Press in 2014.

Michael Schmitt is the Charles H Stockton Professor of International Law and Director, Stockton Center, Naval War College; Professor of Public International Law, Exeter University; and Senior Fellow NATO Cooperative Cyber Defence Centre of Excellence. He directed the *Tallinn Manual on the International Law Applicable to Cyber Warfare* project and is also directing the follow-on Tallinn 2.0 project.

Sean Watts is Professor of Law at Creighton University Law School and an Instructor at the United States Military Academy at West Point in his capacity as a Lieutenant Colonel in the US Army JAG Reserve. From 2010 to 2012 he participated in drafting the *Tallinn Manual on International Law Applicable to Cyber Warfare*. He is currently a Senior Fellow at the NATO Collective Cyber Defence Center of Excellence, engaged in drafting an update to the Tallinn Manual. From 2009 to 2011 he served as a defense team member in *Gotovina* et al at the International Criminal Tribunal for Former Yugoslavia. Prior to teaching, Professor Watts served as an active-duty US Army officer for fifteen years in legal and operational assignments.

Christopher S Yoo is the John H. Chestnut Professor of Law, Communication, and Computer & Information Science at the University of Pennsylvania, as well as director of its Center for Technology, Innovation & Competition. He is the author of *The Dynamic Internet: How Technology, Users, and Businesses Are Transforming the Network* (AEI Press, 2012), *Networks in Telecommunications: Economics and Law* (Cambridge Univ. Press, 2009) (with Daniel F. Spulber) and *The Unitary Executive: Presidential Power from Washington to Bush* (Yale Univ. Press, 2008) (with Steven G. Calabresi). Yoo testifies frequently before Congress, the Federal Communications Commission, and the Federal Trade Commission.

Table of Cases

Table of Legislation and Executive Orders

Table of Treaties and Conventions

List of Abbreviations

AAA	Anti-Aircraft Artillery
AI	artificial intelligence
ARPA	Advanced Research Projects Agency
CERL	Center for Ethics and the Rule of Law
C2W	command-and-control warfare
CNCI	Comprehensive National Cybersecurity Initiative
CO	Cyberspace Operations
CCDCOE	Cooperative Cyber Defence Centre of Excellence
DARPA	Defense Advanced Research Projects Agency
DDoS	directed denial of service
DNS	Domain Name System
DoS	denial-of-service
DoD	Department of Defense
EBO	effects-based operations
ECM	electronic countermeasures
EEZ	exclusive economic zone
ELINT	electronic intelligence
EIW	economic information warfare
EPIC	Electronic Privacy Information Center
EW	electronic warfare
FBI	Federal Bureau of Investigation
GGE	Group of Governmental Experts
GPS	Global Positioning System
HSPD	Homeland Security Presidential Directive
IA	information assurance
IAEA	International Atomic Energy Agency
IBW	intelligence-based warfare
ICANN	Internet Corporation for Assigned Names and Numbers
ICJ	International Court of Justice
ICRC	International Committee of the Red Cross
ICS	industrial control system
ICT	information and communications technology
ICTY	International Criminal Tribunal for the former Yugoslavia
IDF	Israel Defense Forces
IETF	Internet Engineering Task Force
IGE	International Group of Experts
IHL	International Humanitarian Law
ILC	International Law Commission
IO	information operations
IP	internet protocol
IT	information technology
ITU	International Telecommunications Union
JDAM	Joint Direct Attack Munition
JWT	just war tradition

KISA	Korea Internet Security Agency
LOAC	Law of Armed Conflict
NATO	North Atlantic Treaty organization
NGO	non-governmental organization
NMS-CO	National Military Strategy for Cyberspace Operations
NSA	National Security Agency
NSPD	National Security Presidential Directive
OCS	offensive counter space
OMB	Office of Management and Budget
OTP	Office of the Prosecution
PCIJ	Permanent Court of International Justice
PLA	People's Liberation Army
PPD	Presidential Policy Directive
PSYW	psychological warfare
RPV	remotely powered vehicle
SIGINT	signals intelligence
SAM	surface-to-air
SCADA	Supervisory Control and Data Acquisition
TCP	Transmission Control Protocol
TPIM	Terrorism Prevention and Investigation Measures
UAV	unmanned Aerial vehicle
UK	United Kingdom
UN	United Nations
UNCITRAL	United Nations Commission on International Trade Law
US	United States
WMD	weapons of mass destruction
WTO	World Trade Organization

PART I

FOUNDATIONAL QUESTIONS OF CYBERWAR

1

The Nature of War and the Idea of "Cyberwar"

Larry May

I. Introduction

War is morally and legally problematic in that it is an institution that involves the intentional killing and disabling of humans. Rules or laws of war have been established over the millennia that are designed to restrict the activities of war to make war and its effects more humane while not making war impermissible. War is triggered when a state's territorial integrity or sovereign immunity are breached. Today, other forms of armed conflict are recognized, such as civil war, which lack these triggers. But the naming of a conflict as a "war" still is very important for determining which rules or laws are relevant.

In this chapter I discuss the rich literature on the nature of war in the Just War tradition, where there was an attempt to distinguish war from other conflicts that did not involve a coordinated series of lethal attacks. I argue that cyberwar is not sufficiently like normal cases of war or armed conflict to justify the relaxing of the rules and laws concerning intentional killing. And I argue that there are good reasons to think of cyber attacks more like embargoes than like the type of lethal attacks that war has historically involved. I admit that cyber attacks could rise to the level of being wars or armed conflicts, but that there are not likely to be many, if any, such attacks today. And yet, there are very serious consequences for assimilating cyber attacks to the war paradigm that will make the world a less safe place.

Specifically, in section II, there is a brief historical survey of definitions of war. In the third section, I argue that cyber attacks do not rise to the level of what has historically been called war, and also this is true of contemporary accounts of war and armed conflict as well. In section IV, it is argued that a different paradigm is needed for cyber attack that is not premised on the idea that large numbers of people will be permissibly killed or disabled. In cyber attacks, the aim is to disable machines not humans, and so the type of rules governing cyber attacks need not resemble those governing war or armed conflict. In the fifth section, I consider problems with assimilating cyber attacks to the war paradigm. And in the sixth section I look especially at problems of proportionality that arise in discussing cyber attacks as war.

II. The Historical Definitions of War

Francisco Suarez, writing at the end of the sixteenth century, called attention to the various meanings of the word "war" but indicated that only one was proper:

> An external contest at arms which is incompatible with external peace is properly called war, when carried on between two sovereign princes or between two states.

> When however, it is a contest between a prince and his own state, or between citizens and their state, it is termed sedition. When it is between private individuals it is called a quarrel or duel. The difference between these various kinds of contest appears to be material rather than formal.[1]

So, while Suarez recognized that the term war is used for several different forms of contest, war is only used properly to refer to contests at arms between states.

Alberico Gentili, also writing at the end of the sixteenth century, employs a very similar definition to that offered by Suarez, but Gentili offers a few more thoughts on the nature of war that may help us understand the idea of cyberwar:

> War is a just and public contest of arms. In fact war is nothing if not a contest, and it is a contest of arms, because to wage war in one's mind and not with arms is surely cowardice, and not war...And although much is accomplished in war without the use of arms, yet there is never a war without the preparation of arms... Furthermore, the strife must be public; for war is not a broil, a fight, the hostility of individuals.[2]

For Gentili, even though much of war does not concern arms, there is no war without at least the preparation of arms.

To think of war as a contest of arms, or of weapons, seems to leave open the door for cyberwar since the use of cyber attacks can be characterized as a contest involving a certain kind of *weapon*, a destructive computer program placed clandestinely into a foreign computer system. What is less clear is whether it makes sense to think that there could be a *contest* involving the use of computer programs. The difficulty is that the use of computers to destroy property is not the sort of action that has been characterized by contests of arms. The contest of arms that Suarez, Gentili, and many others said was characteristic of war is the mutual use of arms that are directed not at property but at the lives of opposing soldiers.

Hugo Grotius can also be consulted on the topic of what constitutes arms. Here is how he understands the idea of arms that has been so important in the definition of war:

> Now, even as actions have their inception in our minds, so do they culminate in our bodies, a process that may be called "execution". But man has been given a body that is weak and infirm, wherefore extracorporeal instruments have also been provided for its service. We call these instruments "arms". They are used by the just man for defense and [lawful] acquisition, by the unjust man, for attack and seizure. Armed execution against an armed adversary is designated by the term "war".[3]

Writing in 1605, Grotius here gives the term "arms" a very broad definition that could seemingly encompass the use of computer programs, which after all are extensions of

[1] Francisco Suarez, "On War" (*Disputation XIII, De Triplici Virtue Theologica: Charitate*) (c 1610), in *Selections from Three Works*, translated by Gladys L Williams, Ammi Brown, and John Waldron (Oxford: Clarendon Press, 1944) Disputation XIII, 800.

[2] Alberico Gentili, *The Law of War (De Jure Belli)* (1588–1589), translated by John C Rolfe (Oxford: Clarendon Press, 1933) 12.

[3] Hugo Grotius, *On the Law of Prize and Booty (De Jure Praedae)* (1605), translated by Gwladys L Williams (Oxford: Clarendon Press, 1950) Ch II, 30.

a person's body in some sense. And these programs certainly allow people to "execute" their decisions in the sense of putting these decisions into concrete action.

A bit later in his life, in 1625, Grotius came to a different conception of war that is worth considering:

> Cicero defined war as a contending by force. A usage gained currency, however, which designates by the word not a contest but a condition; thus war is the condition of those contending by force, viewed simply as such.[4]

Here two things are worthy of note. Grotius expands the domain to take into account contests of force, not merely ones that employ arms. And he also stresses that war is not merely a contest but a "condition." He does not elaborate but it seems relatively clear that he thinks of war as a state of affairs that continues for a time, not merely a single act of contesting as would be true in a duel or a jousting match, for instance.

To see war as a condition or state of affairs is to see war as perhaps a series of contests, or battles, isolatable from each other but bound together in that they are all in the service of a particular cause. In this sense, cyberwar would only be war if it involved more than merely a single attack spurred by a computer program. But of course there is nothing in principle to prevent there being a series of cyber attacks that collectively constituted a cyberwar. Yet, war normally involves two parties each employing arms in a continuing state of affairs.

In this respect one needs to think about the implicit idea that war is a *public* condition—all of the views of war distinguish the public conflict, which is called war, from the private act of conflict, which goes by the name sedition or even duel. Part of what is involved here is that that war is openly declared and the element of surprise is thereby limited. This gives the other side a fair chance to try to repel the attacking forces. Yet, the best known examples of "cyberwar" have been clandestine, with nothing like a declaration of war or even initially an admission that the destructive computer program has come from a particular state—indeed the United States and Israel are widely believed to be responsible for launching the 2010 program Stuxnet against the computers at a nuclear facility in Iran. As of this writing Stuxnet has still not been acknowledged as coming from a United States and Israel coalition.

Again, though, cyber attacks do not have to be clandestine. But until they are public acts it is once again hard to see them as instances of war properly so called. And it is hard to imagine a computer virus or worm being successfully used against a foreign computer system if it is acknowledged in advance. So, the prohibition on clandestine warfare that runs throughout the Just War tradition seems not to sit well with the idea of cyberwar.

At the end of the seventeenth century, Samuel Pufendorf had yet another variation on the definition of war that brings our discussion much closer to contemporary usage:

> Some states expressly denote a relation toward other men than do others, since they signify distinctly the mode in which men mutually transact their business.

[4] Hugo Grotius, *On the Law of War and Peace (De Jure Belli ac Pacis)* (1625), translated by Francis W Kelsey (Oxford: Clarendon Press, 1925) Bk I, Ch 1, Sec 2, p.33.

> The most outstanding of these are *peace* and *war*. . .. Now peace is that state in which men dwell together in quiet and without violent injuries, and render their mutual dues as of obligation and desire. War, however, is a state of men who are naturally inflicting or repelling injuries or are striving to extort by force what is due to them.[5]

By "state" here Pufendorf means a condition in which persons find themselves in various moral relations. On the basis of the states that pertain to humans they have obligations. War is a state of men, of persons in plural not singular. War is the state of men inflicting or repelling injuries, or attempting to do so. As a result, war creates obligations on the people who are in this state.

Pufendorf defines war in a way that makes the inflicting and repelling of injuries the key component. Cyberwar could merely be the type of war that involves injuring or attempting to injure someone by means of malicious computer programs. But there is a problem with this idea—cyberwar does not aim at inflicting injury on persons but on other computers. And most importantly, cyber attacks are not aimed at inflicting or repelling injuries on "common soldiers" who Pufendorf says are those who are authorized by their states to inflict injuries on the common soldiers of the enemy state.[6]

In my view, Pufendorf is right to say that the term "war" has been reserved for the recourse to violent force by one state against another state, where the violence is primarily directed at the soldiers of the enemy state. The rules and laws of war that have been constructed since the seventeenth century have been designed to limit the horrors of war that mainly concern the killing of civilians and the suffering of soldiers. In section III, I explain in more detail why I think that so-called cyberwar is not the kind of war that is primarily of interest to those who have written on the rules and laws of war.

III. The Rules of War and "Cyberwar"

War has historically focused on public contests that involve arms and especially those contests that involve the attempt to kill or wound soldiers of the opposing state. So-called cyberwar can be seen as a kind of contest of force and perhaps even of arms, but such attacks are rarely public and the point of cyber attacks is not to kill or wound soldiers but to destroy property. Of course, the destruction of property, such as the computer programs that control centrifuges in a nuclear power plant or the electric power grid that supplies power to military installations, certainly can have as their foreseeable secondary effects that civilians and perhaps also soldiers will suffer. But the primary aim is not publicly to kill or wound soldiers and this, in my estimation, is an important difference between cyber attacks and war.

[5] Samuel Pufendorf, *On the Law of Nature and Nations (De Jure Naturae et Gentium)* (1672), translated by Charles H Oldfather and William A Oldfather (Oxford: Clarendon Press, 1934) 9.

[6] Pufendorf, *On the Law of Nature and Nations* (n 5) 11.

Henry Wheaton, writing in 1836, presents what is often thought to be the modern notion of war:

> A contest of force between independent sovereign States is called a public war. If it is declared in form, or duly commenced, it entitles both the belligerent parties to all the rights of war against each other. The voluntary or positive law of nations makes no distinction, in this respect, between a just and unjust war. A war in form, or duly commenced, is to be considered, as to its effects, as just on both sides. Whatever is permitted by the laws of war to one of the belligerent parties is equally permitted to the other.[7]

Recognizing a contest of force as a war means that there are various rights, not recognized in other situations, which pertain. Most importantly, the rights of war allow for the intentional killing and wounding of people (soldiers) that would normally be forbidden.

Recognizing a contest as a public war involves rules designed to minimize suffering but not to restrict intentional killing and wounding. This is because war is thought to be different from any other sphere of life. Before designating a contest or condition as one of war, care must be taken since such a designation carries with it significant permissions. War, especially of the defensive variety, is not outlawed but seen as at least a necessary evil. Once this is recognized, then the rules of war are framed so as to make war as humane as possible, given that killing and wounding will still go on as a matter of right for the parties.

When an attack is raised to the level of being called a war, this means that that attack is treated according to the special rules governing war and armed conflict. And this means that the killing or wounding of people is not proscribed but is instead accepted, at least for the direct attacks on soldiers and even for many indirect attacks on civilians. The main task of the laws and customs of war, as understood today under international humanitarian law, is to contain the violence in the most humane way that still allows for large scale killing and wounding. And the rationale is that when states resort to war as a means of self-defense, successful execution of the war requires permission for its soldiers to kill or wound enemy soldiers.

However, the question is why in "cyberwar" we would want to allow for large scale killing and wounding, settling instead only to contain some of the violence that results from cyber attacks rather than to treat them as non-war attacks that must conform to a stricter standard. It is not clear that cyber attack, to be successful as a matter of self-defense of states, needs to be granted the permission to kill or wound in order to effect its purpose. Indeed, cyber attacks seem to be most successful when the main destruction is already limited to property and only incidentally harms soldiers or civilians. Since there is no intention to kill or wound enemy soldiers that is an essential component of cyber attacks, the war convention's equal permissions to kill and be killed makes little sense.

[7] Henry Wheaton, *Elements of International Law* (1836), edited with notes by George G Wilson (Oxford: Clarendon Press, 1936) 295–6.

One reason to assimilate cyber attacks to war is that the recent examples of cyber attacks seemingly violated the territorial integrity of a sovereign state. Such a violation typically triggers the categorizing of an act as an act of war. Acts of war are those that breach the territorial integrity or sovereign immunity of a state. The laws of war are in part aimed at protecting sovereign states from infringements on their sovereignty. The trigger for an act of aggression has been the crossing of a state's borders by uninvited foreign troops. In a sense, cyber attacks appear to follow this model since the computer worm or virus does cross the state's borders and is most certainly uninvited.

Yet, there are reasons to doubt that cyber attacks are enough like uninvited border crossings by foreign troops to count as acts of war. Three factors are important in creating the consensus over the centuries that borders of sovereign states should not be crossed by uninvited foreign troops. First, there is a concern that crossing of borders by foreign troops will undermine the self-determination of citizens of the attacked state. Second, the uninvited foreign troops, as constituted by armed and dangerous soldiers, risk the killing of the soldiers of the attacked state. Third, uninvited foreign troops can cause great disruption of the normal workings of a state and also risk the physical well-being of the citizens of the attacked state. Each of these issues is taken up in the following paragraphs.

First, cyber attacks do not cross borders in the way that foreign troops do. The main worry about territorial integrity concerns the threat to the self-determination of a state that follows upon a physical invasion by foreign troops. The kind of attacks that involve computer programs do not have the same kinds of worries normally associated with armed border crossing. Cyber attacks can be quite disruptive of the normal functioning of a state, and can certainly lead to deaths as a side effect of those attacks. But cyber attacks only cross borders in a metaphorical sense of the term "cross." While there is some physical material that crosses borders in a cyber attack, no uninvited person traverses the border and sets foot in the state in question.

Second, because in the launching of a cyber attack there is no intention directly to kill enemy soldiers there is another reason not to extend to cyber attacks the label of war with its corresponding rules allowing for killing but seeking only to render the killing more humane. The reciprocal permission to kill and wound on both sides of a public war does not need to be extended to cyber attacks that seek to undermine a state by non-lethal means. Cyber attacks can occur, pretty much unhindered, whether or not the permission to kill and be killed is extended to troops on either side. There are good reasons to see cyber attacks as similar to the use of economic sanctions rather than foreign troops.

Third, the kind of disruption of services that cyber attack can achieve is insufficient to count as an act of war. Think about embargoes in this respect. Embargoes involve the attempt to stop trade between two or more states by the act of one state. Embargoes are not considered acts of war, but economic blockades are since they involve the physical intrusion into the territorial waters of a state by the uninvited military forces of a foreign state. As discussed in section IV, cyber attacks are more like embargoes than economic blockades. Like embargoes, cyber attacks can cause great disruption and result in deaths within the target state. Yet, like embargoes, there is no physical presence of foreign troops in the target state. But when embargoes are contemplated we do not treat them according to the standards of humanitarian law, where we say that there is an equal right to kill combatants or that the only worry is about the collateral killing of civilians. Instead, embargoes are treated according to the standard moral conceptions not the special moral rules of war. Cyber attacks should be treated similarly, as shown in section IV.

The rules or laws of war are different from the rules concerning everyday life. Except in the rare case of attacks that take on a self-defense response, people in everyday life do not make themselves liable to be attacked by others merely because they put on uniforms, carry arms openly, and follow a chain of command. The rules of war are exceptional rules for very exceptional circumstances. And whether one agrees with these rules being extended equally to soldiers who fight on the just as well as the unjust sides of a war, the rules of war are only justified because of the exceptional, and I would say emergency, circumstances of state aggression.

Cyberwar may very well call for its own special rules and laws, although I shall give reasons to doubt this, but the use of the rules of war for this type of attack is simply not warranted. Cyber attacks may be so disruptive, as when they target the electrical grid of a state, that special rules for how morally and legally to deal with such attacks may need to be crafted. But I do not see why the rules and laws designed for war should be applied to the case of cyber attacks given how different from war cyber attacks seem to be. Section IV explores what kind of model of rules is best suited to cyber attacks.

IV. A Paradigm Shift for Conceptualizing Cyber Attacks

We need a paradigm shift from seeing cyber attacks as war to seeing them as similar to embargoes. The rules of war are, by design, meant to apply only to situations that are exceptional. In such situations it is thought that the normal rules and laws of society need to be adjusted, in some cases quite radically. Specifically, situations of war involve the clash of armies almost always claiming to act in national self-defense, each aiming at inflicting massive destruction on the other side so as to incapacitate and render harmless the threat of enemy armies. It is true that some cyber attacks may have similar rationales, namely national self-defense. But the point of cyber attacks is not to incapacitate as many enemy soldiers as possible. And because of this significant difference with situations of war, cyber attacks should not be assimilated to the war paradigm. Indeed, I argued in section III that there is much dissimilarity between cyber attacks and war or armed conflict.

A better paradigm for cyber attacks than the war paradigm is one closer to the rules of everyday life where killing is not seen as justified and even sometimes required. In everyday life, there is a near absolute prohibition on intentional killing, even if the person targeted is an enemy soldier. Indeed, soldiers have not done anything to make them liable to be killed when a cyber attack is launched. Or if there are some soldiers who have the liability to be attacked it will be the very few soldiers that it takes to create and install a computer worm or virus in a foreign computer system. Soldiers cannot kill other soldiers in everyday life since all killing is seen as the same, and subject to the same near absolute prohibition.

Some theorists have recently argued that war itself should be assimilated to the standards of everyday life, specifically to the standards of individual self-defense in the face of lethal attacks.[8] But even these theorists do not mean the moral and legal rules of everyday life but the moral and legal rules concerning emergencies when a person is attacked and can only save his or her own life by mounting a lethal counter-attack. The kinds of assaults that cyber attacks constitute do not pose such stark choices about the retaliatory use of lethal violence. Indeed, the best way at the moment to confront a cyber attack is by counter-cyber attacks rather than anything lethal at all. Indeed, calling these counter-cyber attacks "attacks" at all seems inappropriate since the acts may be purely defensive in the sense that they merely put up a shield that blocks the computer virus or worm from entering computer systems in the target state. In addition, there is not really any need for a counter-attack, at least not in the way attacks are normally understood. Purely defensive action that risks no one's life or even property is the main way that cyber attacks are confronted and stymied.

The better paradigm for cyber attacks is one based in the intentional infliction of harm on property and other aspects of the infrastructure of a state, including but not limited to the state's ability to engage in successful economic transactions with other states, both sending goods out of the state for export sale as well as bringing goods into the state for import sales. Cyber attacks can risk life and limb of the citizens of the state that is attacked. But this is certainly also true of economic embargoes. In Iraq after the first Gulf War comparable harm to the civilian population occurred due to embargoes than due to the war itself.

Cyber attacks interfere with the ability of a state to maintain communication between humans in the state and the various modes of infrastructure that are run by computers. And cyber attacks also disrupt the ability of a state to communicate with people in other states for mutually beneficial matters. The inflow and outflow of information in a society is disrupted, but such disruption is better seen on the paradigm of economic exports and imports rather than on the model of troops attacking each other or engaging in lethal attacks against foreign troops that are aggressing against them.

Insofar as cyber attacks and embargoes interfere with the sovereign functioning of a state, they are certainly not to be taken lightly. But in the Just War tradition as well as contemporary international law, embargoes are treated very differently from war or armed conflict—and this seems completely right to me. Wheaton distinguishes between embargoes that affect a state's own output commerce and embargoes that physically restrain goods delivered by ship as input. The seizure of ships as a means to enforce an embargo verges into an act of war insofar as the intention is to take property of a state. But embargoes that are enforced by civil authority alone are not to count as acts of war and are governed by different rules than are acts of war.[9]

In contemporary international law, embargoes are generally considered acts short of war. But trade embargoes, for instance, are restricted in various ways. Today international legal practice has recognized certain embargoes as illegal, namely those that

[8] See Henry Shue, "Do We Need a 'Morality of War'?" in David Rodin and Henry Shue (eds), *Just and Unjust Warriors: The Moral and Legal Status of Soldiers* (Oxford: Oxford University Press, 2008) 87.

[9] Wheaton, *Elements of International Law* (n 7) 312–13. Wheaton here refers favorably to Vattel.

affect the free access to medicines, medical supplies, and certain basic foods that are necessary for the prevention of unnecessary suffering by civilian populations. Some of the restrictions on embargoes resemble restrictions on acts of war, but the overall regime is different in that embargoes are not generally seen as able to inflict loss of life on soldiers and other combatants. The long-standing US embargo of goods to and from Cuba was condemned by a series of UN declarations—but this was because of the effects of the embargo, especially on medical care in Cuba, rather than because the embargo was seen as an act of war.

I believe that cyber attacks should be regarded as similar to embargoes in respect to most of the points just covered in the preceding paragraphs of this chapter. Embargo regimes do not allow for killing of soldiers and other combatants, indeed killing is generally condemned when caused by embargoes just as would be true for other normal types of killing outside of the context of war or armed conflict. So, since it is physical destruction rather than killing or disabling soldiers that is the point of cyber attacks, we should not assimilate cyber attacks to the paradigm of war or armed conflict.

One objection to my view could build on the analogy with embargoes to argue that cyber attacks are aggressive measures that, when seen in embargoes, can constitute acts of war. Indeed, it is the physical intrusion into the territory of the targeted state that can turn embargoes from civil administrations into acts of war. One could argue that cyber attacks normally are aggressive in this twofold sense: there is invasion of the territory of the targeted state, and there is an attempt to harm significantly the physical property of the targeted state. So, just as aggressive acts of embargo are counted as moving into the war paradigm, cyber attacks should be similarly treated in the war paradigm as well.

My response to this objection is first to point out that the main worry about aggressive embargoes is that they will indeed involve killing of civilians as a directly expected goal of the embargo—namely to make the civilian population suffer. Yet, cyber attacks do not have civilian suffering as a directly anticipated or intended effect. It is true that some cyber attacks—that target the computers responsible for the water supply or electrical grid of a state—are likely to effect civilian death and suffering. And in such cases, calling cyber attacks acts of war may indeed be warranted. But this is different from seeing the regime of cyber attacks as best approached though a paradigm of war and armed conflict. Rather, when cyber attacks are closely linked with civilian death and suffering only then should we employ the war paradigm, and here as an exception to the normal way that cyber attacks should be understood.

A second objection is that cyber attacks are more serious assaults on the sovereignty of a state than I have recognized. Increasingly, most of the major services that states provide for their citizens are controlled by computers that can be subjected to malicious worms and viruses. The disruption of basic services of a population is also a disruption of the sovereign ability of a state to maintain the peace and security of its population. The sovereignty of the state is challenged in much the same way that an invasion challenges the sovereignty of a state, and leads to the charge that an act of war has occurred.

My response to the second objection is to admit that cyber attacks can have just the wide-ranging effect that has been indicated. But once again I would point out that the necessary response to cyber attacks is not to kill enemy soldiers but rather to defend the cyber networks of the targeted state. And I concede that when the effects of cyber attacks do have an impact on the civilian population, then just as in cases of embargoes, this gives us reason to shift the paradigm. I next elaborate on my reasons to worry about the wholesale assimilation of cyber attacks to the model of war and armed conflict.

A third objection is that in some cases cyber-attacks have at least as an indirect goal the taking of human life. There is a difference with standard examples of war or armed conflict in that the causal chain in cyber attacks between the launching of the cyber attack and the killing of humans is more attenuated than in the normal case of war. But since there are still goals of killing humans by the launching of cyber attacks, perhaps at least in some cases cyber attacks would fall under the rules and laws of war and armed conflict.[10]

My response to this objection is generally to agree that insofar as cyber attacks have as their aim, even if not their main aim, the taking of human life then there is more reason to think of these attacks on the paradigm of war than if these lethal aims are not present. But there are two reasons to believe that this condition will not often be met in the relevant way. First, the rules of war are primarily aimed at events that involve the intentional killing of soldiers, not merely any taking of human life. It is very unclear that a cyber attack would be aimed at soldiers' lives since the attacks on the electrical grid or water supply would aim at killing humans indiscriminately not because people are soldiers or combatants. Second, the targeting of civilians is problematic in war either because the civilians are directly targeted when they cannot easily defend themselves, or indirectly targeted (as collateral damage) where the loss of life is disproportionate. Here again it is unlikely that cyber attacks will be launched that directly target civilians. And while it is more likely that cyber attacks could be seen as similar to armed attacks in terms of the indirect attacks (as collateral damage) of civilians and noncombatants there is an additional problem. The possibility of collateral damage is not normally seen as sufficient to trigger the war paradigm's lack of restraints on soldiers killing and being killed. Nonetheless, there is nothing in principle that would rule out seeing some future cyber attacks as meeting the conditions for fitting under the war or armed conflict paradigm.

V. The Problems of Assimilating Cyber Attacks to a War Paradigm

One of the ways to see some of the potential problems with assimilating cyber attacks to a war paradigm versus an economic embargo paradigm can be seen in the different departments of the government in charge of each. In the United States wars are the exclusive purview of the Department of Defense (which used to be called the War Department), whereas embargoes are primarily handled through the Department of State (where matters of diplomacy are its prime purview). When consigned to the defense department cyber attacks are under the direction of generals and lower

[10] I am grateful to Jens Ohlin for pressing this objection concerning an earlier draft of this chapter.

ranking officers who have grown up in a military culture that sees intentional killing and wounding as acceptable, indeed nearly required, parts of their mandate. In the Department of State, by contrast, intentional killing and wounding is not acceptable behavior, just as is true in most of the rest of ordinary life.

If cyber attacks are put under the purview of generals in the Pentagon there will be little to limit the development of cyber attack technology, especially development toward a more lethal means of attack than what is currently on offer. If instead, cyber attacks are treated as we treat economic embargoes there will be significant restraint on developing cyberweapons that are lethal. As part of the normal regime of how killing is viewed in ordinary life, cyber attacks would not be developed alongside other lethal means of attack. This is one of the main worries in assimilating cyber attacks to military attacks that are designed to effect maximal destruction of lives of soldiers and other combatants. Cyber attacks, if governed by the state department, would then be a part of the diplomatic means available for pursuing state interest.

In this scenario, we can think of cyber attacks as acts short of war that states should contemplate prior to the stage where states begin contemplating recourse to the violence of war. But once cyber attacks are assimilated into the war paradigm, there will not be the advantage of having a potent alternative to war and armed conflict that can be employed by negotiators and other officials who are focused on peaceful settlement of disputes.

Another problem with assimilating cyber attacks to military attacks is that the states against which they are used will be more likely to respond militarily rather than merely to try to block the attacks with counter-cyber means or to organize diplomatic resistance in the world community. This is the scenario that has developed since it was made known that the Stuxnet virus used against the centrifuges in Iran was designed and launched by the United States and Israel. Iran has vowed to kill Americans and Israelis for what it regards as the act of war perpetrated by the use of the Stuxnet virus. The response that surely would be best for world peace is either merely to build better cyber defenses or to launch retaliatory cyber attacks, but not to move to the plane of killing and threats of killing that is the normal domain of escalating armed conflict. In addition, as has been seen as of February 2014, three and a half years after the Stuxnet attack, Iran is still using the supposed military attack as a reason to claim that it needs to defend its sovereignty by any means necessary.

Curtailing or cabining cyber attacks in the way in which economic embargoes are controlled would alleviate the risks to world peace that are the effect of cyber attacks understood as just one of many military means to attack a country's sovereignty. One question to pose at this point is what then is the right way to view such cyber attacks as that involving Stuxnet if not as a military attack on Iran's sovereignty. I see cyber attacks as deterrents and means of pressuring states rather than offensive weapons, once again similar to the employment of economic sanctions.

Cyber attacks can be deterrents, as in the way they were used against Iranian centrifuges. Neither the United States nor Israel launched the Stuxnet virus so as to invade and conquer Iran, but only to set back Iran's nuclear missile capability. And from all accounts, the plan succeeded. In addition, there was the implicit message that other cyber attacks would be launched if Iran tried to rebuild its nuclear weapons capability.

As is well known, the cost of cyber attacks is so small compared to the damage that can be done that states would be well advised to accede to the pressure and simply stop the prohibited activity that has brought upon itself the cyber attack.

Cyber attacks can also be used as a means of pressuring states to act more peaceably or stop acting in threatening ways to their own citizens or to citizens of neighboring states. If cyber attacks are used against the banking industry of a bellicose or oppressive state, for example, this could significantly increase the pressure on the target state to cease these illegal activities and rejoin the community of peaceable states. And the disruptions could be intensified until it became unpalatable for the target state to continue policies disfavored by the international community. The advantage of deterrence and pressure is muted, I believe, if cyber attacks are merely one form of military attack in a state's arsenal. The world will be a safer place also if states do not think of their cyber capabilities in military rather than in diplomatic terms. And we will all be better off if we do not extend to cyber attacks the immunity from sanction if people are killed as a result of these cyber attacks.

Throughout this chapter, I have argued that cyberwar is badly named as a form of war. Rather I have argued that cyber attacks should be understood as a type of international pressure that is similar to the way that economic embargoes have been used over the centuries. This means that there is an alternative paradigm readily available for how cyber attacks are understood that does not force us to invent something out of whole cloth. And while the parallels between embargoes and cyber attacks is not a perfect one, I have given reasons for thinking that the embargo paradigm is much better than the war paradigm for understanding cyber attacks.

VI. Proportionality and "Cyberwar"

Cyber attacks, in my view, fall into the category of attacks that are short of armed attacks. The main reason for this is that the cyber attacks are not aimed at the taking of lives, and so the countermeasures that involve the taking, or the risk of taking, of lives should not be assumed to be justified. Outside of the war context, proportionality basically is governed by human rights concerns rather than humanitarian concerns. And this means that the use of armed force is completely restricted only to those situations of emergency.

Countermeasures short of armed conflict, especially those that do not respond to armed conflict, are strictly controlled in international law. It is not assumed that armed actions will be considered proportionate merely if collateral damage is minimized. The higher standard employed is the one of proportionality in human rights law. In this respect, the killing of soldiers as a countermeasure to a cyber attack that is not aimed at the killing of soldiers would normally be considered disproportionate.

When cyber attacks are taken out of the "war or armed conflict" category and instead considered as "actions short of armed conflict" the response to them is not drawn in terms of Article 51 actions. Article 51 of the UN Charter enshrined the idea that states had "an inherent right of self-defense."

> Nothing in the present Charter shall impair the inherent right of individual or collective self-defense if an armed attack occurs against a member of the United Nations,

until the Security Council has taken measures necessary to maintain international peace and security.

If cyber attacks are not assimilated to the model of "armed attack," then states are not regarded as having the right to take armed action as a countermeasure. Nearly all armed action taken as a countermeasure to a sub-armed attack is very likely to be considered disproportionate.

The rules of humanitarian law, which are fairly permissive concerning countermeasures taken in response to armed attack, are replaced by rules of human rights law that are much more restrictive than those of humanitarian law. If cyber attacks are considered more like embargoes than like armed attacks, there will be very serious consequences for what is thought proportionate and what disproportionate as a countermeasure.

Several judgments of the International Court of Justice, most especially the Oil Platform Case and the Nicaragua Case, set out very different standards of proportionality that must be met when a state is responding to an attack that falls short of an armed attack. Suffice it here to say that it will matter greatly if cyber attacks are seen as fitting into the category of attacks short of armed attacks. In such a case we do not assume that justified killing will take place and that the main issue is to diminish cruelty and collateral damage. Instead, the main issue is to prohibit killing altogether, since the intentional taking of life not in response to a threat to life violates the human rights of the persons attacked.

VII. Conclusion

As a final point it should again be noted that there could be cyber attacks that rise to the level of armed attacks, although this does not seem to be the case, or even likely, at the moment. Throughout this chapter I have argued that cyber attacks, because they do not involve "arms" in any sense of that term, are to be treated more like embargoes than like war or armed conflict. If it should turn out that cyber attacks risked the lives of soldiers or innocent civilians, then my assessments would have to change. But before cyber attacks are assimilated to the model of war, intentional risk of killing would have to be shown as one of the main aims of that instance of cyber attack.[11]

[11] For more discussion of this issue, see Michael Newton and Larry May, *Proportionality in International Law* (Oxford: Oxford University Press, 2014).

2

Is There Anything Morally Special about Cyberwar?

James L Cook[1]

I. Introduction

The literature on the ethics and law of cyberwar traces a broad pattern: ethicists are still discussing whether the just war tradition (JWT)[2] can deal adequately with cyberwar[3] while legal experts seem largely to agree that the law must evolve even if the particulars of that evolution are up for vigorous debate. So among philosophers the primary question is *whether* to revise the JWT; among lawyers the debate is over *how* to adapt the law to accommodate cyberwar.

This chapter joins the philosophers' debate, which broadly conceived takes two approaches to the question of whether the JWT can accommodate cyberwar. One approach sees cyberwar as so new that it bursts the mold of the received JWT. The other approach assumes the JWT pays more attention to effects and intentions than to means; in so far as cyber is new only as a means of war, the JWT continues to apply. For similar reasons the JWT did not need rewriting *because* humanity had reached any specific technological way station between sling and hydrogen bomb. The tradition did evolve in that span but for socially complex reasons that included many other factors besides advances in weapons technology.[4]

While I am among those who think the received JWT can accommodate the intentions and effects of cyberwar with no foundational modification—for reasons explained more fully in section II—I nonetheless also believe that cyberwar holds a unique ethical status at least for the historical moment. Three factors contribute to making cyberwar morally special—a growing sense of cyber's ubiquity, uncontrollability, and what I call its

[1] Disclaimer: the views stated in this chapter are solely the author's. They do not necessarily represent positions of the US government, the US Department of Defense, or the US Air Force.

[2] One finds both "just war *theory*" and "just war *tradition*" in the literature, often enough at the hand of the same author. For instance, Brian Orend calls just war theory a "tradition" (along with realism and pacifism), but later talks about both "just war theory" and "just war tradition" without explicitly distinguishing one from the other. See Brian Orend, "War" (2005) *Stanford Encyclopedia of Philosophy*, sec 1, "The Ethics of War and Peace," *Stanford Encyclopedia of Philosophy*, sec 1 (accessed July 2, 2014). In this chapter I prefer "tradition" because it connotes something that evolves as a product of culture as well as scholarship—a connotation "theory" would not capture quite as well.

[3] Randall R Dipert, "The Ethics of Cyberwarfare" (2010) 9 *Journal of Military Ethics* 384–410; James L Cook, "'Cyberation' and Just War Doctrine: A Response to Randall Dipert" (2010) 9 *Journal of Military Ethics* 411–23.

[4] Bernard Brodie and Fawn M Brodie, *From Crossbow to H-Bomb:The Evolution of the Weapons and Tactics of Warfare* (Bloomington: Indiana University Press, 2nd revised edn, 1973) 3–29.

neoreality (sections III, IV, and V).[5] Because of these three factors our culture finds it ever easier to view cyberwar as a constant and near-universal threat to good people and the ways they go about their daily lives.

Although this relatively new and still-evolving perception of cyberwar requires no specific, foundational change to the received JWT, students of just war would be wise to revive what I call *the ethics of threat*, the sixth and concluding section. The ethics of threat is a moral framework that grew up within the JWT and was motivated by the advent of nuclear weapons and the Cold War. This framework comprises a hodgepodge of views that emerged from the work of many thinkers and movements. Perhaps treaties and other official protocols provide the best evidence of efficacy.[6] It would be equally difficult to measure the concrete impact of the Union of Concerned Scientists, the US National Conference of Catholic Bishops, and many other groups that grappled with the nuclear threat. Substantial common ground emerged even though these groups' respective views did not always coincide.

Between 1945 and 1989, but sometimes drawing on much earlier sources, popular, intellectual, and political culture in the US during the Cold War described or imagined the worst nuclear scenarios. Works of the imagination, self-interested political policies, and hard science all urged strict controls on the development, testing, and use of nuclear technology. The disintegration of the Soviet Union (officially in 1991 but well underway by 1989), Saddam Hussein's reputed nuclear ambitions (from the beginning of his presidency in 1979 until his ouster by coalition forces in 2003), and the acquisition of nuclear weapons by Pakistan (1998) and North Korea (confirmed 2006) motivated hasty rethinking of scenarios that seemed most relevant at the Cold War's height.[7] What if terrorists obtained a nuke or enough radioactive material to make a dirty bomb? Could a minor nation-state build or buy a nuclear weapon yet fail to behave like the rationally self-interested actors the nuclear powers had long considered themselves and each other to be?[8]

[5] Not so long ago "cyber" was used only as an adjective or a prefix. A brief period of semantic transition followed. See Richard Stiennon, "Cyber Confusion: What Is the Airforce Talking About?" *ZDNet.com*, January 24, 2008, at <http://www.zdnet.com/blog/threatchaos/cyber-confusion-what-is-the-air-force-talking-about/512> accessed July 2, 2014. Here and throughout the chapter I use "cyber" as a noun even though a standalone nominative sense has not made it into the *Oxford English Dictionary*. Even the best of dictionaries can lag usage.

[6] This view is implied by a mid-1960s analyst of the Pugwash movement: "Any attempt to evaluate the Pugwash Conferences is beset with difficulties. There is much about Pugwash that is intangible and unrecorded. Many conversations take place privately or under informal or unanticipated circumstances. The effect of such dialogues is extremely difficult to judge.... Pugwash has been credited, in varying proportion, with originating or promoting a number of disarmament proposals, including the Test Ban Treaty, the banning of weapons of mass destruction in orbit, automatic seismic stations, minimum deterrent and others." Leonard E Schwartz, "Perspective on Pugwash" (1967) 43 *International Affairs* (Royal Institute of International Affairs 1944–) 498, 509–11.

[7] Among unclassified sources, US congressional hearings are an especially good barometer of post-Cold War concerns. See, for example, "U.S. Nonproliferation Strategy: Policies and Capabilities," Hearing before the Subcommittee on Oversight and Investigations of the Committee on International Relations, House of Representatives, One Hundred Ninth Congress, Second Session, July 20, 2006, Serial No 109–198, (Washington DC: US Government Printing Office, 2006). On the origin and magnitude of post-Cold War proliferation issues see David E Hoffman, *The Dead Hand:The Untold Story of the Cold War Arms Race and its Dangerous Legacy* (New York: Doubleday, 2009) 379–423, 475–83.

[8] A documentary film focusing on John F Kennedy's and Lyndon B Johnson's Secretary of Defense, Robert S McNamara, illustrates why this assumption might have been wrong and why it might not have

However, once Saddam proved not to have weapons of mass destruction, the US public's attention turned more to the conventional struggles in Iraq and Afghanistan. The newest nuclear player, North Korea, and the aspiring nuclear power, Iran, provided staccato reasons to attend again to nuclear threats, but how long could the associated stories loom on front pages and at the top of thirty-minute TV news summaries when competing with the so-called war on terror, a global recession, and the home team's exploits?

In short, the ethics of threat spurred originally by the threat of nuclear war has lost momentum over the last decade. That is not to say that since 9/11 US or other Western cultures have been heedless of the dangers nuclear weapons pose, but only that the United States and other nations have been understandably distracted by the sizable wars in Afghanistan and Iraq. The Cold War ethics of threat was based on the belief that nonproliferation and deterrence were both possible. Acquisition aside, the US administration between 2000 and 2008 arguably lost focus on deterrence.[9] Another sign that the ethics of threat has itself been under threat is the alleged inability of US policy-makers to perceive nuclear hazards holistically rather than piecemeal.[10] The prospect of cyberwar may and should revive that ethics and take it in a new direction, I argue in this chapter. Among other effects, the threat of cyberwar reinvigorates an important question posed during the waning years of the Cold War: may one morally *threaten* to do what would be immoral if one in fact did it?

II. Definition and Nature of Cyberwar: Applicability of the JWT

How should we define "cyber" so we can go on to define "cyberwar"? The US Air Force offered the following definition of cyberspace in NSPD-54: "The interdependent network of information technology infrastructures [that] includes the Internet, telecommunications networks, computer systems, and embedded processors and controllers in critical industries."[11] Although this definition may suffice for some purposes, it fails to capture two important aspects of cyber as I hear the term used in common parlance in today's US Air Force. First, it misses the information and directives that are an

mattered even if it was essentially correct. See Errol Morris (director), *The Fog of War: Eleven Lessons from the Life of Robert S McNamara* (Sony Pictures Classics, 2003). McNamara expresses his surprise that Cuba's President Castro favored launching Russian nuclear weapons at the United States even though doing so would have meant Cuba's near-annihilation and the end of his revolution. McNamara concludes that US ignorance of the true status quo during the Cuban missile crisis was so profound that "dumb luck" was all that allowed the United States and the USSR (The Union of Soviet Socialist Republics) to avoid war. But this example is an outlier: Cold War nuclear policy presupposed world leaders who were rational actors determined to preserve their nations rather than proto-suicide bombers ready for the wholesale destruction of their own people in a nuclear-fueled Götterdämmerung.

[9] Peter Beinart, *The Icarus Syndrome: A History of American Hubris* (New York: HarperCollins, 2010) 382–3 ("In 2002, George W. Bush declared the concept [of deterrence] dead. In the post-9/11 world, he insisted, threats of nuclear retaliation were not only useless against Al Qaeda; they were useless against anti-American dictatorships with WMD.").

[10] Claudia Rosett, "Iran Could Outsource Its Nuclear-Weapons Program to North Korea," *The Wall Street Journal*, June 20, 2014.

[11] National Security Presidential Directive 54, January 8, 2008, REDACTED, at <http://fas.org/irp/offdocs/nspd/nspd-54.pdf> accessed July 2, 2014.

integral part of cyber at any moment. The second problem with NSPD-54's definition (and despite the word "includes") is that the phrase "information technology infrastructures" arguably misses what could become a critical feature of cyber and the battleground of cyberwar: "the Internet of Things."[12] In this sense of "thing," for example, every US government official's or military general's private car, if it has Bluetooth or even GPS (Global Positioning System), is part of the Internet of Things and, therefore, vulnerable to cyber attack. As our sense of what constitutes the internet evolves so must our understanding of cyber and its potential military uses.[13]

This intuitive approach to understanding cyber as something more than mere infrastructure—an approach most easily understood through concrete examples rather than abstract definitions—seems to work well enough to have earned the endorsement even of US military trainers. For instance, the Air Force, which takes as its mission to "Fly, fight, and win in air, space, and cyberspace," does not even try to define cyberspace in a computer-based training module that is mandatory for all Air Force members, uniformed and civilian. Through this online training on information assurance (IA) the Air Force aims to accomplish an essential mission—to protect its official data and Air Force members' personal data in cyberspace. Rather than attempt to define cyberspace, the module simply presents threat scenarios to information security and asks the trainee what steps would ensure data security and integrity. An intuitive understanding of the context suffices.

That is all to the good given the difficulty of defining "cyber" and its cognates in a way more specific than the working concept of the ethics of threat already sketched.[14] The term comes from an ancient Greek word for a steersman, the person who controls a vessel. Since he is within the ship, the function of the steersman is also somewhat reflexive—he cannot help but steer himself along with the vessel. This double sense of control—of the ship but also of self—will prove relevant to the perception of uncontrollability discussed in section IV.

Whatever else a more specific and definitive account of cyber's nature would tell us, we know that in war cyber is a means and not an end in itself. It is a tool useful for achieving effects in accordance with a policy-maker's or military commander's intentions. That corresponds nicely to the JWT, which primarily addresses effects and intentions; the JWT has relatively little to say about the means used. (This claim should require no argument. It is self-evident when one examines the nature of the jus ad bellum, in bello, and post bellum principles as well as the JWT's ontogeny.[15])

[12] Kevin Ashton, "That 'Internet of Things' Thing," *RFID Journal*, July 2009, at <http://www.rfidjournal.com/articles/view?4986>accessed July 9, 2014.

[13] Jared Serbu, "Air Force Looks to Revamp Its Definition of Cyberspace," *Federal News Radio*, October 12, 2012, at <http://www.federalnewsradio.com/395/3075763/Air-Force-looks-to-revamp-its-definition-of-cyberspace>accessed July 9, 2014.

[14] Consider the following definition of cyber operations: "The employment of cyber capabilities where the primary purpose is to achieve objectives in or through cyberspace. Such operations include computer network operations and activities to operate and defend the Global Information Grid." Joint Publication 1-02, *Department of Defense Dictionary of Military and Associated Terms* (November8, 2010, as amended through June 15, 2014), at <http://www.dtic.mil/doctrine/new_pubs/jp1_02.pdf> accessed July 9, 2014. A drawback of this definition for present purposes is that it is not explicitly about cyber*war*; it could apply to espionage, logistical communications, and so on.

[15] Martin L Cook, *The Moral Warrior* (Albany: State University of New York Press, 2004) 21–38; Brian Orend, *The Morality of War* (Peterborough: Broadview Press, 2nd edn, 2013) 190–206, 226–30.

Because the JWT focuses on effects and intentions, the emergence of a new means of war does not by virtue of its novelty or nature force a wholesale change to the tradition. This is increasingly obvious in US discussions of cyberweapons: they are a discernible and relatively new category in the arsenal, and their development and control is largely secret. But debates over if and when to use them take place very much within the JWT.[16] Of course this does not rule out the possibility that the JWT will evolve; by their nature traditions are not wholly static.

So the JWT applies as readily to cyberwar as to any other kind of war, provided we can identify agents, intentions, and effects. This means the JWT will be useful for analyzing some—but certainly not all—aggressive cyber activities. Just war theorists are used to that same pattern as it applies to other means of war. Some activities in the realms of aviation and navigation fall under the JWT's purview because those activities have to do with war; some activities do not. So there is nothing special about cyber as a means of war or "cyberation" as a class of activities some of which fall under the JWT and some of which do not.[17]

Cyberwar may seem different from other kinds of war because often there is no clear sense of sovereignty in cyberspace, either of actors or territory, as there tends to be in more traditional realms of conflict such as land, sea, air, and even space. Absent sovereignty, how could one demonstrate just cause in defense of one's own territory or in righting a wrong that amounts to trespass? These questions forget that in an ad bellum context nation-states and other potential belligerents such as non-state actors *choose* rather than *discover* casus belli whether in cyberspace or elsewhere. If a Chinese fighter aircraft collides with a US Navy EP-3 reconnaissance aircraft over international waters—another realm where sovereignty is either questionable or moot—forcing the US aircraft to land at a Chinese airfield, the US may choose to view the incident as a casus belli or not, just as China may.[18] The same holds true of cyber attacks: the greyer areas demarcating activities such as spying or vandalism on the one hand from war on the other allow for significant choice. This is not to say the world *should* work that way, but in fact it does.

Arguably there is one minor exception to the rule that cyber as a means need not motivate revision of the JWT, though the exception would affect only the margins of the tradition. If a certain military cyber activity were widely held to be morally repulsive in itself, then the international community might choose to invoke the concept of a means *malum in se*—wicked in itself. Civilized nations have determined that tactics such as combatants' perfidious imitation of medical personnel to escape detection and explosives containing white phosphorous in some contexts fall outside the range of morally and legally acceptable means of waging war. Attributing intrinsic evil to such

[16] A good example is David E Sanger, "Syria War Stirs New US Debate on Cyber Attacks," *The New York Times*, February 24, 2014.

[17] Cook (n 3) 411–23. A similar irrelevance of means seems to apply to cyber belligerence other than cyberwar, for example, cyberbullying. See Lizette Alvarez, "Charges Dropped in Cyberbullying Death but Sheriff Isn't Backing Down," *The New York Times*, November 21, 2013.

[18] Shirley A Kan et al, "China-U.S. Aircraft Collision Incident of April 2001: Assessments and Policy Implications." Updated October 10, 2001, Congressional Research Service Report for Congress, at <http://www.fas.org/sgp/crs/row/RL30946.pdf> accessed June 23, 2014.

means gets us around counterarguments that using these means might conform to other just-war principles—for example, that a certain instance of perfidious deception might be a military necessity which does more good than harm without intentionally hurting any non-combatants.

But it seems highly unlikely that the international community would deem *all* cyber activities to be beyond a moral pale, that is, to be mala in se. Imagine the indiscriminate destruction of all the medical records in a city during a life-threatening epidemic, for instance, or obliteration during severe weather of the industrial control systems (ICSs) of power plants that do not support military operations. Even in these cases cyber attacks would not need to be labeled mala in se since the in bello principles of discrimination and proportionality of means would already condemn them. Besides, one can easily imagine legitimate attacks on databases (e.g., on military personnel records) and ICSs (such as those fronting centrifuges that hostile nations use to refine fissile material).

However, if cyberwar is not so unique as to require important revisions to the JWT, it may nonetheless be morally unique because of three characteristics. The first is ubiquity.

III. Ubiquity of the Cyber Threat

Normally we call something ubiquitous if it is *present* nearly everywhere or very frequently. It is most natural to think of *presence* spatially, but ubiquity can also have a temporal dimension, as when one never finds a moment's respite from something such as an unwanted memory. When it comes to the culture of cyber threats and their present ubiquity in our consciousness, neither space nor time is a flat dimension. A cyber event such as a denial-of-service (DoS) attack on the bank we use may have happened on a given day, and the same goes for other DoS attacks on other banks on that same day. Yet we can and do hold in mind the collection of such events: as a pattern they transcend the individual times of their concrete occurrences. A military cyber expert cannot help but think of the Stuxnet and Flame attacks as categorically similar, two more bricks in a foundation of temporal ubiquity. The spatial aspect of cyber events likewise has depth—a capacity for collection within what we call cyberspace—that typifies the social psychology of cyber.[19]

For experts in various aspects of cyber—programmers and networkers, for instance—cyber is a realm of unknowns, of mystery. And that mystery in turn magnifies cyber threats in a way that one might explain in Darwinian terms: surely it must be adaptive to speculate about and fear what lies out of sight but within reach of one's own position—behind that bush, say, or just within the shadows of that cave—and so it is no wonder if military planners and even laypeople have developed a sense that the next phone caller may be a phisher, the next email attachment may contain malware,

[19] William Gibson's novel *Neuromancer* (New York: Ace, 1984) is the source of the term "cyberspace" according to some. See, for example, Greg Rattray, *Strategic Warfare in Cyberspace* (Cambridge: MIT Press, 2001) 17. The *Oxford English Dictionary* entry for "cyber" suggests Gibson coined the term two years earlier in his short story "Burning Chrome."

and the next military apocalypse may be cyber-borne. Post-Edward Snowden, whose revelations about National Security Agency procedures and leaks of thousands of classified documents surprised many around the world,[20] one is now less sure that phoning in a pizza order is a private matter; it is harder to think in terms of safe private versus exposed public spheres of life. Motive is a key term because it gets at causal linkage. The senses of mystery and threat decrease whenever we know enough or can find out enough about causality to figure out what has happened or what is happening. Hence the prevalence of the so-called attribution problem[21] in the literature on cyber attacks: if we cannot figure out who did what, or who has been where, cyberspace is like the far side of the bush or the shadows in the back of the cave—a mystery and a threat. Experts are divided on how tractable the attribution problem is and unfortunately the salient details are classified.[22] And it is not just a matter of what *information* one can collect while maintaining anonymity.[23] In the wake of Edward Snowden's revelations, even heads of state are trying to figure out how to hide their cyber activities from what some see as a hyper-intrusive United States. For example, the president of Brazil envisions dramatic new ways of circumventing US-based servers.[24]

In short, cyber as pervasive mystery and threat is snowballing through our cultural consciousness. Particularly the dangerous aspects of cyber have our full attention. Perhaps we are our own worst enemies in this regard.[25] Besides augmenting our sense of cyber's ubiquity, constant reports about abuses of cyber may also contribute to our perception of cyber's uncontrollability (the subject of section IV) and, therefore, to the moral uniqueness of cyberwar even within the JWT. Whether this sense of ubiquity and uncontrollability lasts depends largely on the tempo of real or potentially damaging cyber attacks, especially ones carried out by traditional participants in war such as nation-states.

[20] For a sympathetic journalist's take on the Snowden affair, see Glenn Greenwald, *No Place to Hide: Edward Snowden, the NSA, and the U.S. Surveillance State* (New York: Metropolitan Books, 2014).

[21] By "attribution" I mean the identification of the attacker, not the categorization of the attack itself. On the distinction between "who-attribution" and "what-attribution" in cyberwar, see Susan Bremer, "'At Light Speed': Attribution and Response to Cybercrime/Terrorism/Warfare" (2007) 97 *Journal of Criminal Law and Criminology* 379, 457. For a conscientious account of attribution issues, see Martin C Libicki, *Cyberdeterrence and Cyberwar* (RAND Monograph Series) 39–73 (on the relationship between attribution and cyberdeterrence).

[22] The unclassified literature can only deal in generalities: "Traditional arms control regimes would likely fail to deter cyberattacks because of the challenges of attribution, which make verification of compliance almost impossible." William J Lynn III, "Defending a New Domain: the Pentagon's Cyberstrategy" (2010) 89 *Foriegn Affairs* 100.

[23] From a cultural perspective, phenomena such as Tor software that helps navigate Onionland will inevitably affect our sense of how threatening cyberweapons can be. Onionland is a sort of Old West Robber's Roost in the virtual landscape where even the most ardent trackers may lose the scent because in theory Tor allows users to move anonymously. Brad Chacos, "Meet Darknet, the Hidden, Anonymous Underbelly of the Searchable Web," *PCWorld*, August 12, 2013.

[24] Amanda Holpuch, "Brazil's Controversial Plan to Extricate the Internet from US Control," *The Guardian*, 20 September 2013.

[25] Tuchman's Law, devised by the historian Barbara Tuchman, holds that "[t]he fact of being reported multiplies the extent of any deplorable development by five- to tenfold." Barbara W Tuchman, *A Distant Mirror:The Calamitous 14th Century* (New York: Alfred A Knopf, 1978) xviii. Tuchman also asserts that "[i]n individuals as in nations, contentment is silent, which tends to unbalance the historical record." *A Distant Mirror*, 210.

That raises an obvious question: to what degree can military uses of cyber be regulated so as to make cyber attacks less likely and less damaging? Guesses range from the relatively optimistic to the nearly apocalyptic. In 2012 then US Secretary of Defense Leon Panetta staked out the dire end of the spectrum and the sense of teetering on the edge of a cyber abyss:

> A cyber attack perpetrated by nation states [or] violent extremists groups could be as destructive as the terrorist attack on 9/11. Such a destructive cyber-terrorist attack could virtually paralyze the nation....The collective result of these kinds of attacks could be a cyber Pearl Harbor; an attack that would cause physical destruction and the loss of life. In fact, it would paralyze and shock the nation and create a new, profound sense of vulnerability.[26]

In my view, Secretary Panetta's remarks represent the prevailing cultural sense, at least in the United States, that cyber attacks pose a clear, present, and hugely serious danger.

Perceptions matter greatly in the JWT to the extent that applying the tradition's principles requires knowing one's own side's aim and guessing another nation's intention. If another Chinese fighter jet collides with another US surveillance aircraft,[27] the question of whether either side has a just cause to attack the other depends in large measure on *perceived* intention. Maybe the collision was just an accident, the result of momentary inattention or a mechanical failure. In bello, perceptions of intention similarly matter a great deal. Did military necessity require destruction of the ad hoc Iraqi column fleeing Kuwait at the end of operation Desert Storm?[28] Ethically speaking, the answer depended on coalition *perceptions* of the Iraqi forces' intentions to reconstitute themselves into combat-effective units and on *perceptions* of Saddam's intention to redeploy the forces for future unjust ends; policy-makers, military leaders, and the general public are more likely to see hostile intentions in a climate of perceived ubiquitous cyberthreat. This is not to argue that objective facts of the matter do not exist, but only to point out that in cyberwar, as in other kinds of conflict, dire perception can undercut rosier objectivity.

IV. Uncontrollability of Cyberspace and Cyber Activities

Closely related to the multidimensional temporal and spatial *ubiquity* of cyberspace and cyber threats is their perceived *uncontrollability*. In part this uncontrollability

[26] Leon E Panetta, "Remarks by Secretary Panetta on Cybersecurity to the Business Executives for National Security, New York City" (2012), at <http://www.defense.gov/transcripts/transcript.aspx?transcriptid=5136> accessed July 10, 2014. For a measured but hopeful vision, see James A Lewis, "Multilateral Agreements to Constrain Cyberconflict" (2010) 40 Arms Control Today 14–19.

[27] Shirley A Kan et al, "China-U.S. Aircraft Collision Incident of April 2001: Assessments and Policy Implications," Updated October 10, 2001, Congressional Research Service Report for Congress, at <http://fas.org/sgp/crs/row/RL30946.pdf> accessed July 14, 2014.

[28] For one of the more incendiary accounts, with emphasis on the roles of perception and deception, see Joyce Chediac, "The Massacre of Withdrawing Soldiers on the "Highway of Death'," in *War Crimes: A Report on United States War Crimes Against Iraq to the Commission of Inquiry for the International War Crimes Tribunal*, by Ramsey Clark et al (The Commission of Inquiry for the International War Crimes Tribunal, 1992.

derives in an obvious way from cyber's ubiquity. Suppose only a small percentage of mosquitoes or ticks or some other pest served as vectors for dangerous diseases, and suppose such insects and arachnids were so rare that one might never see a mosquito except in a museum display case with a pin through its thorax. In that case we would not think much about mosquitoes. But when the beasties seem constantly about and in great numbers, especially in locales where they are statistically most likely to be carriers of dangerous diseases, one would naturally think that the mosquito population was out of control. Cyber is as prevalent in the lives of most of us these days as mosquitoes are in the lives of those in a village on a tropical river.

How is cyber different from ideas that were dangerous in the time of the printing press or oral traditions? No doubt the ideas we might call memes—"the units [that] are the smallest elements that replicate themselves with reliability and fecundity"[29]—can be dangerous and even cause kinetic harm through a human agent, for example, ideological extremists, no matter the means of their spread. In this regard, cyber is no different in kind from the printing press or other tools of communication. But in degree, cyber is shockingly more potent even though it accelerates the spread of ideas in a way analogous to what Gutenberg's invention achieved. A difference in kind may be that cyber needs no human intermediary even if it requires a human creator at the start of its genealogy. Through a sermon or a printed tract, a Reformation theologian might sow a seed that eventually would motivate iconoclasts to destroy statuary and other images. Stuxnet *directly* destroyed Iranian centrifuges once it was set loose. One can imagine malware that becomes "intelligent" in accordance with the "sense-think-act" paradigm used in robotics[30] and then creates memes that require no renaissance printer to mediate between author and reader/actor. The independence of human mediation would seem to support a sense of cyber's uncontrollability.

However, there may be another—and somewhat paradoxical—contributor to this sense. Cyberspace and the things that go on in it are artifacts. They are *made* things (e.g., file servers, hubs, routers, cables, signals consciously produced and perpetuated within the physical space by physical things such as charged particles). There are people who believe cyber or "the virtual" is somehow non-physical. Yet when such a person suffers a problem with their computer at work, odds are they will still call their local information technology (IT) office to work the physical issue rather than a local cyber shaman to intercede with cyber gremlins in some non-physical realm. Much of what happens in cyber has a dynamic character that we value and that has no clear analog among other artifacts: we want our computer programs to adapt *without our intervention* to new data ranges and formats that arise; we want our cyber-controlled artifacts in the Internet of Things—smart gadgets such as thermostats, for instance—to react creatively when necessary and *without our intervention*. At the far end of the spectrum of dynamic cyber things is artificial intelligence (AI)[31] that in essence frees the artifact

[29] See Daniel C Dennett, *Darwin's Dangerous Idea: Evolution and the Meanings of Life* (New York: Simon and Schuster, 1995) 344.

[30] Ronald C. Arkin, *Behavior-Based Robotics* (Cambridge: MIT Press, 1998) 130.

[31] An old definition seems particularly apt here: "Artificial intelligence...The aspect of computer science that is concerned with building computer systems that emulate what is commonly associated with human intelligence." See Dennis Mercadal, *Dictionary of Artificial Intelligence* (New York: Van Nostrand

from the intelligence and perhaps even the will of the artificer. But conceivably an artifact could be totally free without ever misbehaving or otherwise thwarting its creator's will. The most masterful artificer is in fact the one that does not remain master and never wants to be master again. There is something of a paradox embedded in that sentence, a paradox of the sort one senses in a statement such as "It would be a surprise if there were no surprises today." The paradox evokes the etymology of the word "cyber" and the concept of a steersman who is in control of something external—a ship—but also of himself.

Arguably this potential for independence makes cyberweapons seem uncontrollable and by extension makes cyberwar morally unique. Just as the nuclear age ushered in anxiety over uncontrollable destruction, previous Western works had studied the decay and extinction of cultures at least as early as Homer's account of the demise of Troy.[32] But the catalyst of decline was usually seen as something as much inherent in all human culture as old age and death are built into the individual human being. On this account, retrospective analysis can identify specific causes of cultural senescence and demise just as an autopsy can point out whether the heart stopped or cancer got out of hand and killed the nonagenarian lying on the slab. Yet the specifics always turn out to be universal. For cultures the killer is greed, ambition, or some other form of hubris. The pace can seem glacial, but its progress is not to be denied.

One might rightly argue that no tool—whether a hammer, nuclear technology, or a cyber thing—is responsible for evil that results from its misuse, especially since most any artifact may also be used to achieve good ends. But even a tool usually used to do good can be wielded for foolishly apocalyptic ends. Perhaps some tools are more dangerous than others—because of their power to destroy indiscriminately, for example, or the ease and anonymity with which they can be used—and hence ubiquitous calls for disarmament since early in the Cold War and now omnipresent worries about the dangers of cyberwar.

Many of the cyberwar scenarios we imagine involve cyberweapons destroying or otherwise manipulating other cyber things—databases, for instance, and program code such as the diagnostic algorithm of an ICS (i.e., the cyber-based front-end control of an industrial process)[33] or the code that runs centrifuges. In fact it is difficult to think of a cyber attack that fails to destroy cyber things even if that destruction is not an end in itself. If we want to destroy something "real" with a cyberweapon, we still must attack cyber things as a means to that end. To destroy centrifuges at Natanz, Stuxnet placed a man-in-the-middle attack to fool operators and otherwise changed ICS code.[34]

Reinhold, 1990) 15. This is a broad charter indeed, since we might say that "what is commonly associated with human intelligence" covers much of what it means to be human.

[32] Homer, *The Iliad*, translated by Richmond Lattimore (Chicago: University of Chicago Press, 1951).

[33] Dave Edwards, "Robust ICSs Critical for Guarding Against Cyber Threats" (2012) 102 *Journal* (American Waterworks Association) 30–3.

[34] For a general account of the Stuxnet attack, see William J Broad, John Markoff, and David E Sanger, "Israeli Test on Worm Called Crucial in Iran Nuclear Delay," *The New York Times*, January 15, 2011. For a more detailed account of the man-in-the-middle tactic, see Geoff McDonald, Liam O Murchu, Stephen Doherty, and Eric Chien, "Stuxnet 0.5: the Missing Link" (*Symantec Corporation White Paper*, February 26, 2013), at <http://www.symantec.com/content/en/us/enterprise/media/security_response/whitepapers/stuxnet_0_5_the_missing_link.pdf> accessed July 10, 2014.

No other weapons seem to be quite like this—a fact that has not received emphasis in the literature so far as I can tell. Kinetic weapons work in a fashion that we might call unmediated and heterogeneous. If one dropped a bomb on or launched a missile at a facility such as Natanz, the destruction would be essentially direct, and the projectile itself would bear little or no similarity to the thing destroyed. Chemical, radiological, and especially biological weapons seem to occupy a middle ground, being sometimes less heterogeneous than kinetic weapons. Cyberweapons always function, at least initially, in a mediated and homogeneous way: they are cyber things designed to affect other cyber things directly even if the ultimate intended effect does not occur in the cybersphere. (This is not to contradict what was said earlier about cyberweapons' independence of *human* intermediaries compared with, say, the memes produced on a printing press.) For example, Stuxnet code directly affected other code to achieve an indirect kinetic consequence. If this distinction among cyber versus other sorts of weapons does not appear intuitively true, one might consider a four-part Aristotelian account: what Aristotle would call a missile's "material cause," the stuff of which it is made, would usually differ from that of its target, as would the formal, final, and probably efficient causes.[35] By contrast, cyberweapons and their immediate targets would seem to share at least a common material cause.

The mediated and homogeneous nature of cyberweapons bolsters their perceived uncontrollability in obvious ways. They are easy to hide and hard to trace:[36] witness how long Stuxnet went undetected and how hard it is to attribute despite what might have been built-in calling cards. We naturally tend to use the terminology of disease in talking about some cyberweapons and their effects—"virus," "infection," "worm" (a parasite), and the like—because we think of cyber attacks as being things that might be able to perpetuate themselves without continued human intervention. Not that any particular cyberweapon *will* do so any more than a given strain of flu will survive over the summer; but malware such as Stuxnet can and has spread beyond its intended target.[37] Armies, navies, air forces, and the space forces of our imagination go into action on human orders, perhaps sometimes acting independently of the highest headquarters' wishes when communications go down, but still under human control at some level, even if that of the individual soldier or seaman or airman as they decide whether to attack, stay put, or retreat. But cyberweapons, like some biological agents, are simply *turned loose*. We assume its authors and users never intended for Stuxnet to attack Siemens machines outside of Iran, but such attacks did in fact occur.

[35] Aristotle, *Physics*, translated by Philip H. Wicksteed and Francis M Cornford (Cambridge: Harvard University Press, 1957), Bk II, Sec 3, 194b–195a, 128–31; *Metaphysics*, translated by Hugh Tredennick (Cambridge: Harvard University Press, 1957), Bk V, Sec 2, 1013a–1013b, 210–14.

[36] Chacos, "Meet Darknet, the Hidden, Anonymous Underbelly of the Searchable Web" (n 44).

[37] See Broad et al (2011) regarding the apparently unintended spread of the Stuxnet worm.

V. Neoreality: The Increasing Prominence of Cyber in Everyday Human Experience

By the late twentieth century it appeared that what had begun as the Defense Advanced Research Projects Agency's (DARPA's) humble attempt to network academic researchers, now the mighty internet, had already defined an era and changed discourse by encouraging us to think of the world in terms of the real and the virtual.[38] DARPA worked with other institutions to create ARPANET, the internet's predecessor, in the late 1960s, proving the concept of a packet-switched network. Simulations of reality predated the internet, of course, but many ways of simulating reality through non-cyber means—in a film studio's special effects departments, for example—became obsolete when digital tools became more sophisticated. These tools relied on immense amounts of data.[39] In itself the distinction between the real and virtual might seem harmless enough; there is no apparent reason to find the bifurcated worldview much more problematic than a pre-cyber outlook, morally speaking, though the doubling of reality cannot help but present legal challenges in the form of privacy violations and the like. Or so one might think.

Baudrillard imagined what would happen—what *is* happening, he would say—if a virtual realm did not just complement the formerly accepted real world but came to supplant or even subsume it in the cultural consciousness. He called this increasingly dominant reality the simulacrum—not synonymous with but certainly augmented by cyber.[40] From a phenomenological point of view, Baudrillard's vision seems very possible: what we experience as primary need not flow wholly from the "real" dimension in the real–virtual distinction. Phenomenologically it is quite possible to devote increasingly more of oneself to the experience of a reality filtered through others' consciousnesses or even through realities that are generated independently of human authorship (a facet of the uncontrollability already discussed). Confirming Baudrillard's observation of the simulacrum's dominance, technology such as Google Glass makes the recording, replaying, and sharing of idiosyncratic experience easier. The record created by Google Glass becomes a kind of super memory—a higher-fidelity sense of the past than would have been possible in the absence of the technology and its perpetuation in cyberspace. One can create the parallel world of a record of one's experiences, and then by returning to the record, experience the experience while simultaneously keeping another parallel record of that present experience of past experience, and so on, like the mirror and eye mutually reflecting one another ad infinitum in an Escher lithograph. This does not

[38] Jaron Lanier often receives credit for originating or popularizing the term "virtual reality." Jennifer Kahn, "The Visionary," *The New Yorker*, July 11, 2011.

[39] Sonya Shannon sees virtual reality techniques not just as tools available to the artist but as a kind of fusion of art and science: "virtual reality can be deciphered as an art in its own right, namely, the art of science made manifest.... If art is the creation of forms expressive of human feeling, then virtual reality expresses a profound human passion to understand nature in terms of logical abstractions and reiterative, verifiable formulas—nothing other than the quest of many a scientist." Sonya Shannon, "The Chrome Age: Dawn of Virtual Reality" (1995) 28 *Leonardo* 379.

[40] Jean Baudrillard, "The Precession of Simulacra," in *Simulacra and Simulacrum*, translated by Sheila Faria Glaser (Ann Arbor: University of Michigan Press, 1994) 1–42.

exactly contradict Heraclitus, but now one can listen and watch as one steps, over and over again, into the recorded facsimile of a river. None of this is utterly new. Memories have been around as long as humans. Neither is the journal, or diary, new and journal entries can, like Google Glass recordings, be expanded indefinitely and self-referentially and reviewed an indefinite number of times.[41] Cyber provides a quantum leap in the possible *volume* of recorded experience, its *continuity*, and the relative *passivity* of recording. These are so striking that one is tempted to heed Baudrillard's fear that the simulacrum may be edging out unfiltered reality as the primary object of human experience.

In war, the moral character of manipulating perception depends on the context, hence the distinction between mere deception (legally wearing camouflage and face paint, say) versus perfidy (illegally pretending that a weapons depot is a hospital by marking it with large red crosses or crescents, for instance).[42] This is not new. Sun Tzu seems to accept the distinction when he discusses various kinds of deception as legitimate aids in war-making, implying some sorts of deception might be illegitimate.[43] In peacetime under threat of cyber violence, however, the simulacrum as Baudrillard conceived it perhaps indicates a different tension—not between deception and perfidy but rather among dominant realities. What primarily engages everyman's attention? Is it an unmediated reality or a world created in or at least filtered through a cyber lens? It is hard enough to sift competing accounts of war's activities, even of good wars and even among nominal allies, even when war is going well[44]; it is much more difficult to parse a cybersphere in which we fear concealed threats are ubiquitous but do not know how to confirm their presence, let alone their precise nature and effect. Cyberwar is not unique compared with other means or intentions of war from the perspective of the JWT, but cyberwar *is* morally special in a cultural sense. At least in the present moment we tend to fear cyber because of the factors already discussed—its ubiquity and uncontrollability. The fact that à la Baudrillard the real and the virtual are increasingly blurred into a simulacrum, or put in less sinister fashion, that we now spend more of our time experiencing cyber things than we used to, means threats to our cyberworlds seem increasingly ominous.[45] The threat is less to

[41] Literary examples include Henry David Thoreau, *Walden* (Princeton: Princeton University Press, 2004) and Michel de Montaigne, *The Complete Essays*, translated by MA Screech (London: Penguin Books Ltd, 2003). Tolstoy wrote multiple perspectives in *Anna Karenina* long before Baudrillard reflected on the multiple views of the same car crash. Leo Tolstoy, *Anna Karenina*, translated by Constance Garnett. Proust's narrator could find in a madeleine, a small French cake, a potent conduit to the past. Marcel Proust, *In Search of Lost Time*, vol. 1, *Swann's Way*, translated by CK Scott Moncrieff (1913) Overture.

[42] Protocol Additional to the Geneva Conventions of August 12, 1949, and relating to the Protection of Victims of International Armed Conflicts (Protocol I), June 8, 1977.

[43] Sun Tzu, *Sun Tzu*, translated and commentary by JH Huang (New York: William Morrow, 1993) 40, 60. (This text is usually entitled "The Art of War," but Huang chose to use Sun Tzu's original, eponymous title.) See Tzu, *Sun Tzu*, 136–7 on the possibility that some kinds of deception are unethical.

[44] A startling instance of internecine disagreement on what happens in war involves the Oscar-winning documentary film of the end of World War II, *The True Glory*. See Frederic Krome, "*The True Glory* and the Failure of Anglo-American Film Propaganda in the Second World War" (1998) 33 *Journal of Contemporary History* 21–34.

[45] For example, in the opinion of former Secretary of Defense Panetta (n 26); of former security advisor and now analyst Richard Clarke, q.v.*Cyber War The Next Threat to National Security and What to Do About It* (London: Harper Collins 2010); of virtually any government interested in protecting its national

a means of pleasantly passing the time or working efficiently (the former uses of the internet) than to the dominant stuff of life. Because of the homogeneity and mediated character of cyber attacks (discussed in section IV), cyberwar would threaten the simulacrum even if defensive measures such as air-gapping power plants succeed and become the norm. Moreover, the increasing dominance of the simulacrum in our perception makes it simultaneously more important in our lives and any threat to it that much more intimidating. If the extent of cyber in our culture amounted to a video game or two, or an obscure network used only by scientists at research universities, the threat of cyberwar would not loom so large in the collective imagination. But cyber plays an enormous and increasingly widely recognized role in our culture—as discussed in section III on ubiquity—so the threat of cyberwar intrudes more and more into our thinking about the present and future.

That makes cyberwar a topic of serious moral concern, but in fact cyber affects us even more profoundly than most of us probably realize. Our lives would not necessarily be intolerable if some proportion of cyber infrastructure were devastated—if information transfer capacity were wiped out by an attack on servers, say, affecting air traffic control, banking, voice and data communications, and so on. Even smaller scale destruction—for example, of individual power plants' ICSs—would be inconvenient to millions and life-threatening to a subset of those. But life would go on. Still, the deaths and disruption could occur on a scale we associate with war rather than crime or vandalism, so it is reasonable that we should ask ourselves how to prevent or at least control the death and suffering that cyberwar could cause. To that end we have a precedent that several analysts have recognized.

VI. Reviving the "Ethics of Threat"

Why "reviving"? Because of numerous, worrisome indications that the world is drifting away from its hard-won, Cold War era ethics of threat, and because of the fact that we have barely begun to apply Cold War methods to cyberwar. In the realm of weapons of mass destruction (WMD), we see evidence of inattention to controlling the development and proliferation—for example, in Iran's alleged attempts to acquire nuclear weapons,[46] North Korea's nuclear saber rattling,[47] and Russia's alleged violations of the 1987 Intermediate-Range Nuclear Forces Treaty.[48] But our focus here is on the cybersphere. We have seen that taken together the ubiquity, uncontrollability, and neoreality of cyber in our cultural consciousness magnify the threat that cyberwar poses. Not that cyberwar is merely a threat; there have been actual attacks that the victims could have chosen to call casus belli—for example, Stuxnet and a Reagan era attack on automation facilitating the

industries, q.v. Ellen Knickmeyer, "After Cyberattacks, Saudi Steps Up Online Security," *The Wall Street Journal*, August 26, 2013, at <http://blogs.wsj.com/middleeast/2013/08/26/after-cyberattacks-saudi-steps-up-online-security/> accessed 10 July 2013.

[46] James M Lindsay and Ray Takeyh, "After Iran Gets the Bomb: Containment and Its Complications" (2010) 89 *Foreign Affairs* 33–49.

[47] Anna Fifield, "North Korea Conducts New Drills, Raising Tensions with South Korea," *The Washington Post*, July 14, 2014.

[48] Michael R Gordon, "U.S. Says Russia Tested Missile, Despite Treaty," *The New York Times*, January 29, 2014.

transport of Soviet-produced natural gas.[49] However, those attacks were relatively isolated. Others—the DOS assaults on Estonian banking,[50] for instance, or the cyber-based espionage that netted China the plans of "more than two dozen major [US] weapon systems"[51]—seem unlikely to be deemed acts of war given cultural traditions distinguishing espionage from war. But if cyberweapons have never been a standalone means of waging war, they are a potential facilitator of war and, therefore, a constant threat. Presumably there is not much that can be done about the use of cyber as a peripheral tool of war besides trying to use existing protocols—for instance, to slow the militarization of space.

The *threat* of cyberwar is another matter. As a looming menace, cyberwar shares at least two characteristics with the nuclear threat that has shaped our culture since Hiroshima—ubiquity and uncontrollability, qualities of cyber more generally. From duck-and-cover drills for school children to bunkers deep under backyards, the crescendo of nuclear consciousness mirrored the acme of perceived threat.

Implied in the JWT's ad bellum principle of last resort is the notion that threatening war can be a means of preserving peace. Of course the threat must be accompanied by diplomatic measures meant to address grievances or other factors that could lead to war. Similarly, the principle of just cause, including the notion of when preemptive war might be necessary, must take implicit account of fear and threat. Other treatments of war's causes are more explicit in their recognition of threat and fear. For instance, Hobbes's 1651 taxonomy of war's causes (following Thucydides)—"competition, diffidence [i.e., suspicion or fear], and glory"[52]—confirms the observation that a casus belli is chosen against a backdrop of cultural norms. When it comes to fear, for example, perception—the fear itself, in Franklin D Roosevelt's phrase[53]—can trump any fact of the matter that is supposedly objective in the sense of transcending culture. The same is true of Hobbesian diffidence and glory.

Even if the efficacy of threat is implied in just war principles such as last resort, cultural (social-psychological) facts—facts of perception—are not explicitly mentioned.[54] Perhaps the JWT's focus on effects and intentions has left little room for attention to the *perception* of threat and the real harm that threats alone can do to their recipients. That might be because much of the JWT developed in eras characterized by a more or less simple ontological realism—before Cartesian dualism, before idealism's noumenal-phenomenal split, before psychology appealed to a divisible inner reality, and certainly before phenomenology emphasized the primacy of experience rather than

[49] Tim Weiner, *Legacy of Ashes* (New York: Anchor Books, 2008) 447–8.

[50] For details of the attack, see Jason Richards, "Denial-of-Service: The Estonian Cyberwar and Its Implications for U.S. National Security," *International Affairs Review*, at <http://www.iar-gwu.org/node/65> accessed July 11, 2014. For the reason the cyber attack in Estonia was an important precedent, see joywang (blogger identity), "The 2017 Estonian Cyberattacks: New Frontiers in International Conflict," *Harvard Law School Weblogs*, December 21, 2012, at <https://blogs.law.harvard.edu/cyberwar43z/2012/12/21/estonia-ddos-attackrussian-nationalism/> accessed July 11, 2014.

[51] Ellen Nakashima, "Confidential report lists U.S. weapons system designs compromised by Chinese cyberspies," *The Washington Post*, May 27, 2013; Associated Press, "Chinese man charged with hacking into US fighter jet plans," *The Guardian*, July 11, 2014.

[52] Thomas Hobbes, *Leviathan* (New Haven: Yale University Press, 2010) Part I, Ch 13.

[53] Franklin Delano Roosevelt, First Inaugural Address.

[54] See, for example, Brian Orend, "A Sweeping History of Just War Theory," in *The Morality of War* (Peterborough: Broadview Press, 2nd edn, 2013) 9–32.

theoretical models in establishing what constitutes reality. What mattered most to the just war reflections of Augustine and Aquinas, for instance, had little explicitly to do with social psychology, even though there is something in the prospect of a Christian Roman empire beset by hostile armies that smacks of threat and the psychology of threat, and even though matters of individual psychology are prominent (as in the "mournful" stance of Augustine's Christian warrior). So the ontogeny of the JWT might explain why the ad bellum and in bello criteria make no direct mention of threat or fear.

During the Cold War, however, numerous voices pointed out the unprecedented power of atomic weapons and wondered about the psychological effects of the looming danger. So it was natural for those most attuned to the JWT to turn their attention toward fear and threat to a degree that was somewhat new in just war dialogue. Groups of scientists (most prominently the Union of Concerned Scientists), intellectuals, and artists coalesced around fear of nuclear Armageddon and a range of potential second-order effects such as nuclear winter. Especially emphatic was the US Catholic bishops' episcopal letter of 1983.[55] The bishops pose the key question, "May a nation threaten what it may never do?"[56] For instance, may a nation explicitly or implicitly (e.g., through publicized development and deployment) threaten to use nuclear weapons to devastate another nation's or non-state group's infrastructure—even if only its electrical grid, water supply, sanitation systems, transportation networks, and so on?

Obviously this was not the first era in which such questions had been posed in one form or another. Thucydides' account of the Athenians' threat against Melos raises the questions of whether it is indeed just a threat, and whether a threat to do something horrendous (albeit then culturally acceptable) would itself be immoral. In 1599 Shakespeare revisits the gates of Harfleur c. 1414 so his Henry V may threaten rape, pillage, and general mayhem unless the city capitulates.[57] Later Henry threatens to murder prisoners at Agincourt unless the French forces either attack or yield the field.[58] But the bishops' late Cold War statement is still hugely important historically because it goes against the cultural grain in a very self-conscious way. In their careful consideration of the arms race, its origin, and its purpose, the bishops admit that having a huge, hair-triggered nuclear arsenal paradoxically may prevent nuclear war and might even prevent other conflicts (while draining precious resources that could be used for the common weal). Yet the bishops still worry over the culture of threat—not just the indiscriminate nature of nuclear weapons if used but the indiscriminateness of their all-penetrating presence in society and the angst they cause. It is not just the attack that matters; it is also the continual threat of attack that makes life anywhere on the planet akin to sitting on Damocles' throne and, therefore, always under the sword.[59]

Much fine theoretical work was done in the face of the nuclear threat. Gregory Kavka, for instance, shared the US Catholic bishops' concern about the morality of

[55] The National Conference of Catholic Bishops, *The Challenge of Peace: God's Promise and Our Response: A Pastoral Letter on War and Peace by the National Conference of Catholic Bishops*, May 3, 1983.

[56] The National Conference of Catholic Bishops, *The Challenge of Peace* (n 55) paras 137, 221.

[57] William Shakespeare, *The Life of King Henry V*, Act III, Scene 3.

[58] Shakespeare, *Henry V*, Act IV, Scene 7.

[59] National Conference of Catholic Bishops, Bishops, *The Challenge of Peace* (n 103) paras I.C.1., 106.

threatening to target mass numbers of non-combatants.[60] How could it make moral sense to threaten nuclear counterattack that would violate the principle of discrimination? Can we even coherently intend to cause harm of such magnitude? Kavka's answers to these questions turn on a utilitarian comparison of alternatives. As regrettable as it is that one must maintain a large nuclear arsenal and threaten to use it after suffering a nuclear first strike, the risks of unilateral disarmament or of demonstrating an unwillingness to use nuclear weapons are too great. So long as the United States seriously sought bilateral disarmament, Kavka suggested, the threat of nuclear counterattack was the least of evils.[61] David Gauthier's pragmatic justification for nuclear retaliation took a somewhat different tack, allowing that actual use of nuclear weapons might result in so much destruction that one could not justify it on utilitarian or pragmatic grounds. But according to Gauthier, one might still reasonably threaten to counterattack with nuclear weapons if the risk of doing so were small enough and if the threat otherwise created the best state of affairs. So it is *not* the case that "threat behavior is never rational"[62]—a point he had already made eight years earlier.[63] These and other potent ways of thinking through the ethics of threat continue to be refined. Claire Finkelstein, for example, points out a possible way of making Gauthier's pragmatic defense of plans (of which threat is a species) more coherent by rethinking his threat-assurance dichotomy.[64] This seems an especially promising tack in thinking about cyberwar, where the number of simultaneous threats and their results if actually carried out far outnumber the comparatively simple schematics of nuclear threats, attacks, and counterattacks.

The cyber-angst that seems now to be growing has much in common with the culture of foreboding that characterized the Cold War and continues to raise its head as North Korea rattles sabers and silos. Perhaps any cure will also have much in common with Cold War era measures, especially arms control agreements.[65] Given the potential dangers of cyberwar indicated in common catchphrases such as "cyber Pearl Harbor," diplomacy needs to catch up and, with luck, overtake cyberwar technologies by creating protocols for limiting some cyber activities. The danger is the automaticity of technology and of the processes by which that technology is used. If those processes could somehow be brought to heel by existing treaties the world would be a safer place from cyber as well as nuclear attack.

The "if" is the rub, of course. Effective treaties governing cyberweapons presuppose a sustained, rational multilateralism among tech-savvy players. Although I have used the network of treaties and protocols during the Cold War as an example of how a new ethics of threat could be built up around cyberweapons, we now know that the world nearly teetered into a nuclear exchange between a superpower and a superpower

[60] Gregory S Kavka, *Moral Paradoxes of Nuclear Deterrence* (Cambridge: Cambridge University Press, 1987).

[61] Kavka, *Moral Paradoxes of Nuclear Deterrence* (n 102) 79–99.

[62] David Gauthier, "Assure and Threaten" (1994) 104 *Ethics* 690, 713.

[63] David Gauthier, *Morals by Agreement* (Oxford: Oxford University Press, 1986) 99–100.

[64] Claire Finkelstein, "Pragmatic Rationality and Risk" (2013) 123 *Ethics* 673–99.

[65] That is the most important message (along with a call for attention to cyber defense) of Richard Clarke and Robert Knake, *Cyber War The Next Threat to National Security and What to Do About It* (New York: Harper Collins, 2010) sec 8. For the US Department of Defense's intention to emphasize cyber defense (because cyber deterrence depends on accurate attribution, currently a technical problem), see Timothy Farnsworth, "Pentagon Issues Cyber Strategy" (2011) 41 *Arms Control* 37–8.

surrogate.[66] Former Secretary of Defense McNamara implies that the near-nuclear exchange between the United States and Cuba was of a piece with the irrational parochialism that led to the failure of the League of Nations in general and the failure of the US Senate to ratify Article X of the Covenant of the League (the article establishing a mutual defense agreement among member states in case of attack).[67] A dark interpretation of McNamara's reflections might suggest that Wilsonianism simply is not palatable to self-interested nations (and perhaps equally odious to non-state groups), as evidenced by World War II and a near nuclear exchange between Cuba and the United States. If that is the case then the Cold War ethics of threat has not worked; the world has simply been lucky.

Assuming effective treaties *might* be put in place, however, the common sense and good will of rational agents would work best proactively rather than reactively. Hence, the forcefulness of Clarke and Knake's call for an arms control like approach to cyberweapons. The first step toward helpful changes like treaties or modifications in the realm of international customary law is the moral condemnation of indiscriminate threat exemplified by the US Catholic bishops' 1983 letter. The next step would be concrete resolutions such as an agreement to use Internet Protocol version 6 and other technologies to make it more difficult to launch cyber attacks *anonymously*.[68]

Having found an apparent analogy between cyberweapons and nuclear weapons, it makes sense to look for disanalogies too, in order to ensure we do not unwittingly stretch the analogy further than it should go. One important difference between the nuclear weapons during the Cold War and cyberweapons as commonly conceived is that nuclear weapons were by definition weapons of mass effects and, therefore, necessarily indiscriminate. By contrast, at least some cyberweapons might be discriminate. During the Cold War nuclear weapons were to be employed for decisive effects measured in big numbers—to kill many, to destroy a lot. Unlike the highest-profile military technologies of our decade—cyber and unmanned aerial vehicles (UAVs)—potential actors in war, especially nation-states, will not be tempted to resort to nuclear weapons rather than diplomacy to settle disputes. Put another way: nuclear weapons and other WMD tend to drive the threshold for war *up* rather than down. By contrast cyber and UAVs arguably drive the threshold of military action down. That lower bar can tempt actors away from diplomacy and toward military attack. Cyberweapons can be expensive—certainly Stuxnet must have been costly to develop—but even the most elaborate cyberweapons would not *necessarily* be devastating in their effects. Secretary Panetta's "cyber Pearl Harbor" is unsettling, but I have yet to hear the term "cyber winter." So if over the next decade or so cyber attacks turn out to be as discriminate as Stuxnet—not perfect, but still pretty good—it seems reasonable to think the specter of cyberwar will disappear from the public imagination. We might begin to believe a post-cyber attack world would be less convenient and less entertaining, but still nothing like whatever a nuclear holocaust would leave behind.

[66] Robert S McNamara and James G Blight, *Wilson's Ghost: Reducing the Risk of Conflict, Killing and Catastrophe in the 21st Century* (New York: Public Affairs, 2001) 188–91.

[67] McNamara and Blight, *Wilson's Ghost* (n 66) 6–9.

[68] See Libicki *Cyberdeterrence and Cyberwar* (n 37) 43 n 5; Patrick Lin, Fritz Allhof, and Neil Rowe, "Is It Possible to Wage a Just Cyberwar?" *The Atlantic Monthly*, June 5, 2012.

This disanalogy between the Cold War context and our cyber era may make it difficult to move forward from moral condemnation of indiscriminate uses of cyberweapons to legally binding limitations such as cyber treaties. But, no matter. A diplomatic ounce of prevention is still worth trying, and there seems to be no better alternative than Clarke and Knake's prescription. Perhaps we will find the disanalogy of effect less important than it might now seem. After all, relatively late in the Cold War there emerged so-called "tactical nuclear weapons" designed for small-yield blasts compared with their strategic big brothers. These tactical nuclear weapons were, therefore, supposed to be capable of greater discrimination if used properly and in the right context. Rather than aiming at a large metropolitan area for a so-called countervalue strike, a tactical nuke (or a series of them) might be used to stop or delay a column of armor charging through accommodating terrain toward a high-value objective. Diplomacy of the time—diplomacy being by definition a pursuit that seeks out and exploits nuance—could deal separately with tactical and strategic nuclear weapons. Presumably the same would be true of cyber weapons and attempts to limit their development, proliferation, and use: even if they are not necessarily devastating means of war, at least they can be destructive enough to merit limitations spelled out in treaties.

Though elaborate cyberweapons such as Stuxnet are bound to be expensive, their overall cost of development, maintenance, and deployment would appear to be significantly lower than that of most other weapons. (Some biological weapons may unfortunately prove to be quite cheap as well.) Certainly it should be less expensive to develop a cyber capability than a nuclear arsenal and the means to deploy it. That suggests the cyber battlefield could have many more players than most other military arenas. This is another aspect of cyber's ubiquity and uncontrollability discussed in this chapter. Never mind, then, that nuclear weapons may be more dangerous in fact than most cyberweapons in most contexts. Cyberweapons are potentially so numerous, and the possibility of cyber attack is, therefore, so great, that it makes good sense to put in place protocols that limit cyber-borne violence.

Further disanalogies will make reviving the ethics of threat still more difficult and require additional adjustments. Strategic nuclear weapons are not just relatively massive in their effects and cost relative to some cyberweapons; nuclear weapons are also hard to hide, and their use requires a constellation of purposeful, sustained development efforts to produce not just the warheads but also the means of delivering them, storing them, and so on. By contrast, cyberweapons are highly portable and easily hidden.

Furthermore, many *offensive* cyberweapons will emerge more or less automatically from the hard work of cyber defense. The process of combing through lines of software to find one's own vulnerabilities may all but design offensive weapons to use against other nations and groups using the same software. Of course the degree of effort required and the price will depend on the nature of the offensive cyberweapon. We can guess that weapons aimed at simply rendering a cybersystem inoperable by destroying it or using up its resource will be easier to produce and deploy than cyberweapons which try to penetrate well beyond a user interface without destroying it. Stuxnet is a good example: the malware did not destroy the functionality where it entered but instead manipulated the ICS so that the centrifuges, not the ICS itself, would be destroyed. The Stuxnet attack also included a "man-in-the-middle" tactic to fool operators into believing the centrifuges were running normally.

Finally, it is unclear whether and how deterrent threats might work in the cyber realm.[69] Traditional means of delivering nuclear weapons—for example, silos, runways to accommodate long-range penetrating bombers, submarine docks—are far more difficult to hide than the means of launching cyber attacks.[70] Counterforce cyber attacks are, therefore, proportionately more difficult to deter than their nuclear analogs.[71] Countervalue cyber attacks could range from the mild (cellphone service interrupted) to the life-threatening (power out across an urban area), but at their worst they seem less dramatic than a nuclear attack on, say, a large urban area. They are also less likely to be effective if threatened in advance. This link between efficacy and secrecy in cyberweapons is especially evident in "zero-day" vulnerabilities and corresponding weapons.[72] Air-gapping a power plant to defend against malware is easier than hardening a city against blast, fire, electromagnetic pulse, and the other results of a nuclear detonation. One sort of air gap is a so-called sneakernet, in which data are transferred via standalone storage devices such as thumb drives, keeping the data source and recipient wholly isolated from each other.[73] The most expensive part of such a system is the cost of using humans to transfer the storage device across the air gap. Within the air-gapped complex, however, data connections are optimized. The result is slow data transfer but relatively cheap, somewhat reliable security. By comparison, Iraq's Saddam Hussein spent $60 million in "the early 1980s" for a nuclear detonation-proof bunker.[74] So clearly much of the practical diplomacy of the Cold War, as well as the theoretical work of thinkers such as Kavka and Gauthier, might not apply easily to a cyber future where arsenals emerge organically and secretly, and where threats are made only in the most general terms. For example, Gauthier never assumes we can always carry out a plan or make good on a threat; in fact, "[t]his is an unrealistic supposition in most situations."[75] But some of the good Cold War era work on the ethics of threat should be immediately applicable. In any case, there seems to be no better alternative to lessen the real danger and perceived threat of cyberwar.

[69] Libicki, *Cyberdeterrence and Cyberwar* (n 37) 8.

[70] Libicki, *Cyberdeterrence and Cyberwar* (n 37)16.

[71] On the range of asymmetric threats to cybersecurity in the US, see *Department of Defense Strategy for Operating in Cyberspace*, July 2011, 3, at <http://www.defense.gov/news/d20110714cyber.pdf> accessed July 11, 2014.

[72] See Ryan Gallagher, "Cyberwar's Grey Market," *Slate*, January 16, 2013, at <http://www.slate.com/articles/technology/future_tense/2013/01/zero_day_exploits_should_the_hacker_gray_market_be_regulated.html> accessed July 12, 2014.

[73] Cory Janssen, "Air Gap," *Technopedia*, at <http://www.technopedia.com/definition/17037/air-gap> accessed July 12, 2014.

[74] Jaime Holguin, "Saddam's Nuke-Proof Bunker," *CBS News*, March 30, 2003, at <http://www.cbsnews.com/news/saddams-nuke-proof-bunker/> accessed July 12, 2014. For the downsides of air gaps as a means of security, see PW Singer and Allan Friedman, *Cybersecurity and Cyberwar: What Everyone Needs to Know* (Oxford: Oxford University Press, 2013) 63.

[75] David Gauthier, *Morals by Agreement* (Oxford: Oxford University Press, 1986) 200.

VII. Conclusion

In our present culture, cyber's perceived ubiquity, reputation for uncontrollability, and apparently increasing role in our daily lives make cyberweapons morally different from other tools of war. Yet cyber is not unprecedented in the prominence of its cultural profile, as I have tried to show by comparing the cyberthreat with nuclear threats. That analogy, imperfect though it is because of several differences—for example, in the magnitude of effect, degrees of possible discrimination, cost of developing and maintaining military capability, and likely number of players—should help us transition from a descriptive study to a robust normative agenda that includes backing away from some kinds of cyberweapons as nations have from other weapons (e.g., poison gas, expanding bullets, white phosphorous, and even firearms in one instance[76]). For example, nations could mandate that some cyberweapons be engineered such that their effects would be immediately reversible,[77] although such a rule could not apply to cyberweapons like Stuxnet that achieve a kinetic effect. The nature of the mandate and the best means of enforcement would be matters of debate.[78] We cannot know how a robust ethics of threat will look, say, ten years from now. Cyber could have advanced enough to require treaties and other protocols we cannot imagine from our present vantage point. What we do know is that the numerous Cold War groups who worked for limitations on nuclear weapons and their use did not let threat paralyze them; they did what they could do, with energy and optimism. We should follow their lead as the cybersphere expands and the threat of cyber attacks grows.

[76] On the prohibition on poison gas, see Declaration (IV, 2) concerning Asphyxiating Gases, The Hague, July 29, 1899. See also Protocol for the Prohibition of the Use of Asphyxiating, Poisonous or Other Gases, and of Bacteriological Methods of Warfare, Geneva, June 17, 1925. On limitations on the use of expanding bullets, see Declaration (IV, 3) concerning Expanding Bullets, The Hague, July 29, 1899. Although not prohibited per se by international humanitarian law, white phosphorous as an incendiary weapon is controlled under the Protocol on Prohibitions or Restrictions on the Use of Incendiary Weapons (Protocol III), Geneva, October 10, 1980. For a brief period Japan officially outlawed all firearms. See Jared Diamond, *Guns, Germs, and Steel: The Fates of Human Societies* (New York: WW Norton, 1999) 257–8. However, the Law of Armed Conflict may be difficult to apply to cyberwar even if international legal authorities agree that cyberwar falls under LOAC. See Michael N Schmitt (ed), *The Talinn Manual on the International Law Applicable to Cyber Warfare* (Cambridge: Cambridge University Press, 2013) 68–70, para 9.

[77] Lin et al, "Is It Possible to Wage a Just Cyberwar?" (n 111), Sec 6.

[78] For an argument against immediately trying to establish formal treaties governing cyberweapons, see Jeffrey TG Kelsey, "Hacking into International Humanitarian Law: The Principles of Distinction and Neutrality in the Age of Cyber Warfare" (2008) 106 *The Michigan Law Review* 1427, 1449–50.

3

Cyber Causation

Jens David Ohlin

I. Introduction

The law of war has generally ignored causation. This is not because causation is irrelevant for jus in bello and jus ad bellum, but more likely because causation is usually uncontroversial during war. To put it simply: one side drops a bomb and the building blows up; the individuals inside are injured or killed. The law of war produces a dense fog of conceptual and theoretical problems, but causation is not one of the problems generating any particular urgency. The one exception is the problem of over-determination, or the idea that multiple causes could each be sufficient to generate a particular result, as in a firing squad with multiple bullets. Although the law of war is generally unconcerned with the problem of over-determination, various doctrines of international *criminal* law are potentially undermined by over-determination, or the possibility that the crime (or some similar crime) would have occurred even in the absence of the defendant's actions. Although international courts have not offered a satisfactory answer to this dilemma, the issue is at least on the radar of international criminal law scholars.[1]

In general, though, the law of war has usually remained blissfully ignorant of the problems of causation—problems that rise to the level of full-blown obsession in domestic tort and criminal law. Specifically, doctrines in these disciplines, with varying degrees of coherence, have grappled with the required causal connection for civil liability or criminal punishment, and whether (or when) third-party actions are sufficiently disruptive to negate liability. But the law of war has so far ignored this problem because it had the luxury to do so.

This chapter argues that the increasing threat and deployment of cyber-weapons will force (or should force) the law of war to develop a sophisticated and nuanced account of causation. Section II will explain in greater detail why causation is largely irrelevant (or at the very least uncontroversial) to the basic structure of traditional international humanitarian law (IHL). Section III will then introduce various scenarios of cyber attacks that will trigger immense pressure on IHL to develop an account of causation. I will place less emphasis on which account of causation is *abstractly* correct and will confine my remarks to supporting the proposition that cyber attacks implicate the concept of causation in previously unseen ways and place immense pressure on the

[1] See, for example, J Stewart, "Overdetermined Atrocities" (2012) *Journal of International Criminal Justice* 1189–218; CF Stuckenberg, "Problems of 'Subjective Imputation' in Domestic and International Criminal Law" (2014) 12 *Journal of International Criminal Justice* 311.

adjudication of causation. The result is the emergence of a primary research agenda for IHL at both levels: theory (scholarship) and codification (via state practice and potential treaty provisions). Finally, section IV will explain why some traditional theories of causation cannot be reflexively and uncritically grafted into IHL. Simply put, IHL demands a level of publicity and transparency that generates a significant asymmetry as compared to other fields of domestic law, where the fact-finding machinery of domestic courts is more suited to parsing complex causal phenomena. By deploying George Fletcher's famous distinction between the pattern of subjective criminality and the pattern of manifest criminality, I will show that the former is appropriate for the criminal law's extensive fact-finding system, while IHL must be built, out of necessity due to the lack of fact-finding resources, from the raw materials of the pattern of manifest criminality. Cyberwar presents an especially acute case of this general phenomenon within IHL; the causal processes of a cyber attack and its downstream consequences are difficult to chart, thus suggesting that the law governing cyberwar should place a premium on transparent rules that, like the pattern of manifest criminality, can be applied by a reasonable third-party observer.

II. Causation and Traditional IHL

The basic building blocks of IHL are relatively simple: The principle of distinction forbids attacking forces from deliberatively targeting civilians; only soldiers and military assets may be the object of attack.[2] The principle of proportionality sanctions collateral damage of civilians and civilian assets, unless the damage to civilians is disproportionate to the anticipated military advantage.[3] The principle of necessity licenses the killing or disabling of military personnel that hasten the resolution of the war; alternate formulations or interpretations of the principle limit the killing of soldiers to situations where they cannot be reliably captured or disabled as a means of bringing them hors de combat.[4] The principle of humanity requires that attacking forces take precautions to reduce civilian casualties and to reduce unnecessary suffering

[2] See Knut Ipsen, "Combatants and Non-Combatants," in D Fleck (ed), *The Handbook of International Humanitarian Law* (Oxford: Oxford University Press, 2nd edn, 2008) 79, 80.

[3] Protocol Additional to the Geneva Conventions of August 12, 1949, and relating to the Protection of Victims of International Armed Conflicts (Additional Protocol I), June 8, 1977, Article 51(5) ("Among others, the following types of attacks are to be considered as indiscriminate…(b) an attack which may be expected to cause incidental loss of civilian life, injury to civilians, damage to civilian objects, or a combination thereof, which would be excessive in relation to the concrete and direct military advantage anticipated.").

[4] For debate concerning the correct interpretation of the principle of necessity, *compare* Ryan Goodman, "The Power to Kill or Capture Enemy Combatants" (2013) 24 *European Journal of International Law* 819 (arguing for a least-harmful-means interpretation of the principle of necessity) and Nils Melzer, "Keeping the Balance between Military Necessity and Humanity: A Response to Four Critiques of the ICRC's Interpretive Guidance on the Notion of Direct Participation in Hostilities" (2010) 42 *NYU Journal of International Law & Policy* 831 (2010), with MN Schmitt, "Wound, Capture, or Kill: A Reply to Ryan Goodman's 'The Power to Kill or Capture Enemy Combatants'" (2013) 24 *European Journal of International Law* 855–61, JD Ohlin, "Recapturing the Concept of Necessity" (manuscript on file), and GS Corn, LR Blank, C Jenks, and ET Jensen, "Belligerent Targeting and the Invalidity of a Least Harmful Means Rule" (2013) 89 *International Law Studies* 536 (concluding that the least-harmful-means interpretation of necessity is not lex lata and also impractical).

amongst civilian and combatants alike.[5] Specific prohibitions outlaw particular methods of warfare or particular weapons: perfidy, treachery, landmines, chemical weapons, and so on. There is little room—or need—in this basic structure for the concept of causation.

The one major exception to this assessment is the rule permitting attacks against military objectives and prohibiting attacks against civilian objects. Causation matters for this legal norm. How are military objectives and civilian objects defined? According to Article 52(2) of Additional Protocol I:

> attacks shall be limited strictly to military objectives. Insofar as objects are concerned, military objectives are limited to those objects which by their nature, location, purpose or use make an effective contribution to military action and whose total or partial destruction, capture or neutralization, in the circumstances ruling at the time, offers a definite military advantage.[6]

Article 52 codifies a causal standard in two senses: the perspective of both sides of the conflict.[7] From the perspective of the defending force, the object must make some contribution to their military action. From the perspective of the attacking force, destroying the enemy object must confer a military advantage on the attacking force. The requirements are conceptually linked and may be viewed as two sides of the same coin: destroying the target confers a military advantage on the attacking force precisely *because* the target played some role in the enemy's military operations.[8]

The standard involves a causal contribution because the question remains: what level of effective contribution is required and how close a connection must there be between the destruction of the target and the military advantage conferred on the attacking force? Consider the following examples, all of which are minimally controversial under the Article 52 standard:

1. The air force bombs a munitions factory, since the ordinances manufactured there make an effective contribution to the enemy's military campaign. Without weapons, the enemy will be unable to fight.
2. The army uses artillery shells to destroy a mine and the miners within it. Metals from the mine will be used to manufacture weapons; without the metal, the enemy army will soon deplete its ammunition.

[5] GS Corn, "Principle of Humanity," *Max Planck Encyclopedia of Public International Law* (Oxford: Oxford University Press, 2013) ("The principle protects combatants from unnecessary suffering, and individuals who are no longer, or never were, active participants in hostilities by mandating that they be treated humanely at all times... The principle of humanity provides an essential counter-balance to the equally fundamental principle of military necessity.").

[6] Additional Protocol I (n 3) Article 52.

[7] Y Dinstein, *The Conduct of Hostilities Under the Law of International Armed Conflict* (Cambridge: Cambridge University Press) 84–6.

[8] M Sassoli and AA Bouvier, *How Does Law Protect in War? Case, Documents and Teaching Materials on Contemporary Practice in International Humanitarian Law* (ICRC, 1999) 161 ("In practice, however, one cannot imagine that the destruction, capture, or neutralization of an object contributing to the military action of one side would not be militarily advantageous for the enemy; it is just as difficult to imagine how the destruction, capture, or neutralization of an object could be a military advantage for one side if that same object did not somehow contribute to the military action of the enemy."), cited in Dinstein (n 7) 85 n 28.

3. The navy attacks a farm and a food processing plant; food arguably makes an effective contribution to the military effort since soldiers need food. Destroying the food confers a military benefit, since hungry and weak soldiers are ineffective.
4. The air force bombs an electrical grid power station, which provides electrical capacity to both civilian and military installations. Destroying the electrical station will hamper the ability of the enemy forces to adequately maintain command and control over their infantry on the ground.
5. Special forces destroy the enemy's state-run television station in order to eliminate the propaganda machine deployed by its ultra-nationalist president.

The Additional Protocol provides shockingly little guidance on how to resolve these cases. Nonetheless, the broad outlines of an answer can be inferred by applying the standard announced in Article 52. The key issue is the closeness of the causal connection along two axes: the advantage that the installation confers on the enemy and the advantage conferred on the attacking force by destroying it. In the first case, the munitions factory is one-step removed from regular military assets. Destroying an enemy tank or artillery position confers an immediate and direct benefit because the enemy is then unable to deploy the asset in its own attack. The munitions factory is one-step removed from this level of causation, since the factory *produces* the very assets that will one day be used in the military campaign. Although the causal connection between the factory and its military significance is attenuated, it is only one-step removed and most lawyers would agree that the causal connection is sufficiently close along both axes to render it a legitimate target.[9]

The mine example is more attenuated. The metals are necessary for the production of military assets, though only after one considers a two-step causal story: destroying the mine inhibits the factory production of weapons that in turn inhibit the capacity of the enemy forces. Even so, the *shape* of the causal connection is sufficiently direct to render the mine a legitimate target, despite the fact that the significance of the causal connection is further downstream than the munitions factory example.

The food example is much more controversial and strains the limit of our causal intuitions. The problem is not so much the remoteness of the causal connection, since food arguably stands in an analogous position to metals as two causal steps removed from the degradation of the fighting force. In that sense, the causal remoteness is not very far. One can say with sufficient predictability that food contributes to the enemy's military readiness, and that its destruction would confer a military benefit. Rather, the issue is with the nature and shape of the causal connection. The degradation of military effectiveness is accomplished *by* removing something that constitutes a basic human right: the right to food. That suggests an impermissible causal step, though it is not easy to articulate the nature of the objection here. One argument might be that it treats soldiers and their basic human need to sustenance as a mere means; the soldiers' right to be free from hunger is impermissibly hijacked by the goal of degrading the enemy's fighting capacity. This intuition seems plausible though it proves too

[9] See Janina Dill, *Legitimate Targets: Social Construction, International Law and US Bombing* (Cambridge: Cambridge University Press, 2014).

much, since soldiers are always subject to being used as a mere means to an end: killing them in order to destroy the enemy's fighting capacity. A more plausible expression of our anxiety over destroying food is that it hijacks an element of fighting which is utterly extrinsic to fighting and that human beings (soldiers and civilians alike) share. That is the source of our intuition that the shape of the causal connection is somehow impermissible. The Lieber Code specifically prohibited the poisoning of wells even if it would frustrate the ability of the enemy forces to fight.[10] The source of our intuition that food and agriculture should be largely immune from military targeting is an ancient intuition.[11]

The fourth example, bombing the electrical power grid, is similarly problematic. The electrical power grid constitutes a dual-use infrastructure, the type of installation that NATO famously targeted during the bombing campaign against Serbia. In that case, the power grid serviced both military and civilian customers.[12] At the very least, the military capacity of the power grid was presumably a legitimate target, though this too involved at least a two-step causal relation since the electricity—like the metal for ammunition—was a necessary ingredient in getting military assets to function properly. The real question was the status of the civilian electrical capacity. Under one argument, the civilian capacity of the grid was collateral damage, incidental to the grid's military use, and, therefore, to be governed by the principle of proportionality, and only permissible if the damage to the civilian power capacity was not disproportionate to the military advantage obtained by destroying the military capacity of the power grid. But under another theory, even the civilian capacity of the grid was a legitimate target, since it was a necessary element in the capacity of the Serbian forces to conduct their military operations.[13] The conceptual connection here was that the military campaign required substantial popular buy-in from the civilian population; if civilians suffered through the discomfort of power failures and other disruptions of daily life, their support for Milosevic's military campaign would wane. Consequently, there would be a definite military benefit for NATO in attacking these civilian installations if they were destroyed; in the absence of internal political support, Milosevic would be (and in fact

[10] Lieber Code, Article 70 ("The use of poison in any manner, be it to poison wells, or food, or arms, is wholly excluded from modern warfare. He that uses it puts himself out of the pale of the law and usages of war."). See also Hague Regulation 23(a). But see Dinstein (n 7) 219, who argues that "a food-producing area may be bombarded, if the purpose is to forestall the advance of enemy troops rather than to prevent the enemy from growing food for civilian consumption").

[11] Compare with Dinstein (n 7) 150 ("It is almost universally agreed that if the supplies in the truck consist of foodstuffs—although destined for use by combat troops in the contact zone—the driver would not be regarded as directly participating in hostilities"), citing N Boldt, "Outsourcing War—Private Military Companies and International Humanitarian Law" (2004) 47 *German Yearbook of International Law* 502, 522.

[12] Stephen T Hosmer, *The Conflict over Kosovo: Why Milosevic Decided to Settle When He Did* (RAND, 2001) 98 (noting Serb complaints about the NATO airstrikes against electric power transformer yards, bridges and tunnels, and television and radio transmission facilities).

[13] But see Henry Shue and David Wippman, "Limiting Attacks on Dual-Use Facilities Performing Indispensable Civilian Functions" (2011) 35 *Cornell International Law Journal* 559, 565 ("state practice suggests that governments are uncomfortable with the notion that the civilian function of a dual-use facility can be ignored. For example, in its attacks on the Federal Republic of Yugoslavia's electrical power facilities, NATO forces sometimes used special carbon-graphite filaments designed to disrupt power temporarily, in part to minimize long-term incidental harm to civilians").

was) forced to capitulate and come to the negotiating table earlier than he would otherwise have done. The fifth example (bombing the TV station) falls in an intermediate position between more traditional targets and the electrical grid as pure inconvenience. In the case of the TV station, it provides an essential element in the politician's propaganda machine, and its destruction prevents the political leadership from achieving the very civilian buy-in that it so desperately needs, and in that respect its destruction may make an effective contribution to the military campaign.[14]

This is a controversial argument, in both theory and practice. NATO's bombing strategy during the Serbian campaign occasioned substantial criticism and even triggered a preliminary investigation from the ICTY Office of the Prosecution (OTP) (though the OTP ultimately declined to launch a full investigation).[15] The argument proves too much: almost *any* kind of horribleness inflicted on the civilian population would trigger political instability that would force political authorities to capitulate to the enemy's political demands, including bombing hospitals and schools to sow political dissension with the hope that the populace will blame its political leaders (as opposed to hardening its resolve against the attackers). Clearly this is impermissible and suggests that the causal argument has gone awry. The counter-argument is that bombing hospitals and schools is impermissible for other reasons—specific jus in bello prohibitions—and not because they fail to make an effective contribution to military action.

However, it seems that the causal requirement of the law of war is violated in cases of bombing targets whose military contribution is purely based on the civilian pressure applied against the political leadership (based on their unhappiness over unpleasant conditions). In such situations, the causal connection is not just attenuated—reaching to the level of a three-step process—but it also takes an impermissible shape. Although the causal chain ends with an effective military contribution, it runs through the political process to get there: applying pressure against civilians who in turn place pressure on political leaders who in turn are more likely to compromise or withdraw from military engagement. Or the attack prevents politicians from using propaganda against their civilian population, thus causing them to lose popularity and making them less likely to enjoy complete discretion in military affairs (and, therefore, more likely to compromise). This causal connection is impermissible because its route is more political than military. It certainly is *effective*—that is not the problem. Indeed, if effectiveness were the only criterion, the law might actually favor such interventions because they cut quickly to the deciding factor.[16]

[14] For a discussion of the Kosovo campaign, see Hosmer, *The Conflict Over Kosovo* (n 12); Charles J Dunlap, Jr, "Kosovo, Casualty Aversion, and the American Military Ethos: A Perspective," (2000) 10 *Air Force Journal of Legal Studies* 95; Charles J Dunlap, Jr, "The End of Innocent: Rethinking Noncombatancy in the Post-Kosovo Era" (2000) 28 *Strategic Review* 9, 14 (arguing that air power needs to be deployed in a broader category of targets against "societies with malevolent propensities").

[15] See Final Report to the Prosecutor by the Committee Established to Review the NATO Bombing Campaign Against the Federal Republic of Yugoslavia, para 3; Andreas Laursen, "NATO, The War Over Kosovo, and the ICTY Investigation" (2002) 17 *American University International Law Review* 765, 789 (discussing military function of dual-use targets); WJ Fenrick, "Targeting and Proportionality during the NATO Bombing Campaign against Yugoslavia" 12 *European Journal of International Law* 489.

[16] See Dill (n 9) x (describing and rejecting the "logic of efficiency").

Applying pressure to the civilian population would be the quickest and most efficient way of forcing the enemy's capitulation. But the whole point of the causal requirement is to impose a second layer of constraint over and above effectiveness and efficiency. In this case, the detour through the political process threatens to undermine the very principle of distinction since it turns almost any civilian target into a military one by virtue of the role its destruction might play in forcing an enemy population to sue for peace.

As a final point, it should be noted that the two prongs of the standard might be satisfied though unconnected with each other. In other words, the reason for the asset's contribution to its own military might be different from the effective contribution that would flow to the attacker from its destruction. Consider the following example taken from the Additional Protocol Commentary: In 1944, the Allies attacked German controlled assets in the Pas de Calais region in occupied France.[17] The assets, including bridges, fuel, and airstrips, were of obvious benefit to the German war effort. However, their destruction was not important to the Allies because it deprived the Germans of adequate transportation.[18] Rather, their destruction misled the Germans as to the location of the eventual allied invasion.[19] Due to the bombing, the Germans were more likely to think that the invasion would take place near Pas de Calais, rather than Normandy. In this case, the assets contributed to German military operations, though their destruction was advantageous to the Allies not because of their elimination but because of the deception engendered by their destruction. In this situation, the two elements of the causal prong are unmoored from each other.

Unfortunately, the law of war has done little to resolve these issues with a general theory of causation, leaving more questions than answers. The uncertainty surrounding NATO's bombing campaign in Serbia, and the general lack of legal contestation over its strategic deployment, suggests that causation is woefully under-theorized in the law of war. While criminal and tort law feature richly textured theories of causation, the law of war deploys concepts of causation with little sophisticated theory to add much needed content to simple concepts employed in treaties. Nor can the disparity be attributed to the case law development inherent in tort and criminal law, and the comparative lack of positive case law in the law of war. Many controversies in the law of war are rarely adjudicated in court and prosecutions before international tribunals are few and far between and rarely hinge on issues of causation. The bigger issue is that theorists of tort and criminal law are obsessed with causation, but the major publicists and scholars of jus in bello have given only fleeting attention to doctrines of causation, preferring to expend intellectual resources elsewhere.

[17] See M Bother, KJ Partsch, and WA Solf (eds), *New Rules for Victims of Armed Conflicts: Commentary on the Two 1977 Protocols Additional to the Geneva Conventions of 1949* (Leiden: Martinus Nijhoff, 2nd edn 2013) 366.

[18] Bother et al, *New Rules for Victims of Armed Conflicts* (n 17) .

[19] See T Crowdy, *Deceiving Hitler: Double-Cross and Deception in World War II* (New York: Osprey, 2011). The incident is also discussed in Dinstein (n 7) 86; I Henderson, *The Contemporary Law of Targeting* (2009) 53; Jeanne Meyer, "Tearing Down the Façade: A Critical Look at the Current Law on Targeting the Will of the Enemy and Air Force Doctrine" (2001) 51 *Air Force Law Review* 143, 173.

It seems clear, as section III will demonstrate, that the future of cyberwarfare will place immense pressure on the field's failure to develop an adequate account of causation.

III. Proximate Cause and Cyber Attacks

It is common knowledge that cyberweapons have already been used in strategic conflict. The Stuxnet virus deployed by the US and Israel is only the most obvious example; future conflicts between state parties will inevitably include at the very least the *threat* of cyber deployment, even if cyber attacks are not actually performed.[20] In terms of regulating these attacks, the primary uncertainty with regard to the legal landscape is whether a cyber attack would trigger the application of IHL to the conflict in question, and whether the cyber attack would trigger a right of kinetic self-defense in response. Although there are many legal uncertainties engendered by cyber attacks, this would undoubtedly count as the most acute.

In Rule 13, the *Tallinn Manual* adopts a "scale and effects" standard inspired and drawn from the ICJ *Nicaragua* judgment, which "identified scale and effects as the criteria that distinguish actions qualifying as an armed attack from those that do not."[21] Under this criterion, attacks that injure or kill persons or damage property would satisfy the requirement, according to the *Tallinn Manual*.[22] Consequently, cyber intelligence gathering and cyber theft do not meet the criteria, according to the *Tallinn Manual*, since their scale and effects would not be damaging enough.[23] Although this conclusion seems clear with regard to cyber spying, it is unclear why cyber theft would not qualify since it has the exact same effect as any conventional damage to property, that is, it deprives the owner of its possessory interest in the property. In any event, any scale and effects standard must confront the inescapably difficult tipping point at which the scale and effects passes the threshold into an armed attack. In particular, the *Tallinn Manual* takes no position on whether a series of smaller cyber attacks could be aggregated to form, for legal purposes, a single incident sufficiently large to meet the scale and effects test.[24] This issue is already contested for conventional attacks, and the issue is no less difficult in the cyber context.

However, the most difficult issue is not scale or effects but the required relationship between the effects and the attack itself. The *Tallinn Manual on the International Law Applicable to Cyber Warfare* offers the following analysis of the problem:

> A further challenging issue in the cyber context involves determining which effects to consider in assessing whether an action qualifies as an armed attack. The

[20] David W Sanger, "Obama Order Sped Up Wave of Cyberattacks Against Iran," *The New York Times*, June 1, 2012.

[21] See MN Schmitt (ed), *The Tallinn Manual on the International Law Applicable to Cyber Warfare* (New York: Cambridge University Press, 2013) 54; ICJ *Nicaragua* judgment, para 195.

[22] *Tallinn Manual* (n 21) 55 ("[T]he parameters of the scale and effects criteria remain unsettled beyond the indication that they need to be grave. That said, some cases are clear. The International Group of Experts agreed that any use of force that injures or kills persons or damages or destroys property would satisfy the scale and effects requirement.").

[23] *Tallinn Manual* (n 21) 55.

[24] *Tallinn Manual* (n 21) 56.

> International Group of Experts agreed that all reasonably foreseeable consequences of the cyber operation so qualify. Consider, for example, the case of a cyber operation targeting a water purification plant. Sickness and death caused by drinking contaminated water are foreseeable and should therefore be taken into account.[25]

The *Tallinn Manual* standard essentially grafts a proximate cause standard onto the law of war. If the kinetic effect of a cyber attack is reasonably foreseeable, then the effect is sufficiently causally related to the cyber attack to trigger a traditional right of self-defense.

The *Tallinn Manual*'s proximate cause standard is a good first step, though it does not answer the questions raised by more complicated scenarios.[26] Consider the following case: a cyber attack disrupts a state's stock market, not just shutting it down but triggering substantial financial losses for both institutional and individual investors. The attack has a destabilizing effect on the local economy, producing widespread unemployment and a loss in discretionary income. Subsequent waves of the cyber attack target the state's currency, causing it to fluctuate widely on the global market. A final wave targets ATM machines and other retail-level financial institutions, causing widespread panic that people's savings have simply disappeared. As a result of the panic, the attack might trigger price gouging, widespread looting and even general lawlessness resulting in physical violence, death, and murder on the streets.

Although this cyber attack seems far-fetched, it is certainly possible. The real question here is how to evaluate the legal consequences of the attack if it produces widespread death and destruction as a local economy plunges into chaos. At first glance it would appear that the relevant effects are far too attenuated from the cyber attack. The source of this intuition is that much of the relevant effects are the product of intervening human actors: free agents engaged in price gouging, looting, and physical violence on the streets. Unlike the hypothetical bomb which directly kills its intended target, a cyber-weapon may kill its intended target by triggering a long chain of events that snake their way in and out of free human agents who engage in self-interested economic activity, ultimately resulting in death and property destruction on the streets. The cyber attacker might not just *foresee* these consequences but might also *intend* them, even though they will require the intervention of other human agents.[27]

The law of war has generally eschewed these causal accounts (involving intervening human agents) because such attacks are not terribly significant in traditional armed conflict scenarios; the most efficient attacks are those that directly injure their intended targets. Consequently, the law of war has been able to ignore these problems

[25] *Tallinn Manual* (n 21) 57.

[26] Just to clarify, this should not be considered a defect of the *Tallinn Manual*, which provides an excellent overview of the major issues of cyber regulation.

[27] On the intent-foresight distinction in targeting, see JD Ohlin, "Targeting and the Concept of Intent" (2013) 35 *Michigan Journal of International Law* 79, 123. For a distinction of the intent-foresight distinction in general, see J Finnis, "Intention and Side-Effects," in *Collected Essays: Intention and Identity* (Oxford Scholarship Online, 2011) 173, 180; M.P. Aulisio, "On the Importance of the Intention/Foresight Distinction," (1996) LXX *American Catholic Philosophical Quarterly* 189. The philosophical literature on the Doctrine of Double Effect has for quite some time concentrated on the issue of intentional attacks and collateral damage. See also FM Kamm, *The Moral Target: Aiming at Right Conduct in War and Other Conflicts* (New York: Oxford University Press, 2012).

of causation. But with the coming age of cyber attacks, these more attenuated attacks with intervening human agents might become the weapon of choice. Attackers may trigger a chain of causation and then watch as the very self-interested economic behavior unfolds exactly as they predicted (and hoped) it would.[28] The results might be deadly and impermissibly infringe the territorial integrity and sovereignty of the target state. Evaluating these cyber attacks requires resolving the thorny problem of intervening human agents within a general theory of causation. Although traditional military attacks rarely pose this problem, the same cannot be said for cyber attacks. Appealing to the proximate cause standard of reasonable foreseeability does not—by itself—resolve the question of whether the actions of intervening human agents break the causal chain initiated by a cyber attack.

The section IV will canvas the most likely theories for resolving the question of intervening human agency and argue that domestic law theories are ill suited for importation into the law of war. Domestic law can afford to craft principles of causation around seemingly obscure causal phenomena because domestic law can rely on the extensive ex post fact-finding embodied in tort and criminal trials. In contrast, the law of war is comparatively weaker in ex post fact-finding and works best when it sticks with legal standards that can be transparently and publicly applied by all parties to the conflict without extensive and labor-intensive fact finding.[29]

IV. Theories of Intervening Human Agency

The default legal rule is that intervening human agency will break the chain of causation that establishes proximate cause.[30] However, this rule is analytically deficient for obvious reasons: the intervention of human actions down the line does nothing to erase the contributions of actors at prior points in time. In a sense, all events are caused by a plethora of overlapping human agents at multiple points in time, and adopting a simplistic rule that points to the last-in-time human agent as the "proximate cause" is clear and tidy, but hardly satisfying. So the law has struggled to comprehend when intervening human actions break the causal chain and when they do not. The hornbook answer in criminal law and tort law is that a defendant is still responsible even if a subsequent actor is negligent, as long as that future negligence was reasonably foreseeable to the first party.[31] So the defendant who recklessly places the victim in the hospital is still guilty of manslaughter even if it is the treating physician's negligence that results in the victim's death.[32] Negligent medical care is foreseeable, the law says,

[28] See *Tallinn Manual* (n 21) 106 (Rule 30, stating that a "cyber attack is a cyber operation, whether offensive or defensive, that is reasonably expected to cause injury or death to persons or damage or destruction to objects").

[29] I also make this point in "Targeting Co-Belligerents," in C Finkelstein, JD Ohlin, and A Altman (eds), *Targeted Killings: Law & Morality in an Asymmetrical World* (Oxford: Oxford University Press, 2012) 60, 79–80. But see Su Wei, "The Application of Rules Protecting Combatants and Civilians Against the Effects of the Employment of Certain Means and Methods of Warfare," in F Kalshoven and Y Sandoz (eds), *Implementation of International Humanitarian Law* (Leiden: Martinus Nijhoff,1989) 390 (arguing that the difficulty with enforcing rules of battle based on "uncertainty of the facts" needs to be resolved by creating fact-finding procedures).

[30] *Compare with.* D Hodgson, *The Law of Intervening Causation* (Burlington: Ashgate, 2008).

[31] See, for example, J Dressler, *Understanding Criminal Law* (Lexis, 6th edn, 2012) 192.

[32] See, for example, *State v Shabazz*, 719 A.2d 440 (Conn. 1998).

so the defendant is responsible for the entire causal chain, including the death. The same result applies to the fleeing felon chased by police officers, one of whom drives recklessly and kills an innocent third party.[33] The felon is still guilty of manslaughter, even though the police officer's recklessness was later in time, since it is foreseeable that a police officer "may act negligently or recklessly to catch the quarry."[34]

However, these situations all deal with future recklessness or negligence. When dealing with intervening *intentional* actions, our intuitions are more likely to think of them as breaking the causal chain and negating proximate cause. Consider the following well-known example from the criminal law: a driver is reckless in causing an automobile accident on the highway.[35] The victims survive, exit their vehicle, and transverse the highway to the relative safety of the highway median.[36] Upon realizing that they left their engine running, they cross back across the highway and return to the vehicle to turn the ignition off and deploy their hazard lights. As they are dealing with the vehicle, they are struck and killed by a third vehicle that does not realize that the vehicle is disabled until it is too late to stop.[37] Is the driver of the first vehicle (who caused the first accident) guilty of manslaughter? Or did the victim's decision to renter the road break the causal chain that leads back to the first accident? Clearly, the first accident was a but-for cause of the death, in the sense that the resulting death would not have occurred but-for the first accident, which lead to the victim's car being disabled in the roadway. However, the ultimate death was also attributable to the victim's free and voluntary decision to reenter the road. Although this may have been foreseeable, these cases of voluntary intervention are more difficult to resolve.

One standard rubric to answer these cases is to distinguish between responsive and coincidental causes, where the former does not break the causal chain but the latter does.[38] Under this theory, the responsive cause does not break the causal chain because it is taken in direct response to the first cause, while the coincidental cause bears no relationship to the first cause and can, therefore, be considered a superseding cause.[39] However, it is often difficult to distinguish between responsive and coincidental causes in particular cases. In *Rideout*, was the victim's decision to reenter the road a responsive or coincidental cause? In one sense it seems responsive because it was taken in direct response to the first accident caused by the defendant. On the other hand, the fact that a third vehicle happened to be barreling down the road at precisely the moment that the victims were dealing with their stranded vehicle seems like a coincidence.[40]

The distinction between responsive and coincidental causes sounds analytically convincing but breaks down when applied in difficult cases. Even the court in *Rideout* refused to adopt the distinction or, in fact, any specific theory to explain when

[33] See also *People v Brady*, 29 Cal. Rptr. 3d 286 (Ct. App. 2005).
[34] *People v Acosta*, 284 Cal. Rptr. 117 (Ct. App. 1991).
[35] See *People v Rideout*, 272 Mich. App. 602, 727 N.W.2d 459 (Mich. App. 2006).
[36] *People v Rideout* (n 35). [37] *People v Rideout* (n 35).
[38] See J Dressler, *Understanding Criminal Law* (Lexis, 3d edn); Wayne R LaFave, *Criminal Law* (4th edn., 2003) 334–345.
[39] Dressler, *Understanding Criminal Law* (n 31) 191–3.
[40] See *People v Rideout*, 272 Mich. App. 602, at 610.

intervening human agents negate proximate cause.[41] Indeed, the court in *Rideout* disclaimed the very possibility of developing a crystal clear account of proximate cause that could be applied in this context, arguing that the most that can be hoped for are guideposts to guide our intuitions. The judges quote Dressler for the following pessimistic assessment of our capacity to develop such a theory:

> One early twentieth century scholar observed that all efforts to set down universal tests that explain the law of causation are "demonstrably erroneous." [Jeremiah Smith, Legal Cause in Actions of Tort, 25 Harv. L.R. 223, 217 (1912).] There are no hard-and-fast rules for determining when an intervening cause supersedes the defendant's conduct. However, there are various factors that assist the factfinder in the evaluative process.[42]

Another famous example of this pessimism is *State v Preslar*.[43] The defendant threatened the life of his spouse, who out of necessity wandered into the "freezing night" to seek protection from her violent spouse. She walked 200 yards to her father's home but spent the night outside instead of knocking on the door or otherwise entering his dwelling (presumably not wanting to disturb him or because she was embarrassed).[44] The victim froze to death. Did her voluntary decision to sleep outside break the chain of causation first initiated by her abusive husband? Again, the principle of responsive and coincidental causes does little to resolve this marginal case, since in one sense it was a response to her husband's initial aggression but in another sense her refusal to bother her father was coincidental.[45]

The most that can be hoped for in such situations is a set of criteria that guide our application of the proximate cause standard to the facts at hand. In cases where the intervening cause is an intentional act (as opposed to a negligent or reckless act), then it will be much more likely to break the causal chain and negate proximate cause. As Hart and Honore put the point, a "free, deliberate, and informed intervention of a second person, who intends to exploit the situation created by the first, but is not acting in concert with him, is normally held to relieve the first actor of criminal responsibility," though this is not always the case.[46] Hart and Honore note that the outcome depends on whether the first actor's conduct was sufficient to produce the result in the absence of the second individual's conduct or whether the second individual's action was necessary to bring the result to fruition.[47] In any event, the most that can be hoped for is a set of legal criteria to aid and structure legal decision-making, not an ironclad standard that produces automatic outcomes.

Recognizing the inherent limits of producing a "naturalistic" theory of causation that will always answer such questions, some scholars have adopted a more normative

[41] *People v Rideout* (n 40).

[42] *Rideout*, 272 Mich. App. 602, at 611–612, quoting Dressler, *Understanding Criminal Law* (4th edn) at 193.

[43] *State v Preslar*, 3 Jones (48 N.C.) 421 (1856).

[44] *State v Preslar* (n 43).

[45] For a discussion of this case, see Manuel Cancio Meliá, "Victims and Self-Liability in Criminal Law: Beyond Retributive Negligence and Foreseeability (Without Blaming the Victim)" (2008) 28 *Pace Law Review* 739, 750; "Proximate Cause" (1925) 39 *Harvard Law Review* 149, 182 n 93.

[46] HLA Hart and T Honore, *Causation in the Law* (Oxford: Oxford University Press, 2nd edn, 1985) 346.

[47] Hart and T Honore, *Causation in the Law* (n 46).

approach: simply decide whether the original cause is sufficiently proximate to make legal responsibility "fair" under the circumstances.[48] And so, for example, the Model Penal Code suggests that liability attaches if "the actual result involves the same kind of injury or harm as the probable result and is not too remote or accidental in its occurrence to have a [just] bearing on the actor's liability or on the gravity of his offense."[49] Instead of establishing a naturalistic proxy with normative consequence, why not simply cut straight to the normative level and ask the fact-finder to determine whether the causal connections are sufficiently close, as a normative matter, to make the defendant responsible for the resulting incident?[50] This view is popular in German law where it is called *Objektive Zurechnung* (normative attribution).[51] Proponents of normative theories of causation view this as intellectually more responsible because it avoids crude proxies that fail to capture our moral intuitions regarding their significance.[52] Better instead to drill down to normative criteria at the first level of the analysis and ask whether it is fair to hold the defendant responsible in such situations.

So it is clear that normative causation avoids the artificial and implausible line drawing that appears to be inevitable in accounts of naturalistic causation.[53] But the real question is whether a normative account of causation could be applied in the law of war context in order to link the particular results to the attacking force's original attack. Why not simply ask whether it is normatively correct (fair, just, etc.) to attribute the resulting consequences to the attacking state when deciding whether it meets the "scale and effects" test for defining an armed attack?

Consider again the cyber attack scenario described already involving the attack on the financial system that produces civil unrest. Under the approach of normative attribution, one would simply ask whether the results of the attack—including the resulting murders and looting caused by the damage to the financial system—are sufficiently close that they can justly be attributed to the attacking state. Weighing in favor of this attribution is that the attacking force foresaw and even intended these results, but the fact that free human agents also caused the results weighs against the attribution. With these competing elements, it is unclear what the correct "normative" result would be.

[48] But see Uwe Murmann, "Problems of Causation with Regard to (Potential) Actions of Multiple Protagonists" (2014) 10 *Journal of International Criminal Justice* 283 (arguing that naturalistic causation is one element of normative attribution as opposed to a potential replacement, and therefore naturalistic notions of causation still have a rule to play within theories of normative attribution).

[49] MPC Section 2.03.

[50] See also Hart and Honoré (n 46) 4; H Morris, *Freedom and Responsibility: Readings in Philosophy and Law* (1961) 283–5 (arguing that causal responsibility and legal responsibility are not co-extensive); Stewart (n 1) x.

[51] See the discussion in Stewart (n 1) n 59.

[52] See also the discussion in Guyora Binder, *Felony Murder* (Stanford: Stanford Univeristy Press, 2012) 208 ("It seems we cannot exorcise normative judgment from the attribution of causal responsibility… Causal attribution is normative because it depends on the values expressed by action").

[53] See Hendrik Kaptein, "Criminal Causation," in Christopher Berry Gray (ed), *Philosophy of Law: An Encyclopedia* (London: Routledge, 1999) 93–4 ("Attempts at supplementing condition theories of causation by neutral space-time criteria like immediacy or proximity fail: a victim of murder may die because doctors were negligent, but the criminal offense is still the relevant cause. Conceptions of criminal causation must be normative in one or another sense.").

The problem with normative attribution is that it suffers from vagueness.[54] Naturalistic causation may be more determinate, at least if supplemented by bright-line rules and doctrines to narrow the category of causes that count as proximate cause. But what naturalistic causation gains in clarity, it loses in accuracy; its lines seem arbitrary.[55] But when one shifts to normative attribution, it becomes obvious that clarity is no small virtue. In the criminal context, there are plenty of scholars who reject normative attribution because its vagueness and indeterminate contours are simply too unpredictable to serve as a guide for criminal prosecutions. If one takes the Holmesian Bad Man perspective on the law (designed to give clear guidelines for which conduct will trigger liability and give potential violators the chance to conform their conduct to the requirements of the law), then it is clear that normative attribution may be deeply problematic. The only true predictive answer to the question is to ask what the fact-finder would decide in an ex post evaluation of the causal question—a notoriously difficult prediction to make.

If normative attribution is problematic and controversial in the criminal context, then it is much more so in the international context. The law of war depends on publicity and transparency.[56] International law has few fact-finders capable of issuing decisions on matters of law in real-time—human rights NGOs and UN investigative commissions would be the notable exceptions. Indeed, there is also a lacuna of fact finders who issue binding ex post decisions. Although isolated questions of law may reach the ICC or the ICJ, these are the exceptions that prove the rule. In determining the threshold question of whether an armed attack has occurred, the participants themselves must determine whether the standard has been met and whether an armed response is legitimate under Article 51 of the UN Charter. Ultimately the rest of the world must evaluate the situation based on publicly available information and determine whether the defensive force was legitimate or not, and make quick decisions regarding the lawfulness or illegality of the defensive force. These are not questions that can be decided in the courtroom. Then-presidential candidate John Kerry was openly mocked for referring to the court of world opinion, but his comment displayed an accurate portrait of international law's deliberative machinery, where major questions of jus ad bellum and jus in bello are determined and enforced by the participants themselves and third-party observers in a non-judicial setting.[57]

One might object that international law is developing ever-increasing levels of enforcement and independent adjudication in the form of independent tribunals.[58] The recent growth of the ad hoc criminal tribunals and the ICC provide greater independent oversight of wartime conduct. Reprisals during armed conflict (punitive action undertaken by the participants themselves) are generally disfavored under the

[54] Kaptein, "Criminal Causation," (n 53) ("Central to such normative notions of causation are vague distinctions between conduct as such and conduct as cause of consequences."). For a good discussion of this issue, see Jane Stapleton, "Choosing What We Mean by Causation in the Law" (2008) 73 *Missouri Law Review* 433, 456.

[55] But see Leon Green, *The Rationale of Proximate Cause* (Kansas City: Vernon Law, Book Co. 1927).

[56] See, for example, Stanford International Human Rights & Conflict Resolution Clinic, *Living Under Drones* (Special Report from NYU and Stanford Law Schools) (discussing need for transparency).

[57] See "Kerry forced to explain 'global test' of legitimacy," *The Washington Times*, October 5, 2004.

[58] See M Cherif Bassiouni, *International Criminal Law, Volume 3: International Enforcement* (Leiden: Martinus Nijhoff, 2008).

law of war, although their status with regard to the killing of combatants is somewhat murky.[59] However, one should also not exaggerate the amount of independent enforcement that exists in the international system. The ICC and ad hoc tribunals involve ex post analysis, not real time adjudication that can structure the behavior of the parties to an armed conflict. States continue to withdraw from the plenary jurisdiction of the ICJ, and even in cases where the ICJ enjoys such jurisdiction, the adjudication is virtually historical in nature.[60]

The vast majority of international enforcement is decentralized, carried out by third-party states through reputational effects, formal sanctions, and refusal to engage in bilateral arrangements.[61] These mechanisms are far more powerful than people give them credit, but they are largely external to courtroom litigation and the fact-finding and legal processes that go with it. International enforcement mechanisms require transparency and publicity so that the world community can evaluate the nature of the conflict, decide when transgressors have violated an international legal norm, and impose proper consequences on flagrant violators. Complex assessments of causation that are appropriate in a tort or criminal law courtroom will be insufficient to give participants guidance regarding how to act in the midst of an international crisis.

Consider another example: the law of neutrality and the principle of co-belligerency.[62] Under the standard rules, third-party states must remain formally and practically neutral to a conflict. If they refuse to remain neutral, they can be considered co-belligerents of the party they are assisting, placing them in a state of armed conflict with the opposing party.[63] Decisions regarding the lawfulness of force, therefore, impact not just the original bilateral parties to the conflict, but also potential third parties who must make complex assessments regarding the lawfulness of the original use of force and whether to intervene, and if so, on which side. These determinations must be based on simple rules of self-defense and jus ad bellum that are clear and easily applicable on the international stage: necessity, proportionality, and so on. Normative attribution is simply too obscure to provide a reliable metric for the international community in these situations.

The obscurity of causal processes has long been a problem for international law, which has generally avoided incorporating rules that would require getting into the weeds of causal determinations. Consider the rule regarding invalidation of treaties when procured through coercion of a state.[64] When the Vienna Convention

[59] See F Kalshoven, *Belligerent Reprisals* (Leiden: Brill, 2nd edn, 2005) 344.

[60] Consider, for example, the ICJ's recent consideration of Croatia's and Serbia's cross-complaints of genocide, which were heard more than a decade after the events in question. *Application of the Convention on the Prevention and Punishment of the Crime of Genocide (Croatia v Serbia)* (public hearings from March 3, 2014 to April 1, 2014). See also SD Murphy, "The International Court of Justice," in Chiara Giorgetti (ed), *The Rules, Practice, and Jurisprudence of International Courts and Tribunals* (Leiden: Brill, 2012).

[61] For a discussion and taxonomy of this decentralized system, see O Hathaway and S Shapiro, "Outcasting: Enforcement in Domestic and International Law" (2011) 121 *Yale Law Journal* 252–349.

[62] See M Bothe, "Neutrality, Concept and General Rules," in *Max Planck Encyclopedia of Public International Law* (Oxford: Oxford University Press, 2011).

[63] Bothe, "Neutrality, Concept and General Rules," (n 62).

[64] See O Corten, "Article 52," in O Corten and P Klein (eds), *The Vienna Conventions on the Law of Treaties: A Commentary* (Oxford: Oxford University Press, 2011).

on the Law of Treaties was negotiated, a dispute broke out over how broadly to interpret coercion.[65] Article 52 of the Convention invalidates treaties procured through the threat or use of force in violation of the "principles of international law" embodied in the UN Charter.[66] A further proposal to include within the ambit of coercion a broader set of criteria—economic and political pressure—was defeated by the majority of the delegates.[67] Although the proponents of this broader definition of coercion eventually signed a Declaration articulating their view that political and economic coercion nullified a treaty obligation, this view failed to make its way into the Convention and virtually every commentator states the view simply that international law prohibits military coercion in treaty negotiations.[68]

Why the hostility to economic pressure as a relevant form of coercion? One answer sounds in substance: economically weak states would be able to rely on the proposed rule as a way of getting out of inconvenient treaty obligations; while economically powerful states would not have a similar privilege. So it stands to reason that the developed world would reject the new rule while the developing world might embrace it.[69]

However, there are deeper reasons to reject the new rule. *Implementing* the rule would require that states understand and evaluate economic processes, which are rarely transparent and public.[70] Determining whether a state was laboring under economic coercion would require evaluating why the economic consequences of any particular action unfolded as they did—an analysis that requires not only professional expertise, but more importantly non-public data in the custody of state parties and not generally available to the world community. Although it would be possible for a court to determine, ex post, whether an economic environment was impermissibly coercive towards a state, these are not the type of determinations that could be made on the fly by parties to the conflict and third-party states evaluating the dispute.

Much the same can be said regarding the law of war. International Humanitarian Law is designed to provide a set of rules that the parties to the conflict can self-administer and moreover can *police* when the opposing party has violated these rules. For example, if one party uses illegal combatants because the combatants do not wear fixed emblems recognizable at a distance, then the evidence is fully public behavior that can be recognized by all. If another party to the conflict violates the principle of distinction by massacring civilians in a field, the relevant evidence will be public. If attacking forces declare that no quarter will be given and execute surrendered soldiers rather than granting them prisoner of war status, the relevant events will be publicly manifested.[71]

[65] Corten, "Article 52" (n 64) 1206.

[66] Corten (n 64) 1208.

[67] Corten (n 64). This proposal is often referred to as the Nineteen-State Amendment.

[68] Declaration on the Prohibition of Military, Political, or Economic Coercion in the Conclusion of Treaties.

[69] However, although Britain rejected the proposal, its representative also stated that "of course, there might be cases where flagrant economic or political pressure amounting to coercion could justify condemnation of a treaty." *See* Corten (n 64) 1208, citing Representative of the United Kingdom (Official Records, Summary Records, 1st session, 50th meeting) 283 para 31.

[70] But see Corten (n 64) 1209 ("As it is the case with many rules of international law, certain concepts give rise to controversies that are on some occasions decided by a judge or a third party. In this sense, 'economic or political coercion' is not an exception.").

[71] See George P Fletcher, *Rethinking Criminal Law* (Oxford: Oxford University Press, reprint edn, 2000) 115.

One might object that not all rules in IHL are easily applied. The principle of proportionality requires an analysis of the intended target, its military value, and *expected* collateral damage, and the proportionality relationship between them.[72] Not all of this information is publicly available, which is precisely why the law of war places more emphasis on the principle of distinction and violations of proportionality are almost never prosecuted at military tribunals.[73] But this is not to say that the law of war has no use for private information and complex legal assessment; it certainly does. It is just that the law of war is far more weighted towards what George Fletcher calls the "pattern of manifest criminality" than it is toward the "subjective criminality" that reigns in the criminal law.[74] Principles of subjective culpability and obscure causal phenomena may guide criminal law decision-making, but public international law has a greater need for, and emphasis on, manifest criminality. All of this suggests that the law of war (at least IHL proper as opposed to ICL) cannot be based on assessments of obscure causal requirements. The law of war's failure to develop an adequate account of causation is not based on shortsightedness or academic laziness. Rather, it is based on the structure of the field itself as a creature of public international law.

One simple solution is for the law of war to develop easy-to-apply bright line rules regarding the consequences of cyber attacks. For example, a future convention might simply stipulate that *intended* consequences, regardless of whether they are the product of intervening human agents, should be attributed back to the original cyber attack for purposes of meeting the "scale and effects" standard. However, this rule would place immense pressure on the pattern of subjective criminality because it is based on what the party *intended*. A future convention could, therefore, adopt the opposite rule: future consequences from intervening agents are *always* considered supervening causes that break the causal chain. This would have the virtue of being clear, though it might produce an uncomfortable black hole in the legal regulation. It would permit states to launch cyber attacks that trigger a chain of events that result in property damage and death but at the same time escape the military consequences of the attack since it would not trigger a right of self-defense under Article 51 of the UN Charter. This might provide an incentive for states to engineer causal chains so that they get the benefit of the law and prevent their targets from responding with lawful military force. This result is just as unpalatable.

V. Conclusion

The upshot of this analysis is that deciding on the appropriate legal rule requires consideration of the ease of applying that rule. Moreover, this is not a secondary decision regarding its collateral consequences but rather is *internal* to the question of the appropriate legal rule. The issue of execution and adjudication is intrinsic to legal norm choice. Of course, this is true not just for cyberwar but for all aspects of the

[72] D Saxon, *International Humanitarian Law and the Changing Technology of War* (Leiden: Brill, 2013) 83 ("Even for the best military minds, proportionality is extremely difficult to apply. It has been criticized for being overly vague and difficult to apply consistently.").

[73] This issue is explored in greater depth in Ohlin (n 27) 85–91.

[74] See Fletcher (n 71) 115–18.

law. Even in the basic subjects of domestic law, we are trained to consider questions of public policy. Will the proposed rule open the floodgates to vexatious litigation? Will litigants have sufficient incentives to file a claim or will the remedy be so small that their interests will remain un-vindicated? In international law, these questions take on renewed urgency since the enforcement system is so radically decentralized. Furthermore, the law of war is at the far end of the spectrum even within the domain of international law, since during the heat of battle the parties to the armed conflict (and third-party observers) must take notice of violations even before official fact-finders and neutral international tribunals enter the equation. The demands of clarity and transparency are most acute in the IHL context.

This insight applies with even greater urgency in the context of cyberwar. Since causal processes in the cyber-realm are often non-public, there is a very real danger that the law could evolve a set of legal standards that cannot be adjudicated in the public realm, and that require the robust forensic accounting of a courtroom to unearth. This would be a disaster for the legal regulation of cyberwar, since its norms would then be incapable of being enforced within the current system of public international law.

The law of war should, therefore, be hesitant before it imposes standards of cyber causation that cannot be publicly applied and adjudicated. This might suggest that IHL should impose a rule where all *intended* consequences are attributable to the state that launches the cyber attack. This rule would have the virtue of clarity and would avoid difficult line-drawing exercises. However, it would also invite—indeed demand—heavy reliance on the question of state intention. Indeed, state intention would become the sole criterion of cyber causation. This would impose what Fletcher calls the pattern of subjective criminality—precisely the type of analysis that requires the fact-finding powers of a courtroom. (Indeed, the opposite pattern—the pattern of manifest criminality—is one that produces apprehension in even the most common observer on the street.) So in order to be faithful to the pattern of manifest criminality, the proper result would appear to be the opposite rule: limit cyber causation to situations without third-party intervening causation. Although this leaves some cyber events under-regulated, it is important to remember that the rule would be limited to the identification of consequences that are counted for purposes of an armed attack triggering the right of self-defense under Article 51 of the UN Charter and also potentially triggering the existence of an armed conflict governed by IHL. Under this scheme, a cyber attack with downstream consequences involving intervening human agency might still generate state responsibility under public international law, triggering the more sober and reflective remedies associated with that field of law. But these causally complicated events would not trigger the more immediate legal architecture of Article 51 self-defense and the existence of an armed conflict.

PART II

CONCEPTUALIZING CYBER ATTACKS: THE CIVIL-MILITARY DIVIDE

4

Cyberterrorism and Enemy Criminal Law

Stuart Macdonald

I. Introduction

Since the events of 9/11 commentators have debated the appropriate legal framework for responding to the threat of terrorism. Some, like the Bush Administration, advocated a military response. Others argued that, whilst terrorism should be distinguished from warfare, the gravity of the contemporary terrorist threat justifies exceptional or emergency measures which operate outside of the normal legal framework and/or are temporally limited.[1] Still others urged the importance of a criminal justice based response in which the criminal law is deployed to prosecute suspected terrorists.[2] One of the principal arguments in favor of the latter approach is that it has greater moral authority and is more protective of human rights. The criminal law requires the state to prove its case in open court beyond reasonable doubt and gives the suspect the opportunity to respond to the case against him or her.

As the contributions to this volume show, those responding to the growing threat of cyber attacks also face this choice between different legal frameworks. This chapter contributes to this debate by providing a critical assessment of the UK's criminal justice based response to the threat of cyberterrorism. The chapter will show that, in fact, the different legal frameworks are not mutually exclusive. In recent years the UK has introduced a range of terrorism-related legislation. This has not only significantly extended the criminal law's reach, so that it encompasses both a wide range of preparatory activities and individuals who are only loosely connected to a feared attack. It has also indirectly diminished the procedural rights of suspected terrorists and provides for the imposition of severe sanctions which are rooted in a precautionary approach based on potential future harms. These are all marked departures from the normal standards of the criminal law and so, it will be argued, may be understood as the convergence of the criminal justice and exceptional measures approaches: in other words, as a form of enemy criminal law. The chapter argues that it is contradictory—and,

[1] Some well-known examples include: Bruce Ackerman, *Before the Next Attack: Preserving Civil Liberties in an Age of Terrorism* (New Haven: Yale University Press, 2006); Richard A Posner, *Not a Suicide Pact: The Constitution in a Time of National Emergency* (Oxford: Oxford University Press, 2006); Oren Gross and Fionnuala Ní Aoláin, *Law in Times of Crisis: Emergency Powers in Theory and Practice* (Cambridge: Cambridge University Press, 2006).

[2] See, for example, David Bonner, *Executive Measures, Terrorism and National Security: Have the Rules of the Game Changed?* (Farnham: Ashgate, 2007); David Cole and Jules Lobel, *Less Safe Less Free: Why America is Losing the War on Terror* (London: The Free Press 2007); Conor Gearty, "The Superpatriotic Fervour of the Moment" (2008) 28 OJLS 183; Gary Lefree and James Hendrickson, "Build a Criminal Justice Policy for Terrorism" (2007) 6 *Criminology & Public Policy* 781.

ultimately, self-defeating—to insist on a criminal justice based framework without adhering to the features which give the criminal law its moral authority in the first place.

II. The Inclusion of Cyber Attacks within the UK's Definition of Terrorism

Increasing attention is being paid to the threat of cyber attack. In November 2013 the heads of the FBI, Department of Homeland Security and National Counterterrorism Center told Congress that cyber attacks are likely to eclipse 9/11-style terrorist attacks as a domestic danger over the next decade.[3] These attacks, they predicted, will come from individual actors as well as nation-states.[4] Analysis of the Stuxnet malware—which damaged centrifuges at the Natanz uranium enrichment facility in Iran—has concluded that it could be used as a blueprint for cyber-physical attacks in the future. These would most likely focus on targets which are easier to attack, such as critical infrastructure installations, and importantly would not require nation-state resources.[5]

Warnings of the threat posed by cyberterrorism are not new. In 1997 Barry Collin of the US Institute for Security and Intelligence stated that the potential for multiple casualties and considerable publicity are likely to make cyber attacks desirable to terrorist groups. He gave the examples of contaminating food products through interference with manufacturing processes and the interception of air traffic control systems to engender fatal collisions.[6] Weimann has identified a total of five reasons why terrorists might choose to launch cyber attacks: comparatively lower financial costs; the prospect of anonymity; a wider selection of available targets; the ability to conduct attacks remotely; and, the potential for multiple casualties.[7] Related utility maximization arguments suggest it is inevitable terrorists will employ cyber weaponry if benefits from doing so are likely,[8] and/or if an enemy employs computers and networks as security tools or maintains dominance in this area.[9] Indeed, in a recent survey of the global research community 58% of respondents stated that cyberterrorism poses a significant threat, with a further 12% saying that it may potentially become one.[10]

[3] Spencer Ackerman, "Cyber-Attacks Eclipsing Terrorism as Gravest Domestic Threat—FBI," *The Guardian*, November 14, 2013, at <http://www.theguardian.com/world/2013/nov/14/cyber-attacks-terrorism-domestic-threat-fbi> accessed November 29, 2013.

[4] "Cyberattacks More Serious Domestic Threat to U.S. than Terrorism," *Homeland Security News Wire*, November 20, 2013, at <http://www.homelandsecuritynewswire.com/dr20131120-cyber attacks-more-serious-domestic-threat-to-u-s-than-terrorism-fbi> accessed November 29, 2013.

[5] Ralph Langner, *To Kill a Centrifuge: A Technical Analysis of What Stuxnet's Creators Tried to Achieve* (Arlington: The Langner Group 2013).

[6] Barry Collin, "The Future of Cyberterrorism" (1997) 13 *Crime and Justice International* 15.

[7] Gabriel Weimann, *Cyberterrorism: How Real is the Threat?* (Special Report 119, United States Institute of Peace 2004) at <http://www.usip.org/sites/default/files/sr119.pdf> accessed November 29, 2013.

[8] Steven Simon and Daniel Benjamin, "America and the New Terrorism" (2000) 42 *Survival* 59.

[9] Jerrold M Post, Kevin G Ruby, and Eric D Shaw, "From Car Bombs to Logic Bombs: The Growing Threat from Information Terrorism" (2000) 12 *Terrorism and Political Violence* 97.

[10] Lee Jarvis, Stuart Macdonald, and Lella Nouri, "The Cyberterrorism Threat: Findings from a Survey of Researchers" (2014) 37 *Studies in Conflict & Terrorism* 68.

The UK Government considered the threat of cyberterrorism as part of its review of the definition of terrorism in the late 1990s. As part of this review, it published a consultation paper in 1998, which proposed a new statutory definition of terrorism.[11] This document rejected Lord Lloyd's earlier suggestion that the UK adopt the working definition of terrorism then in use by the FBI.[12] One of the government's criticisms of the FBI's definition was that cyber attacks perpetrated by terrorists fell outside of its scope. The Government stated that such attacks might not only "result in deaths and injuries," but also "result in extensive disruption to the economic and other infrastructure of this country."[13] To illustrate the last of these, the government offered the example of contamination of a public utility system such as a water or sewage works.

The resultant definition can be found in section 1 of the Terrorism Act 2000. For an act to qualify as terrorist, it must satisfy three criteria. First, one of five specified actions must have either been used or threatened. These are: serious violence against a person; serious damage to property; endangering another person's life; creating a serious risk to public health or safety; or, the focus of this chapter, actions which are "designed seriously to interfere with or seriously to disrupt an electronic system."[14] Second, the action (or threat) must have been designed either to influence the government or an international governmental organization, or to intimidate the public or a section of the public.[15] Third, the action (or threat) must have been made for the purpose of advancing a political, religious, racial, or ideological cause.[16] The application of the definition is not limited to the UK; it applies equally to actions outside the UK, to people and property outside the UK, and to foreign governments.[17]

The UK's definition unquestionably has a broad scope.[18] In respect of cyber attacks, it is not limited to attacks on critical infrastructures. Attacks on anything deemed to be an electronic system could potentially qualify (such as a Distributed Denial of Service (DDoS) attack).[19] Moreover, the attack need not actually cause serious interference or disruption. It is enough that it was designed to. This far-reaching definition is expanded still further by the fact that it: applies to threats of cyber attacks as well as actual attacks; applies to cyber attacks which are designed merely to influence, not intimidate, a government; applies equally to all governments, however oppressive; and, contains no exemption for political protest or self-determination.[20]

[11] Home Office, *Legislation Against Terrorism: A Consultation Paper* (Cm 4178, 1998).

[12] Lord Lloyd of Berwick, *Inquiry into Legislation Against Terrorism* (Cm 3420, 1996).

[13] Home Office, *Legislation Against Terrorism* (n 11) para 3.16.

[14] Terrorism Act 2000, s 1(2).

[15] Terrorism Act 2000, s 1(1)(b). If the relevant action involved the use of firearms or explosives it is unnecessary to show that section 1 (1)(b) is also satisfied (s 1(3)).

[16] Terrorism Act 2000, s 1(1)(c).

[17] Terrorism Act 2000, s 1(4).

[18] The breadth of the definition was criticized by the Supreme Court in *R v Gul* [2013] UKSC 64, [2013] 3 WLR 1207 [28]-[29], [33]-[37], [61]–[64]. See also Keiran Hardy and George Williams, "What is 'Terrorism'?: Assessing Domestic Legal Definitions" (2011) 16 *UCLA Journal of International Law & Foreign Affairs* 77, 111–20.

[19] Terrorism Act 2000, s 1(2)(e).

[20] Keiran Hardy and George Williams, "What is Cyberterrorism? Computer and Internet Technology in Legal Definitions of Terrorism" in Tom Chen, Lee Jarvis and Stuart Macdonald (eds), *Cyberterrorism: Understanding, Assessment and Response* (New York: Springer 2014).

For present purposes what is most important about the UK's definition is that it treats cyberterrorism as a subset of the broader category of terrorism, in spite of possible qualitative differences between cyberterrorism and traditional, physical forms of terrorism.[21] By so doing, it grants access to the full panoply of terrorism-related investigative powers, procedures, criminal offences, and sentencing powers in any case involving a cyberterrorist attack that falls within the broadly couched statutory definition. The UK has thus sought to respond to the threat of cyberterrorism by using criminal laws and processes—as opposed to the law of war or emergency powers—just as it has done for traditional terrorism. The remainder of this chapter evaluates this criminal justice based response.

III. Enemy Criminal Law as a Descriptive Concept

The concepts of "enemy criminal law" (*Feindstrafrecht*) and "citizen criminal law" (*Bürgerstrafrecht*) were first advanced by the German professor of criminal law Günther Jakobs.[22] Intended as ideal-types, the aim of these concepts was to draw attention to two contrasting tendencies within the criminal law (as opposed to two isolated spheres of criminal law).[23]

According to Jakobs, since trust alone is not a sufficient basis for individuals in society to engage with one another, the role of law is to enable interaction. But law can only perform this function if members of society believe that legal norms are generally recognized and respected. Citizens (a term which Jakobs uses synonymously with persons[24]) are, therefore, expected to cultivate loyalty to the law. This "anchors the expectations of fellow members of the polity that the law will generally be followed, thereby enabling them to run their lives, if not in total security, at least without constant worry about being wronged."[25] When a citizen commits a crime, the validity of the applicable law is called into question. Punishment is a counter-response, which reinforces the loyalty to the law of both the offender (by forcefully reminding him or her of their duties as a citizen) and members of society in general (by reaffirming the law and making it clear that the conduct is unacceptable).

Citizen criminal law is, therefore, communicative. It assumes that the offender remains a loyal citizen and that the offending behavior was a lapse, which he or she now regards as a mistake, and so continues to address the offender as a "person-in-law."[26] By contrast, enemy criminal law is directed at individuals whose conduct manifests

[21] Lee Jarvis and Stuart Macdonald, "What is Cyberterrorism? Findings from a Survey of Researchers" (2014) *Terrorism & Political Violence*, at <http://www.tandfonline.com/doi/pdf/10.1080/09546553.2013.847827> accessed July 1, 2014.

[22] Jakobs presented his theory at a major conference held in Berlin in 1999: G Jakobs, "Selbstverständnis der Strafrechtswissenschaft vor den Herausforderungen der Gegenwart (Kommentar)," in A Eser, W Hassemer and B Burkhardt (eds), *Die Deutsche Strafrechtswissenschaft vor der Jahrtausendwende* (Munich: Beck 2000).

[23] Günther Jakobs, "Bürgerstrafrecht und Feindstrafrecht" [2004] HRR-Strafrecht 88.

[24] In contrast to some other accounts which also employ the concept of citizenhood: see further Markus D Dubber, "Citizenship and Penal Law" (2010) 13 *New Criminal Law Review* 190.

[25] Daniel Ohana, "Trust, Distrust and Reassurance: Diversion and Preventive Orders Through the Prism of *Feindstrafrecht*" (2010) 73 *Modern Law Review* 721, 724.

[26] Ohana, "Trust, Distrust and Reassurance (n 25) 724.

that they do not respect the validity of the legal system and no longer consider themselves bound by the law. Since these individuals do not provide others with this minimum level of cognitive reassurance, the legal system no longer regards them as citizens or persons but rather as a source of danger. Terrorists are the paradigmatic example of a non-citizen, because the terrorist lacks not only the requisite loyalty to law but also the interest in acting according to it.[27] Other possible examples include sexual predators, members of organized crime, and drug dealers.

It follows that enemy criminal law is concerned not with communication or censure, but with management of risk. The enemy is one who demonstrates by their conduct that he or she can no longer minimally guarantee that they will conduct him- or herself as a loyal citizen. Enemy criminal law, therefore, imposes sanctions not as retrospective punishment for past wrongdoing but prospectively in order to prevent future harms.

As well as the change in discourse (waging war against an enemy), Jakobs identifies three principal features of enemy criminal law. This section of the chapter outlines these features and argues that all three are apparent in the UK's raft of terrorism criminal offences.

(a) Pre-inchoate liability

Few would deny that the criminal law has a preventive, as well as a punitive, function. As Ashworth and Zedner observe, "If a certain form of harmful wrongdoing is judged serious enough to criminalize, it follows that the state should assume responsibility for taking steps to protect people from it."[28] Indeed, as Duff remarks, "a law that condemned and punished actually harm-causing conduct as wrong, but was utterly silent on attempts to cause such harms, and on reckless risk-taking with respect to such harms, would speak with a strange moral voice."[29] Most legal systems accordingly have the general inchoate criminal offences of encouraging or inciting crime, conspiracy, and attempt. These offences have a preventive role, penalizing conduct before any harm actually occurs. What marks out enemy criminal law, however, is that it criminalizes conduct at a far earlier, pre-inchoate, stage.

In the UK, it has been deemed necessary to supplement the ordinary inchoate offences in terrorism cases. Although there have been some high profile convictions,[30] the offences of conspiracy and encouraging crime are notoriously hard to prove. Obtaining admissible evidence of an agreement or words of encouragement within secretive organizations is difficult, particularly given the UK's ban on the use of intercept as evidence in criminal trials.[31] Moreover, even if admissible evidence is obtained

[27] Dubber, "Citizenship and Penal Law" (n 24).

[28] Andrew Ashworth and Lucia Zedner, "Prevention and Criminalization: Justifications and Limits" (2012) 15 *New Criminal Law Review* 542, 543.

[29] RA Duff, *Criminal Attempts* (Oxford: Clarendon Press 1996) 134.

[30] Including Abu Hamza's conviction for soliciting to commit murder and the convictions of seven men on conspiracy charges in the airline liquid bomb plot case.

[31] For discussion of the ban on intercept evidence, see Stuart Macdonald, "Prosecuting Suspected Terrorists: Precursor Crimes, Intercept Evidence and the Priority of Security," in Lee Jarvis and Michael Lister (eds), *Critical Perspectives on Counter-terrorism* (Abingdon: Routledge 2014).

it may lack evidential value (many members of terrorist organizations observe good communications security and disguise the content of their communications) or there may be public interest reasons for not disclosing it (perhaps because it would expose other on-going investigations or reveal sensitive techniques or capabilities).[32] Meanwhile, the offence of criminal attempts has a limited reach. Narrower in scope than the US Model Penal Code's "substantial step" test, the test in the UK is whether the defendant committed an act that was "more than merely preparatory" to commission of the planned offence.[33] In other words, a defendant does not commit a criminal attempt until he or she actually "embarks upon the crime proper."[34] Given the level of risk and severity of the potential harm in terrorism cases, there are strong reasons to (in the words of the UK's Independent Reviewer of Terrorism Legislation) "defend further up the field."[35] This is the role of the pre-inchoate—or precursor—terrorism offences.

There are a large number of terrorism precursor offences in the UK, found predominantly in the 2000 and 2006 Terrorism Acts. These penalize a wide range of preparatory activities, including: fund-raising for terrorist purposes;[36] use or possession of money or other property for terrorist purposes;[37] possession of an article for terrorist purposes;[38] collecting information or possessing a document likely to be useful to a terrorist;[39] training for terrorism;[40] attendance at a place used for terrorist training;[41] and, the catch-all offence of preparation of terrorist acts.[42] In recent years this expanding use of the criminal sanction has received much attention from criminal law theorists.[43] As well as raising concerns about possible overreaching and the impact on human rights and rule of law values, this literature has offered possible principled justifications for these offences.[44] Wörner has suggested that the group-danger rationale that underpins the general offence of conspiracy could also apply to many of the terrorism precursor offences. When a defendant provides weapons, training manuals, or practical advice his or her behavior ceases to be part of their "*internum*" and becomes part of the "*externum*."[45] The defendant's behavior may endanger others and, crucially, they are no longer able to control what happens

[32] Home Office, *Privy Council Review of Intercept as Evidence: Report to the Prime Minister and the Home Secretary* (Cm 7324, 2008).

[33] Criminal Attempts Act 1981, s 1(1).

[34] *R v Gullefer* [1990] 1 WLR 1063.

[35] David Anderson QC, "Shielding the Compass: How to Fight Terrorism Without Defeating the Law" (2013) *European Human Rights Law Review* 233, 237.

[36] Terrorism Act 2000, s 15.

[37] Terrorism Act 2000, s 16.

[38] Terrorism Act 2000, s 57.

[39] Terrorism Act 2000, s 58.

[40] Terrorism Act 2006, s 6.

[41] Terrorism Act 2006, s 8.

[42] Terrorism Act 2006, s 5.

[43] See, for example, AR Duff, L Farmer, SE Marshall, M Renzo, and V Tadros (eds), *The Boundaries of the Criminal Law* (Oxford: Oxford University Press, 2010); AP Simester and A von Hirsch, *Crimes, Harms, and Wrongs: On the Principles of Criminalisation* (Oxford: Hart Publishing, 2011); Ashworth and Zedner, "Prevention and Criminalization: Justifications and Limits" (n 28); GR Sullivan and I Dennis (eds), *Seeking Security: Pre-Empting the Commission of Criminal Harms* (Oxford: Hart Publishing, 2012).

[44] For an alternative perspective that preparatory acts should not, in general, be criminalized but regulated using a system of civil orders, see Daniel Ohana, "Responding to Acts Preparatory to the Commission of a Crime: Criminalization or Prevention" (2006) 25 *Criminal Justice Ethics* 23.

[45] Liane Wörner, "Expanding Criminal Laws by Predating Criminal Responsibility—Punishing Planning and Organizing Terrorist Attacks as a Means to Optimize Effectiveness of Fighting Against Terrorism" (2012) 13 *German Law Journal* 1037, 1052.

next. An alternative, subjectivist, justification is provided by Simester and von Hirsch. They explain that the principal difficulty with imposing criminal liability on "remote harms" (actions which do not themselves directly cause harm to others, such as collecting information, possessing items or raising funds) is that the feared eventual harm is contingent upon some other person or the defendant themself choosing to behave in a particular way in the future. They argue that to hold someone responsible now for the possible future acts of others is contrary to the fundamental right to be treated as autonomous individuals who are distinctively responsible for their own actions, whilst to hold someone responsible now for their own possible future actions is to undermine their autonomy and treat them as being incapable of deliberation and self-control.[46] Criminal liability for remote harms can, therefore, only be justified, they argue, if the defendant "in some sense affirms or underwrites" the subsequent choice to cause harm.[47] They name this the principle of normative involvement: if a defendant endorses the potential future harmful actions of either himself or another, responsibility for the future harm may fairly be imputed to him.

These principled justifications provide a useful yardstick for evaluating the scope of the existing raft of terrorism precursor offences. What is readily apparent is that some of the offences overreach.[48] An example is the offence of collecting information or possessing a document likely to be useful to a terrorist. Not only does this offence not require any proof that the defendant had shared the information or document so that it was no longer under his exclusive control. It also requires no proof whatsoever of a terrorist connection or purpose. As a result, in *R v G*[49] the House of Lords upheld the conviction for this offence of a man who, whilst in custody for non-terrorism offences, collected information on explosives and bomb-making and left it in his cell for a guard to find. The defendant was a paranoid schizophrenic who, it was accepted, wanted to antagonize the prison staff because he believed that they had been whispering about him. In a case like this one, the effect of this offence is to "make a terrorist out of nothing."[50]

It is also important to point out that proof of normative involvement should be regarded as a necessary, but not a sufficient, condition for criminalization. There may be reasons not to enact an offence even if its terms do require proof that the defendant intended to commit, or had normative involvement in, future terrorist acts. An example is the offence of preparation of terrorist acts. A defendant commits this offence if he or she engages in "any conduct" with an intention to commit or assist acts of terrorism.[51] Any form of conduct could potentially be penalized by this offence if carried out with the requisite intention. Simester offers the example of an individual who eats muesli for breakfast as part of a fitness programme in

[46] Simester and von Hirsch (n 43) 80–81. [47] Simester and von Hirsch (n 43) 81.

[48] See further Lord Carlile QC and Macdonald, "The Criminalization of Terrorists' Online Preparatory Acts," in Tom Chen, Lee Jarvis and Stuart Macdonald (eds), *Cyberterrorism: Understanding, Assessment and Response* (New York: Springer, 2014).

[49] [2009] UKHL 13, [2010] 1 AC 43.

[50] Jacqueline Hodgson and Victor Tadros, "How to Make a Terrorist Out of Nothing" (2009) 72 *Modern Law Review* 984.

[51] Terrorism Act 2006, s 5(1).

preparation for a terrorist act.[52] Where the conduct charged is something innocuous, the authorities will need to find some other evidence that the individual performed the act with the necessary intention. This could result in intrusive methods of policing. There is also the danger that the offence will be enforced in a discriminatory manner, with certain groups feeling compelled to forgo some innocent behavior for fear it may be misconstrued. Furthermore, there are numerous other precursor offences which already criminalize various specific forms of preparatory activity. The guidance notes which accompanied the legislation failed to identify any gaps that needed to be plugged.[53] So it is unclear whether this catch-all offence is in fact necessary.

The UK's terrorism precursor offences not only extend the temporal reach of the criminal law. They also encompass a wider range of individuals, penalizing those with an associative or facilitative role as well as potential perpetrators and accessories.[54] It reaches this wider range of individuals in four ways. First and foremost, there are a number of offences which target those with a supporting role, including: membership of a proscribed organization;[55] support for a proscribed organization;[56] encouragement of terrorism;[57] and, dissemination of terrorist publications.[58] Second, these offences not only target acts which facilitate terrorist attacks, but also acts which facilitate the assistance or encouragement of terrorist attacks. So, for example, it is not only an offence for D1 to provide training to D2 with an intention that D2 will use the skills to commit a terrorist act. It is also an offence for D1 to provide training to D2 with an intention that D2 will use the skills to assist someone else (D3) to commit a terrorist act.[59] Third, it is possible to commit many of the terrorism precursor offences in inchoate form. It is an offence, for example, to conspire to engage in conduct that is preparatory to an act of terrorism.[60] Fourth, in certain circumstances the law governing inchoate offences allows one layer of inchoate liability to be layered upon another (so-called double inchoate liability). When these four features are combined, the potential reach of the precursor offences becomes clear. Together, they mean that it is an offence for an individual (D1) to intentionally encourage someone else (D2) to intentionally encourage someone else (D3) to cause someone else (D4) to publish a statement which indirectly encourages someone else (D5) to instigate someone else (D6) to commit an act of terrorism.[61] Ordinarily, individuals who are several steps removed from the harm-causing conduct would be regarded as too remote to fall within the scope of the criminal law.

[52] AP Simester, "Prophylactic Crimes," in GR Sullivan and I Dennis (eds), *Seeking Security: Pre-Empting the Commission of Criminal Harms* (Oxford: Hart Publishing, 2012).

[53] Available at http://www.legislation.gov.uk accessed November 21, 2013.

[54] Lucia Zedner, "Terrorizing Criminal Law" (2012) *Criminal Law and Philosophy*, at <http://link.springer.com/article/10.1007/s11572-012-9166-9> accessed November 21, 2013.

[55] Terrorism Act 2000, s 11. [56] Terrorism Act 2000, s 12. [57] Terrorism Act 2006, s 1.

[58] Terrorism Act 2006, s 2. [59] Terrorism Act 2006, s 6(1).

[60] Criminal Law Act 1977, s 1(1); Terrorism Act 2006, s 5(1).

[61] Serious Crime Act 2007, s 44; Terrorism Act 2006, s 1. Sections 44–6 of the Serious Crime Act 2007 create three separate offences. The form of double inchoate liability described in the text is only available in respect of the s 44 offence of intentional encouragement or assistance (see s 49 and Schedule 3).

(b) The imposition of severe sanctions

The second feature of enemy criminal law is the imposition of severe punishment. The rationale underlying these sanctioning powers is not the retributivist notion of communication and censure, but risk control. As a result, the sentences imposed for preparatory offences may be the same as those imposed for an attempt to cause the harm in question:

> [P]unishment is imposed uniformly, irrespective of the stage of apprehension prior to consummation of the offence, notwithstanding the principle that sanction severity should be commensurate with the blameworthiness of the actor as determined by the actual progress made toward the realisation of the criminal endeavour.[62]

The UK's terrorism precursor offences carry severe sentencing powers. Of the eleven terrorism precursor offences already mentioned in this chapter, two have a maximum sentence of seven years' imprisonment,[63] five a maximum of ten years,[64] two a maximum of fourteen,[65] one a maximum of fifteen,[66] and the other a maximum of life imprisonment.[67] In recent years there have been a number of successful prosecutions for these offences.[68] In order to illustrate the potential severity of the sanctions, three of these cases will be outlined.

First, the case of *R v Worrell*.[69] In this case the police found a significant quantity of racist and right-wing material in the defendant's flat, including books, DVDs, and Nazi memorabilia. The books included manuals on weapons and bomb-making. Officers also found some sodium chlorate, weed killer, matches, lighter fuel, and fireworks. The defendant, therefore, had instructions on how to make an improvised explosive device and some of the materials necessary for their manufacture. At trial he was convicted of possession of articles for terrorist purposes[70] (an offence which, it should be noted, requires a reasonable suspicion that the defendant intended to use the items for terrorism-related activity[71]) and was sentenced to six years' imprisonment.[72] At his appeal against sentence the Court of Appeal acknowledged that the defendant was not part of a conspiracy or terrorist cell, that he had not actually manufactured an

[62] Ohana, "Trust, Distrust and Reassurance" (n 25) 726.

[63] Terrorism Act, ss 1 and 2 (encouragement of terrorism and dissemination of terrorist publications).

[64] Terrorism Act 2000, ss 11, 12, and 58 (membership of a proscribed organization, support for a proscribed organization and collecting information or possessing a document likely to be useful to a terrorist); Terrorism Act 2006, ss 6 and 8 (training for terrorism and attendance at a place used for terrorist training).

[65] Terrorism Act 2000, ss 15 and 16 (fund-raising for terrorist purposes and use or possession of money or other property for terrorist purposes).

[66] Terrorism Act 2000, s 57 (possession of an article for terrorist purposes).

[67] Terrorism Act 2006, s 5 (preparation of terrorist acts).

[68] All successful prosecutions for terrorism-related offences are detailed on the website of the Counter-Terrorism Division of the Crown Prosecution Service. See <http://www.cps.gov.uk/publications/prosecution/ctd.html> accessed November 27, 2013.

[69] *R v Worrell* [2009] EWCA Crim 1431, [2010] 1 Cr App R (S) 27.

[70] Terrorism Act 2000, s 57.

[71] *R v Zafar* [2008] EWCA Crim 184, [2008] QB 810.

[72] He was also convicted on a separate count of racially aggravated intentional harassment, alarm or distress. For this offence he was sentenced to fifteen months' imprisonment to be served consecutively, giving a total sentence of seven years and three months.

explosive device or attempted to do so and that there was no evidence that any attack was actually planned or imminent. The Court nonetheless upheld the sentence of six years, pointing out that these considerations had been taken into account by the sentencing judge and that in cases where a defendant's plan has progressed further the offence is punishable by up to fifteen years' imprisonment. A similar case to this one is *R v Tabbakh*.[73] In this case the police searched the defendant's flat and found three bottles, fertilizer and other chemicals, together with hand-written bomb-making instructions. Although the defendant had collected the correct ingredients, they were of a poor grade and would not in fact have exploded. He also had yet to make or obtain a detonator. At trial he was convicted of preparation of terrorist acts[74] and sentenced to seven years' imprisonment. The Court of Appeal dismissed his appeal against sentence, pointing out that although the bomb was not a viable one the maximum sentence for this offence is life imprisonment.

Second, *R v Karim*.[75] The defendant in this case had come to the UK from Bangladesh to study microelectronics. In 2006 he settled in the country with his wife and son, both of whom were British. In 2007 he began working at British Airways as a graduate IT specialist. In 2009 Karim's younger brother went to Yemen, contacted the notorious Jihadist Anwar al-Awlaki and put him in contact with Karim. Once al-Awlaki discovered that Karim worked for British Airways, he asked him about his knowledge of security and air travel. Karim responded by suggesting either a physical or electronic attack on British Airways computer servers. He also said that he might be able to get a package on-board a plane. When al-Awlaki discovered that Karim wanted to leave the UK and go and fight in Yemen, al-Awlaki told him that he would be of more use if he remained in the UK. At trial, Karim was convicted on four counts of preparation of terrorist acts.[76] The sentencing judge held that the sentence for these offences (twenty-four years) should be served consecutive to the sentence for an earlier period of activity in which Karim had sent money to help mujahideen in Pakistan and Afghanistan, produced a video in support of a terrorist organization, and possessed a computer file containing instructions on making improvised explosive devices.[77] So, in total, Karim was sentenced to thirty years' imprisonment, with a further five year extension to be added to his licence period. At his appeal against sentence Karim's counsel argued that the sentence was excessive because Karim had not "gone far down the road."[78] He had not actually set about doing anything concrete, and he might never have done anything. The Court of Appeal rejected this argument and upheld Karim's sentence. Whilst accepting that Karim might never have gone on to commit the acts he intended to commit, the Court said that this case was "quite different" from other cases which involve "detailed planning by outsiders":

> The gravamen of the case against this appellant was that he was in a position, and was told from the e-mails in January 2010 to remain in position in the front line so he would be able to carry out from the inside acts of terrorism. It seems us to us that someone in that position is someone who has gone very, very far down the route, and

[73] *R v Tabbakh* [2009] EWCA Crim 464, (2009) 173 JP 201. [74] Terrorism Act 2006, s 5.
[75] *R v Karim* [2011] EWCA Crim 2577, [2012] 1 Cr App R (S) 85. [76] Terrorism Act 2006, s 5.
[77] *R v Karim* (n 75) [9]. [78] *R v Karim* (n 75) [27].

> the fact that he has not actually started to put together the paraphernalia for bombing, but has maintained a position where he can act at once, puts him in a category of someone who has overtly committed himself to the probability of committing really serious acts of terrorism. Comparison with the other cases is therefore unjustified, in the sense that it was not necessary to show overt acts, or preparing bombs or the like. It was sufficient that he was a "sleeper"; he had maintained employment where he was in a position to act immediately.[79]

Third, *R v Gul*.[80] The defendant in this case was a law student at a reputable university in London who, it is believed, radicalized himself over the internet. When police searched his home they found videos on his laptop which he had uploaded to various websites including YouTube. The videos included martyrdom videos and ones which showed attacks on Coalition forces in Iraq and Afghanistan by insurgents. At trial he was convicted on five counts of disseminating a terrorist publication[81] and sentenced to five years' imprisonment. Gul's appeal against conviction focussed on whether the insurgents' actions in the videos fell within the UK's statutory definition of terrorism (with the Court holding that they do). Although he also sought leave to appeal against sentence, this was refused by the Court of Appeal.[82] The Court noted the defendant's young age, previous good character and the serious consequences for him for the rest of his life, but stressed the manner in which the videos glorified and encouraged attacks on UK forces overseas. A similar example is the conviction of Craig Slee on four counts of disseminating a terrorist publication.[83] Slee was sentenced to five years for posting videos on Facebook of al-Qaeda beheading captives. The sentencing judge explained that, whilst Slee had no links to any terrorist organizations and no plans to engage in any attack planning, the videos that Slee had distributed had been created in order to encourage people to rally to the terrorist cause.[84]

In their discussion of terrorism precursor offences, de Goede and Graaf suggest that terrorism trials may be understood as a performative space in which "potential future terror is imagined, invoked, contested, and made real, in the proceedings and verdict, as well as through its wider media and societal echoes."[85] As the preceding paragraphs illustrate, this potential future may be one of a multiplicity of possible futures, and need not be a probable future. On the one hand, a repentant Gul expressed regret at his actions,[86] the courts acknowledged that Slee had no terrorist connections or plans to commit an attack, that there was no evidence that Worrell was planning an attack, that Tabbakh's bomb was not viable and that Karim might never have gone on to commit the acts he intended. On the other hand, the courts simultaneously stressed that the videos Gul and Slee posted might have encouraged others to commit terrorist acts, that Worrell had been stopped before he had been able to go further along the road to

[79] *R v Karim* (n 75) [31].

[80] *R v Gul* [2013] UKSC 64, [2013] 3 WLR 1207.

[81] Terrorism Act 2006, s 2.

[82] *R v Gul* [2012] EWCA Crim 280, [2012] 1 WLR 3432.

[83] "Craig Slee Jailed for Posting Beheading Videos on Facebook," *BBC News*, January 18, 2013) at <http://www.bbc.co.uk/news/uk-england-21090488> accessed November 27, 2013.

[84] Information obtained from <http://www.thelawpages.com> accessed November 27, 2013.

[85] Marieke de Goede and Beatrice de Graaf, "Sentencing Risk: Temporality and Precaution in Terrorism Trials" (2013) 7 *International Political Sociology* 313, 314.

[86] "Islamic Terrorist Propaganda Student Mohammed Gul Jailed," *BBC News*, February 25, 2011) at <http://www.bbc.co.uk/news/uk-england-london-12576973> accessed November 28, 2013.

perpetrating a terrorist act, that Tabbakh was doing his best to make a viable bomb and that Karim was a sleeper agent who might have been utilized by al-Awlaki in the future. The imagining of potential futures thus provides a space for the "incorporation of precautionary counterterrorism into criminal law."[87] This is just one example of the broader shift in criminal justice towards a pre-crime society,[88] a society which Zedner describes as one "in which the possibility of forestalling risks competes with and even takes precedence over responding to wrongs done."[89] It is for this reason that Krasmann argues that enemy criminal law is in fact "not about criminal law, it marks rather a new paradigm of security policy."[90]

(c) A reduction in defendants' procedural rights

The third feature of enemy criminal law is a reduction in the procedural rights of defendants. A stark example in the US is the trial of Guantanamo detainees by military commissions. By contrast, since 9/11 the UK Government has not generally sought to introduce modifications to the criminal trial itself. In Northern Ireland, "Diplock Courts" were introduced in 1973 in response to a report submitted to parliament, which addressed the problem of dealing with Irish republicanism through means other than internment.[91] These courts consisted of a single judge, with the right to trial by jury suspended and a number of special rules as to pre-trial processes, evidence, and punishment. But whilst Diplock Courts were common in the 1970s and 1980s in terrorism cases in Northern Ireland, they were abolished by the Justice and Security (Northern Ireland) Act 2007 and replaced with a new system of non-jury trial, which only applies in exceptional circumstances.[92] Moreover, Diplock Courts were not introduced in the rest of the UK. And whilst since 2003 courts across the UK have had a general power to order a non-jury trial in cases where there is a "real and present danger" of jury tampering,[93] to date this power has only been used once. The defendants in this case were charged with the armed robbery of £1.75m from Heathrow Airport, not with terrorism offences.[94]

Whilst the criminal trial itself has not been modified, however, there are two respects in which the procedural rights of suspected terrorists have been indirectly diminished. The first is the wording of many of the terrorism precursor offences. As

[87] Marieke de Goede and Beatrice de Graaf, "Sentencing Risk: Temporality and Precaution in Terrorism Trials" (2013) 7 *International Political Sociology* 313, 327.

[88] Jude McCulloch and Sharon Pickering, "Pre-Crime and Counter-Terrorism: Imagining Future Crime in the 'War on Terror'" (2009) 49 *British Journal of Criminology* 628.

[89] Lucia Zedner, "Pre-Crime and Post-Criminology?" (2007) 11 *Theoretical Criminology* 261, 262.

[90] Susanne Krasmann, "The Enemy on the Border: Critique of a Programme in Favour of a Preventive State" (2007) 9 *Punishment & Society* 301, 302.

[91] Secretary of State for Northern Ireland, *Report of the Commission to Consider Legal Procedures to Deal with Terrorist Activities in Northern Ireland* (Cm 5185, 1972). See further WL Twining, "Emergency Powers and the Criminal Process" [1973] *Criminal Law Review* 406.

[92] See, for example, Henry McDonald, "Brian Shivers Found Guilty of Massereene Murders," *The Guardian*, January 20, 2012, at <http://www.theguardian.com/uk/2012/jan/20/brian-shivers-guilty-massereene-murders> accessed July 1, 2014.

[93] Criminal Justice Act 2003, s 44.

[94] *R v Twomey, Blake, Hibberd and Cameron* [2011] EWCA Crim 8, [2011] 1 WLR 1681.

Tadros has explained, "The fairness of a trial cannot be detached from the fairness of the offences which provide the basis of argument."[95] Not only does the UK's statutory definition of terrorism have a very wide ambit.[96] The conduct identified by the definitions of many of the terrorism precursor offences is also specified in broad terms, such as collecting any information of a kind likely to be useful to a terrorist,[97] providing instruction in the use of any method for doing anything that is capable of being done for terrorist purposes,[98] possession of any property,[99] and even simply "any conduct."[100] Such broad definitions potentially render criminal law safeguards, particularly the beyond reasonable doubt standard of proof, "toothless."[101] Procedural safeguards have little bite when offence definitions are so broad that almost all citizens fall within them.

The breadth of many of the precursor offences has led Edwards to label them "ouster offences."[102] He explains that there is a discrepancy within these offences between the offence definition and the wrong that is being targeted. The offence of encouragement of terrorism,[103] for example, was targeted at extremists who promote a culture of hate, but is broad enough to also encompass North Korean exiles who criticize their native regime and those "like Cherie Blair, who express their ability to understand the actions of Palestinian suicide-bombers."[104] This is not analogous to offences that prohibit all possession of weapons—that is, offences of necessitous over-inclusion—since those offences seek to guide all citizens away from possessing weapons even if their doing so would pose no risk. Terrorism precursor offences, on the other hand, do not seek to guide all citizens away from all of the conduct they encompass. Instead they operate as a "facilitation device."[105] Only some of those who fall within the offence definition will be selected for prosecution. It seems fair to assume that this choice will, to a large extent, be based on whether the individual is deemed to pose a threat to national security. But at trial the issue will be whether the requirements set out in the offence definition are satisfied. The national security considerations that led to the decision to prosecute will sit in the background. So "Even though the pursuit of security is central to the justification for the law itself, it is not open to challenge by the defendant with respect to his particular case."[106] The effect is to deprive the trial court of the opportunity to adjudicate on the actions that the offence is targeting. Whilst this might lighten the prosecutorial burden, it undermines the courts' ability to deliver procedural justice.

The second way in which defendants' procedural rights have been diminished is by the use of Terrorism Prevention and Investigation Measures (TPIMs). Introduced

[95] Victor Tadros, "Justice and Terrorism" (2007) 10 *New Criminal Law Review* 658, 677.

[96] Terrorism Act 2000, s 1.

[97] Terrorism Act 2000, s 58(1)(a). The House of Lords subsequently narrowed the scope of this offence but stating that, whilst the information need not only be useful to a terrorist, it must be such that it calls for an explanation: *R v G* [2009] UKHL 13, [2010] 1 AC 43.

[98] Terrorism Act 2006, s 6(3)(b).

[99] Terrorism Act 2000, s 16(2)(a).

[100] Terrorism Act 2006, s 5(1).

[101] Tadros, "Justice and Terrorism" (n 95) 675.

[102] James Edwards, "Justice Denied: The Criminal Law and the Ouster of the Courts" (2010) 30 *Oxford Journal of Legal Studies* 725, 732.

[103] Terrorism Act 2006, s 1.

[104] Edwards, "Justice Denied" (n 102) 730.

[105] Edwards (n 102) 729.

[106] Tadros, "Justice and Terrorism" (n 95) 688.

in 2011 as a replacement for the Control Order regime, TPIMs are designed for use against individuals who are believed to be involved in terrorism-related activity where there is no prospect of successful prosecution or deportation.[107] Although they are not as onerous as Control Orders,[108] they may still impose a range of obligations and restrictions.[109] These include: restrictions on travel and on places the individual may visit; restrictions on the individual's use of financial services and electronic communication devices; restrictions on whom the individual may associate and communicate with; a requirement to report to a police station at specified times; electronic monitoring of the individual's movements; and a requirement that the individual reside at specified premises overnight.[110] Two conditions must be satisfied for TPIMs to be imposed: first, the Home Secretary must reasonably believe that the individual is, or has been, involved in terrorism-related activity; and, second, the Home Secretary must reasonably consider that TPIMs are necessary in order to protect the public from a risk of terrorism. This latter condition has an obvious resonance with Jakobs' account of enemy criminal law. Instead of focussing on punishing past actions of the individual, the test is forward-looking: are TPIMs necessary to protect the public from offending behavior in the future. As Ohana explains, this implies that the authorities are expected to gauge the individual's capacity and commitment to abide by the law:

> Were the competent authority to find that the actor is suitably disposed to steer himself as a responsible law-abiding citizen, then the making of a preventive order would not be called for: the actor could be trusted to act appropriately, without there being a need to monitor his conduct by setting special restrictions which do not apply to other citizens.[111]

As one would expect given that TPIMs are intended for use in cases where prosecution is not a viable option,[112] the procedure for imposing TPIMs differs from the ordinary criminal process. First of all, TPIMs are imposed by the Home Secretary not the courts.[113] Although the Home Secretary must apply for the courts' permission before imposing TPIMs (save in urgent cases[114]), the courts' function at the

[107] Home Office, *CONTEST: The United Kingdom's Strategy for Countering Terrorism* (Cm 8123, 2011), Ch 4.

[108] The two principal differences are that TPIMs ended the use of forced relocation and they have a maximum duration of two years (save in cases where fresh evidence comes to light). Control Orders lasted for one year, but there was no limited on how many times they could be renewed. The longest period for which someone was subject to a Control Order was fifty-five months (David Anderson QC, *Control Orders in 2011: Final Report of the Independent Reviewer on the Prevention of Terrorism Act 2005* (The Stationery Office 2012), para 3.47).

[109] Terrorism Prevention and Investigation Measures Act 2011, s 3.

[110] Terrorism Prevention and Investigation Measures Act 2011, Schedule 1.

[111] Ohana, "Trust, Distrust and Reassurance" (n 25) 745.

[112] Prosecution may not be considered viable either because there is no realistic prospect of conviction or because prosecution would be contrary to the public interest (perhaps because the evidence against the suspect is of a sensitive nature): Crown Prosecution Service, *The Code for Crown Prosecutors* (CPS Communication Division, 7th edn, 2013).

[113] For an argument that TPIMs should be issued by the courts, see Stuart Macdonald, "The Role of the Courts in Imposing Terrorism Prevention and Investigation Measures: Normative Duality and Legal Realism" (201)] *Criminal Law and* Philosophy, at <http://link.springer.com/article/10.1007%2Fs11572-013-9255-4> accessed November 26, 2013.

[114] Terrorism Prevention and Investigation Measures Act 2011, s 3(5).

permission hearing is simply to determine whether the Home Secretary's decision to issue TPIMs is "obviously flawed."[115] Moreover, the permission hearing may take place in the absence of the individual, without the individual having had an opportunity to make representations to the court and/or without the individual having been notified of the application.[116] Once the TPIMs notice has been served on the individual a review hearing must be held "as soon as reasonably practicable."[117] Here the court reviews the Home Secretary's decision that the conditions for issuing TPIMs were and continue to be met, applying the "principles applicable on an application for judicial review."[118] The court has the power to quash the TPIMs notice or specified measures within it and the power to direct the Home Secretary to revoke the TPIMs notice or modify specified measures within it.[119] In order to ensure that information is not disclosed contrary to the public interest, the court may exclude the individual and his or her legal representative from all or part of the proceedings[120]—although the House of Lords has ruled that Article 6 of the European Convention on Human Rights (the right to a fair trial) requires that an individual is always given "sufficient information about the allegations against him to enable him to give effective instructions in relation to those allegations."[121] During the closed sessions the interests of the individual are represented by a Special Advocate.[122] Before the closed materials are served the Special Advocate may communicate freely with the individual and his legal representative. Once the Special Advocate has been served, however, he may not communicate with either the individual or his lawyer[123] (save in certain limited situations which are rarely utilized in practice[124]). This restriction on communication between Special Advocates and those whose interests they represent has been strongly criticized, with one Special Advocate even suggesting that it renders their efforts "pretty hopeless."[125] Yet, notwithstanding the fact that TPIMs are imposed on a reduced standard of proof and the individual may not have seen all of the evidence against him, breach of a TPIMs notice without reasonable excuse is a criminal offence punishable by up to five years' imprisonment.[126] This hybrid civil-criminal procedure

[115] Terrorism Prevention and Investigation Measures Act 2011, s 6(3)(a).

[116] Terrorism Prevention and Investigation Measures Act 2011, s 6(4).

[117] Terrorism Prevention and Investigation Measures Act 2011, s 8(5).

[118] Terrorism Prevention and Investigation Measures Act 2011, s 9(2).

[119] Terrorism Prevention and Investigation Measures Act 2011, s 9(5).

[120] Civil Procedure Rules, r 80.18.

[121] *Secretary of State for the Home Department v AF & another* [2009] UKHL 28, [2010] 2 AC 269 [59] (Lord Phillips).

[122] Civil Procedure Rules, r 80.20. [123] Civil Procedure Rules, r 80.21.

[124] One of the special advocates has explained: "The position therefore remains that Special Advocates can communicate with the controlled person only with the permission of the court and that applications for permission must be made on notice to the Secretary of State. Such permission is very rarely sought. In practice, it would be very likely to be refused because any question that it would assist the Special Advocates to ask is likely to be one from which part of the closed case could be inferred" (Martin Chamberlain, "Special Advocates and Procedural Fairness in Closed Proceedings" (2009) 28 *Civil Justice Quarterly* 314, 322).

[125] Joint Committee on Human Rights, *Counter-Terrorism Policy and Human Rights (Eighth Report): Counter-Terrorism Bill* (HC 2007-08, 199) para 67.

[126] Terrorism Prevention and Investigation Measures Act 2011, s 23.

(which has been employed in a number of different contexts in the UK[127]) has been likened to a Trojan horse[128] and described as "an ingenious scheme for imposing harsh punishments yet by-passing the appropriate protections at the crucial stage of the proceedings."[129]

TPIMs also have the potential to circumvent the criminal law in a second way: authorities might choose to rely on TPIMs in cases where it would have been possible to prosecute. Although the UK's counterterrorism strategy states that suspected terrorists should be prosecuted "wherever possible,"[130] the Terrorism Prevention and Investigation Measures Act 2011 does not require the courts to review the decision not to prosecute. Instead the Act imposes a requirement that before imposing TPIMs the Home Secretary must first consult with the police about the possibility of prosecution,[131] with an additional obligation to keep the individual's conduct under review with a view to prosecution for the duration of the TPIMs notice.[132] Having examined the cases of the ten men subject to TPIMs at the end of 2012, however, the Independent Reviewer of Terrorism Legislation found no "undue reticence" to prosecute on the part of the police, Crown Prosecution Service or MI5.[133] Indeed, four of the ten men had previously been prosecuted for terrorism-related activity and in each case the jury had chosen not to convict. The Independent Reviewer commented:

> There is certainly an uncomfortable feel to the imposition of TPIMs on acquitted persons. The practice is however troubling not because it constitutes an abuse of the TPIM system, but because it reveals an unpalatable truth: that while it should always be the first and preferable option for dealing with suspected terrorists, the criminal justice system—whose open nature may prevent some relevant national security evidence from being used—is not always enough to keep the public safe.[134]

Interestingly, the Independent Reviewer went on to say that in his experience (corroborated by other studies) the imposition of TPIMs has not generated feelings of resentment amongst Muslim communities, even though all ten of the men that were subject to TPIMs at the end of 2012 were Muslims.[135] This appears to be because TPIMs have so far been used with restraint. He accordingly went on to warn that "the situation could rapidly change if TPIM notices begin to be used on a significantly greater scale, or against less apparently dangerous targets, than has been the case to date."[136]

[127] Ian Dennis, "Security, Risk and Preventive Orders," in GR Sullivan and Ian Dennis (eds), *Seeking Security: Pre-Empting the Commission of Criminal Harms* (Oxford: Hart Publishing 2012).

[128] Editorial, "In Favour of Community Safety" [1997] *Criminal Law Review* 769, 770.

[129] Andrew Ashworth and Jeremy Horder, *Principles of Criminal Law* (Oxford: Oxford University Press, 7th edn, 2013) 43.

[130] Home Office, *CONTEST* (n 107) para 1.16.

[131] Terrorism Prevention and Investigation Measures Act 2011, s 10(1).

[132] Terrorism Prevention and Investigation Measures Act 2011, s 10(5)(a).

[133] David Anderson QC, *Terrorism Prevention and Investigation Measures in 2012: First Report of the Independent Reviewer on the Operation of the Terrorism Prevention and Investigation Measures Act 2011* (The Stationery Office, 2013) 61.

[134] Anderson QC, *Terrorism Prevention and Investigation Measures in* 2012 (n 133)62.

[135] Anderson QC (n 133) Ch 11.

[136] Anderson QC (n 133) para 11.17.

IV. Enemy Criminal Law as a Prescriptive Concept

The concepts of citizen criminal law and enemy criminal law have great value as explicatory and analytical tools. But Jakobs' account of enemy criminal law was also intended to be prescriptive.[137] He argued:

> Whoever does not provide sufficient cognitive reassurance of a law-abiding behaviour, not only cannot expect to be treated as a person by the State, but the State itself should not treat him as such, because if it does so, the State would be harming the right to security to which other persons are entitled. Hence it would be a terrible mistake to demonize what we are calling here 'enemy criminal law'.[138]

Whilst adding that too much enemy criminal law can damage the rule of law, Jakobs argued that if enemy criminal law is carefully disaggregated from citizen criminal law it can in fact preserve the integrity of the latter. This section of the chapter disputes this claim and advances four countervailing considerations, which militate against the use of enemy criminal law. This is not to reject terrorism precursor offences, consequentialist sentencing or adapted, specially protective, criminal trials outright. There is a pressing need for criminal law theorists to evaluate whether, how, and to what extent these may be justified.[139] Rather, the argument is that terrorism-related criminal laws and processes should not be regarded as existing in a separate realm where the ordinary rules and principles do not apply.

First, the empirical basis for the claim that enemy criminal law can secure the (cognitive) requirements for the legal system to exist is uncertain at best. Gómez-Jara Díez uses systems theory to explain that the criminal law not only presupposes the existence of conditions that the criminal law itself is incapable of securing, but that "to the extent that the State uses enemy criminal law to secure citizen criminal law it risks the whole existence of the latter."[140] In a similar vein, Melía states that there is no empirical basis for thinking that the existence of harsh terrorism offences will "deter more or more efficiently than the use of a less draconian criminal law."[141] Similar criticisms have been levelled at the popular balancing metaphor: there is no empirical basis for the over-simplistic assumption that sacrificing liberty will automatically result in enhanced security.[142]

[137] On the distinction between ideal-types and ideals see Stuart Macdonald, "Constructing a Framework for Criminal Justice Research: Learning from Packer's Mistakes" (2008) 11 *New Criminal Law Review* 257.

[138] Quoted in Carlos Gómez-Jara Díez, "Enemy Combatants versus Enemy Criminal Law" (2008) 11 *New Criminal Law Review* 529, 536.

[139] For examples of some of the work done so far, see n 43.

[140] Díez, (n 138) 533.

[141] Manuel Cancio Melía, "Terrorism and Criminal Law: The Dream of Prevention, the Nightmare of the Rule of Law" (2011) 14 *New Criminal Law Review* 108, 114.

[142] Stuart Macdonald, "Why We Should Abandon the Balance Metaphor: A New Approach to Counterterrorism Policy" (2008) 15 *ILSA Journal of International & Comparative Law* 95; Stuart Macdonald, "The Unbalanced Imagery of Anti-Terrorism Policy" (2009) 18 *Cornell Journal of Law & Public Policy* 519; Jeremy Waldron, "Security and Liberty: The Image of Balance" (2003) 11 *Journal of Political Philosophy* 191; Lucia Zedner, "Securing Liberty in the Face of Terror: Reflections from Criminal Justice" (2005) 32 *Journal of Law & Society* 507.

Second is a danger that Jakobs himself adverted to, the possibility of enemy criminal law permeating into and contaminating citizen criminal law. Zedner, for example, has warned that whilst the introduction of exceptional measures is often controversial, "once enacted they become accepted and, over time, percolate down into the everyday criminal law."[143] This has been echoed by Melía:

> Thus, if such draconian measures creep into what has typically been considered legitimate and normal criminal laws, they may generate significant changes in which the logic of enemy criminal law slowly but surely contaminates our system of criminal law until it becomes the norm rather than the exception.[144]

An example in the UK is the expansion in the use of Special Advocates. Introduced by the Special Immigration Appeals Commission Act 1997, Special Advocates were originally only used in appeals in immigration and asylum cases where the Home Secretary's decision was based on national security concerns.[145] In the years since then Special Advocates have been deployed in numerous others contexts—including the Proscribed Organisations Appeal Commission,[146] the Pathogens Access Appeal Commission,[147] Employment Tribunals,[148] and Parole Board hearings,[149] as well as TPIMs review hearings (as explained previously)—culminating in the Justice and Security Act 2013, which provides for the use of closed sessions in *any* civil proceedings before the High Court, Court of Appeal, or Supreme Court where this is required in the interests of national security and the fair and effective administration of justice.[150]

Third, enemy criminal law adopts a relativistic approach to substantive and procedural rights which is at odds with the universality of human rights.[151] On this approach, human rights become conditional. They are not vested in the individual by virtue of their personhood, but have to be earned through loyalty to the law. Since the conditions for singling out citizens are so vaguely defined, this is capable of generating its own form of insecurity: anxiety that one's human rights might be suspended. The alternative is to insist that human rights are absolute:

> Arguably the criminal law is better protected by insisting upon the citizenship status of those against whom criminal proceedings are brought; by maintaining, through the presumption of innocence, that they are law-abiding members of society until proven guilty; and by adhering to the protections of the criminal process even in the gravest case.[152]

[143] Lucia Zedner, "Security, the State, and the Citizen: The Changing Architecture of Crime Control" (2010) 13 *New Criminal Law Review* 379, 394.
[144] Melía, "Terrorism and Criminal Law" (n 141) 112.
[145] John Ip, "The Rise and Spread of the Special Advocate" [2008] PL 717.
[146] Terrorism Act 2000, Schedule 3.
[147] Anti-terrorism, Crime and Security Act 2001, Schedule 6.
[148] Employment Tribunals Act 1996, s 10.
[149] *R (Roberts) v Parole Board* [2005] UKHL 45, [2005] 2 AC 738.
[150] Justice and Security Act 2013, s 6.
[151] Mireille Delmas-Marty, "Violence and Massacres—Towards a Criminal Law of Inhumanity?" (2009) 7 *Journal of International Criminal Justice* 5.
[152] Zedner, "Security, the State, and the Citizen" (n 143) 392.

Closely connected to this is the final danger, that enemy criminal law will undermine the criminal law's moral authority. To apply a diminished level of human rights protection to a specific group of people when many members of that group come from particular ethnic minorities is to risk undermining the legitimacy of government both domestically and overseas.[153] Moreover, as Fletcher argues, the discourse of loyalty and community is exclusionary. He states that enemy criminal law "intensifies the perception of insiders and outsiders," thereby returning us "to the most primitive way of handling criminals—expulsion, excommunication, and banishment."[154] The UK's counterterrorism strategy emphasizes the importance of social inclusion in preventing radicalization.[155] As this suggests, enacting exclusionary laws which generate resentment and ill-feeling is likely to prove counter-productive.

V. Conclusion

In the concept of enemy criminal law the "exceptional measures of the war on terror are legalized and incorporated into criminal law."[156] Using the concept as an analytical aid, this chapter has highlighted: the extensive reach of the UK's raft of terrorism precursor offences; how potential futures and a precautionary desire to mitigate risk lead to the imposition of severe sentences on those convicted of these crimes; and how the procedural rights of those accused of these offences have been indirectly diminished. As concern grows over the possibility of terrorists launching cyber attacks, and policy-makers and legislators assess how best to respond to this threat, it is important to be mindful of the counter-productivity of enemy criminal law. Not only is there a danger that such laws will contaminate other parts of the criminal law and the legal system more generally, but the exclusionary discourse and relativistic conception of human rights are likely to generate resentment and ill-feeling amongst those communities most affected. Ultimately, it is self-defeating to create new offences, procedures, and sentencing powers that undermine the criminal law's moral authority when this moral authority is the very reason for insisting that suspected terrorists should be prosecuted whenever possible in the first place.

[153] David Cole, *Enemy Aliens: Double Standards and Constitutional Freedoms in the War on Terrorism* (New York: New Press, 2003).

[154] George P Fletcher, *The Grammar of Criminal Law*, vol 1 (Oxford: Oxford University Press, 2007) 172.

[155] Home Office, *CONTEST* (n 107) para 5.17.

[156] de Goede and de Graaf, "Sentencing Risk" (n 85) 328.

5

Cyberwar versus Cyber Attack

The Role of Rhetoric in the Application of Law to Activities in Cyberspace

Laurie R Blank

I. Introduction

The word "cyber" has grabbed the world's attention over the past several years: put "cyber" in front of nearly any word and you have a new term for a new millennium. In the media and public discourse, words such as cyber attack, cyberwar, cyber doom, cyber security, and cybercrime sell news and produce entirely new channels for debate and analysis. Cyberspace is, without a doubt, a complex and sometimes impenetrable context that demands new expertise and creative approaches to problems. However, it is also fertile ground for runaway rhetoric—discourse and terminology that can have unintended effects reverberating far beyond the news story or journal article.

In the realm of national security, international relations, politics, and law, cyber activities now play an extensive role in all facets of society, situated along an expansive continuum with information analysis and gathering at one end and hostilities at the other, roughly, and including espionage, surveillance, crime, and other activities along its span:

> Cyber operations can be broadly described as operations against or via a computer or a computer system through a data stream. Such operations can aim to do different things, for instance to infiltrate a system and collect, export, destroy, change, or encrypt data or to trigger, alter or otherwise manipulate processes controlled by the infiltrated computer system. By these means, a variety of "targets" in the real world can be destroyed, altered or disrupted, such as industries, infrastructures, telecommunications, or financial systems. The potential effects of such operations are therefore of serious humanitarian concern.[1]

Technology's constant advances now enable both militaries and civilians to engage in cyber activity to achieve objectives, whether related to protest and revolution, crime, terrorism, espionage, or military operations. Governments and private companies face a nearly constant onslaught of cyber activity seeking to access information, undermine or damage systems, or otherwise gain a financial, political, or strategic advantage of

[1] International Committee of the Red Cross, 31st International Conference of the Red Cross and Red Crescent, "International Humanitarian Law and the Challenges of Contemporary Armed Conflicts," Report 31IC/11/5.1.2, October 2011, at 36.

some kind. Governments and companies alike have established both formal and informal mechanisms for countering these rapidly developing threats and operations in cyberspace, including, for example, US Cyber Command, China's People's Liberation Army General Staff Department's 3rd and 4th Departments,[2] Iranian Sun-Army and Cyber Army,[3] Israel's Unit 8200,[4] and the Russian Federal Security Service's Federal Agency of Government Communications and Information (FAPSI), and other units.[5]

Extensive analysis of the legal, political, and security consequences of cyber operations, cyber capabilities and cyber activities over the past several years has focused on a range of issues, including the application of international law to activities in the cyber arena. At the same time, several high profile events—the cyber operations in Georgia during the 2008 conflict between Russia and Georgia, the Stuxnet virus, the comprehensive computer network operations launched against the Estonian government in the summer of 2007, or the August 2012 computer sabotage of ARAMCO, the Saudi state-owned oil company, for example—have fanned the flames of rhetorical flourish being used to describe events and potential events in cyberspace. A look at news coverage of these issues in recent years demonstrates the growing focus across a range of countries, industries, and disciplines, with the number of news stories mentioning "cyberwar," "cyberwarfare," or "cyber attack" in 2010, 2011, or 2012 more than triple that of any previous year before 2009.[6] The terms "cyberwar," "cyberwarfare," and "cyber attack" are used to describe a broad array of activities, many of which fall far outside the general meaning of the terms "war" or "attack" within international law. This mixing of legal terminology with colloquial discourse can have important ramifications for the application of the law and, as a result, for the protection of persons, preservation of state authority, and stability of the international system. For this reason, understanding the consequences of rhetorical excess is essential to preventing overreach and danger to individual rights in the name of "war" or in response to "attacks."

Counterterrorism policy over the past decade offers a prime example of the impact rhetoric can have on the development and implementation of the law. The rhetoric of the "war on terror" facilitated and encouraged the growth of authority without the corresponding obligation in many cases. For example, the drone campaign in Pakistan, indefinite detention, prosecution of crimes such as conspiracy or material support for terrorism in military commissions and other practices raised significant questions about the application of domestic and international law to counterterrorism

[2] See Larry M Wortzel, *The Chinese People's Liberation Army and Information Warfare* (US Army War College, 2014).

[3] See Tom Gjelten, "Could Iran Wage a Cyber War on the U.S.?" *National Public Radio*, April 26, 2012, at <http://www.npr.org/2012/04/26/151400805/could-iran-wage-a-cyberwar-on-the-u-s>.

[4] Yaakov Katz, "IDF Admits to Using Cyber Space to Attack Enemies," *Jerusalem Post*, June 3, 2012.

[5] FAPSI Operations, GlobalSecurity.org, at <http://www.globalsecurity.org/intell/world/russia/fapsi-ops.htm>. See also "Russia to Create Cyberwarfare Units by 2017," *RIA Novosti*, January 30, 2014, at <http://en.ria.ru/military_news/20140130/187047301/Russia-to-Create-Cyberwarfare-Units-by-2017.html>.

[6] A brief Lexis-Nexis search of major newspapers shows 944, 965, and 839 hits for the term "cyber attack" in 2010, 2011, and 2012 respectively, compared to approximately 200 or fewer for any year before 2009. The same search yields 1,142 results for the first ten months of 2013. The same general pattern holds true for the terms "cyberwar" and "cyberwarfare."

operations, the long-term impact on executive authority and the use of national security as a "trump card" in the face of legal obstacles or challenges. In addition, layering rhetoric on top of the law has affected the application and implementation of key bodies of law, such as human rights law, the law of armed conflict, and various domestic legal regimes relevant to national security and counterterrorism.[7] Over the course of several years, the mix of counterterrorism operations and military operations, and a rhetoric of war that subsumed both, helped lead to minimized rights and magnified executive powers.

Just as the rhetoric of war subsumed a wide variety of counterterrorism measures within the concept of a "war on terror" and ultimately had a profound effect on both law and policy with respect to counterterrorism and war, so the potential for similar consequences in the cyber arena exists as well. Cyber activities can span a continuum of "bad activity" from cyber crime to cyber espionage to cyber terrorism and all the way to cyber attacks and cyberwar. Not all of these acts fit within the paradigm of international law governing the use of force or the law of armed conflict regime, as discussed in greater detail below. Understanding the impact of certain rhetorical choices on potential legal interpretations and applications is, therefore, important. For example, the term "cyber attack" is regularly used in the mass media to denote an extremely wide range of cyber conduct, much of which falls well below the threshold of an "armed attack" as understood in the jus ad bellum or an attack as defined in the law of armed conflict for purposes of triggering the obligations of distinction, proportionality and precautions. Rhetoric that uses a terminology of war, like "cyberwar" or "cyber attack," can create situations in which a state has fewer obstacles to an aggressive response to cyber threats or cyber conduct, stretching or overstepping the relevant legal boundaries. In this way, such rhetoric poses a serious risk of elevating or escalating an apparently hostile action to the status of war or conflict when, in the absence of such rhetoric, it would be more appropriately handled or countered within the criminal system or other non-forceful paradigm.

This chapter analyzes how the interplay between law and rhetoric thus forms an important backdrop to the analysis of the international legal norms that govern how a state can respond to cyber threats or engage in other cyber operations. Rhetoric that opens the door to overly broad responses necessitates an understanding of the relevant legal paradigms, the boundaries between them, and the fundamental principles that guide their application. Use of terms like "war" and "attack" for a much wider array of activities also facilitates a blurring of the lines between relevant and applicable legal frameworks, which can have a detrimental effect on both individual rights and the development of the law. With regard to cyber threats and cyber attacks, both the jus ad bellum and the law of armed conflict help shape the parameters of lawful and effective action, not only by guiding the appropriate conduct when force may lawfully be used or during armed conflict, but also by delineating the dividing line between crime and war and between self-defense and law enforcement.

[7] See, generally, Laurie R Blank, "The Consequences of a 'War' Paradigm for Counterterrorism: What Impact on Basic Rights and Values?" (2012) 46 *Georgia Law Review* 719.

Section II examines the consequences of the use of the terms "cyberwar" or "cyberwarfare" to connote a wide range of actual and potential cyber activity or threats, across a broad spectrum of activity. After setting forth the traditional meaning of "war" and the current framework, which uses the term "armed conflict" rather than "war," this section highlights the consequences of the rhetoric of "war" in the cyber realm, with specific reference to lessons learned from the past decade of counterterrorism. The third section takes a similar approach to analyzing the consequences of the term "cyber attack," focusing on both the jus ad bellum concept of "armed attack" and the law of armed conflict definition of "attack." In particular, the use of "cyber attack" to subsume an extraordinarily broad set of activities has consequences both in terms of the specific understanding of the word "attack" in each legal regime, and in blurring the notion of "attack" into one unspecified and extensive term that conflates two or more legal concepts. Each of these results has significant ramifications for the application of international law, the preservation of the international system, and the protection of persons during times of conflict, thus demanding attention to the rhetoric of cyberspace rather than just the legal implications of cyber activity alone.

II. Cyberwar

An early definition of cyberwarfare, but one still in regular use, is "any operation that disrupts, denies, degrades, or destroys information resident in computers or computer networks."[8] In another study, the authors split cyberwarfare into five general varieties, ranging from the mildest to the most severe: (1) web vandalism, (2) disinformation campaigns, (3) gathering secret data, (4) disruption in the field, and (5) attacks on critical national infrastructure.[9] These definitions use "war" as a descriptive term and a rhetorical term, rather than a legal term. However, the notion of cyberwar encompasses action that could or does fit within the legal conception of war, producing the potential for confusion and conflation. The use of terms that sound like "war" but are in fact much broader in scope than the corresponding legal terms and definitions can have significant consequences for the application of the law, the execution of operations, and the protection of persons and property.

(a) The law of armed conflict's triggers and cyber operations

Wartime triggers the application of a specific legal regime—the law of armed conflict (LOAC). Otherwise known as the law of war or international humanitarian law, LOAC governs conduct during wartime and provides the overarching parameters for the conduct of hostilities and the protection of persons and objects.[10] It authorizes

[8] Walter Gary Sharp Sr, *Cyberspace and the Use of Force* (Aegis Research Corp, 1999) 132.

[9] See Center for the Study of Technology and Society, *Special Focus: Cyberwarfare* (2001) at <http://web.archive.org/web/20061205020720/tecsoc.org/natsec/focuscyberwar.htm>.

[10] The LOAC is set forth primarily in the four Geneva Conventions of August 12, 1949, and their Additional Protocols. Geneva Convention for the Amelioration of the Condition of the Wounded and Sick in Armed Forces in the Field, August 12, 1949, 6 UST 3114, 75 UNTS 31 [hereinafter First Geneva Convention]; Geneva Convention for the Amelioration of the Condition of Wounded, Sick and Shipwrecked Members of Armed Forces at Sea, August 12, 1949, 6 UST 3217, 75 UNTS 85;

the use of lethal force as first resort against enemy persons and objects within the parameters of the armed conflict.[11] It also provides, based on treaty provisions and the fundamental principle of military necessity, for the detention of enemy fighters and civilians posing imperative security risks.[12] Along with these authorities, however, come obligations—such as the obligation to use force in accordance with the principles of distinction and proportionality,[13] the obligation to protect civilians and those no longer fighting from the ravages of war to the extent possible, and the obligation to treat all persons humanely.

The identification of when this legal regime applies is not a rhetorical process, however, or one dependent on rhetorical or political choice. Rather, the application of LOAC rests on an objective analysis of the facts of a given situation to determine whether it meets the threshold for triggering LOAC. Although historically, existing treaty and customary law applied to situations of declared war, the drafters of the 1949 Geneva Conventions eschewed the term "war" in favor of the term "armed conflict." Whereas countries had denied LOAC obligations in the past by claiming that they were not engaged in "war,"[14] a term that once had specific legal connotations, the Geneva Conventions eliminated that particular circumlocution by creating a trigger for law applicability based on the existence of an armed conflict.[15] The difference between the legal analysis necessary to identify a situation of armed conflict—or "war" in the

Geneva Convention Relative to the Treatment of Prisoners of War, August 12, 1949, 6 UST 3316, 75 UNTS 135 [hereinafter Third Geneva Convention]; Geneva Convention Relative to the Protection of Civilian Persons in Time of War, August 12, 1949, 6 UST 3516, 75 UNTS 287 [hereinafter Fourth Geneva Convention]; Protocol Additional to the Geneva Conventions of August 12, 1949, and Relating to the Protection of Victims of International Armed Conflicts (Protocol I), June 8, 1977, 1125 UNTS 3 [hereinafter Additional Protocol I]; Protocol Additional to the Geneva Conventions of August 12, 1949, and Relating to the Protection of Victims of Non-International Armed Conflicts (Protocol II), June 8, 1977, 1125 UNTS 609 [hereinafter Additional Protocol II].

[11] See Geoffrey S Corn, "Back to the Future: De Facto Hostilities, Transnational Terrorism, and the Purpose of the Law of Armed Conflict" (2009) 30 *University of Pennsylvania Journal of International Law* 1345, 1352–3 ("[A]rmed conflict is defined by the authority to use deadly force as a measure of first resort.").

[12] See Third Geneva Convention, (n 6), Article 4 (describing various prisoner of war categories); Fourth Geneva Convention, (n 6), Articles 42, 78 (permitting internment).

[13] The principle of distinction mandates that all parties to a conflict distinguish between those who are fighting and those who are not and that parties only target those who are fighting. In addition, fighters, including soldiers, must distinguish themselves from innocent civilians. See Additional Protocol I, (n 5), Article 48. The principle of proportionality states that parties must refrain from attacks where the expected civilian casualties will be excessive in relation to the anticipated military advantage. See Additional Protocol I, (n 5), Article 51(5)(b).

[14] For example, during World War II, the Japanese claimed that their operations in China and Manchuria were "police operations" and, therefore, did not trigger the law of war. See International Military Tribunal for the Far East, judgment of November 4, 1948, at 490 ("From the outbreak of the Mukden Incident till the end of the war[,] the successive Japanese Governments refused to acknowledge that the hostilities in China constituted a war. They persistently called it an 'Incident.' With this as an excuse[,] the military authorities persistently asserted that the rules of war did not apply in the conduct of the hostilities.").

[15] Common Article 2 of the 1949 Geneva Conventions applies to "all cases of declared war or of any other armed conflict... between two or more [States], even if the state of war is not recognized by one of them." Common Article 2 to the four Geneva Conventions of 1949 (n 6). See, for example, Anthony Cullen, "Key Developments Affecting the Scope of Internal Armed Conflict in International Humanitarian Law" (2005) 183 *Military Law Review* 66, 85 ("[I]t is worth emphasizing that recognition of the existence of armed conflict is not a matter of state discretion.").

colloquial sense—and the rhetorical use of cyberwar provides an important framework for exploring the consequences of expansive "war" rhetoric in the cyber arena.

The 1949 Geneva Conventions endeavor to address all instances of armed conflict[16] and set forth two primary categories of armed conflict that trigger the application of LOAC: international armed conflict and non-international armed conflict. Common Article 2 of the 1949 Geneva Conventions states that the Conventions "shall apply to all cases of declared war or of any other armed conflict which may arise between two or more of the High Contracting Parties, even if the state of war is not recognized by one of them."[17] Common Article 3 of the 1949 Geneva Conventions sets forth minimum provisions applicable "in the case of armed conflict not of an international character occurring in the territory of one of the High Contracting Parties."[18] In both categories of conflict, the term "armed conflict" is used specifically to avoid the technical legal and political pitfalls of the term "war," as already noted.[19] As such, determination of the existence of an armed conflict does not turn on a formal declaration of war—or even on how the participants characterize the hostilities—but rather on the facts of a given situation.

Neither Common Article 2 nor Common Article 3 specifically defines armed conflict. The most common and oft-cited contemporary definition of armed conflict is from the Appeals Chamber of the International Criminal Tribunal for the former Yugoslavia (ICTY) in *Prosecutor v Tadić*, where the tribunal held that an armed conflict exists whenever "there is a resort to armed force between States or protracted armed violence between governmental authorities and organized armed groups or between such groups within a State."[20] The first portion of the definition refers to international armed conflict—any conflict between two states. As the Commentary to the Geneva Conventions explains, "[a]ny difference arising between two States and leading to the intervention of armed forces is an armed conflict within the meaning of Article 2, even if one of the Parties denies the existence of a state of war."[21] The duration of the hostilities or the number of wounded or killed does not impact the characterization as an armed conflict. Any dispute between two states involving their armed forces, no matter how minor or short-lived, thus triggers application of Common Article 2 and the full body of LOAC.

There is general consensus that, although highly unlikely, cyber acts alone could trigger or constitute an international armed conflict. If two states are involved and

[16] Oscar M Uhler and Henri Coursier (eds), *Commentary on Geneva Convention IV Relative to the Protection of Civilian Persons in Time of War* (Geneva: International Committee of the Red Cross, 1958) 26 [hereinafter GC IV Commentary] ("Born on the battlefield, the Red Cross called into being the First Geneva Convention to protect wounded or sick military personnel. Extending its solicitude little by little over other categories of war victims, in logical application of its fundamental principle, it pointed the way, first to the revision of the original convention, and then to the extension of legal protection in turn to prisoners of war and civilians. The same logical process could not fail to lead to the idea of applying the principle in all cases of armed conflicts, including those of an internal character.").

[17] Common Article 2 to the Four Geneva Conventions (n 6).

[18] Common Article 3 to the Four Geneva Conventions (n 6).

[19] GC IV Commentary (n 12) 17–25.

[20] *Prosecutor v Tadić*, Case No IT-94-1, Decision on Defence Motion for Interlocutory Appeal on Jurisdiction, para 70 (ICTY, October 2, 1995).

[21] GC IV Commentary (n 12) 20.

engage in acts amounting to armed hostilities using cyber operations or cyber capabilities, an international armed conflict would exist, even if short and limited in scope. For example, the *Tallinn Manual* suggests that "a cyber operation that causes a fire to break out at a small military installation would suffice to initiate an international armed conflict."[22] Two particular challenges make the identification of an armed conflict difficult in a solely cyber context: whether the damage is sufficient to qualify as "armed" force and whether it is possible to attribute the act to a state or to individuals whose conduct is attributable to a state. Indeed, at this time, "no international armed conflict has been publicly characterized as having been solely precipitated in cyberspace."[23] These legal and practical limitations for characterizing cyber acts as international armed conflict are important to consider when assessing the rhetorical use of the term "cyberwar" or "cyberwarfare" in this chapter.

Non-international armed conflict—a conflict covered by Common Article 3 of the Geneva Conventions—is marked by protracted armed violence between a state and an organized armed group or between two or more organized armed groups. The threshold for non-international armed conflict is thus higher than that for international armed conflict and requires a more comprehensive analysis of the situation at hand. According to the Commentary, no specific test for determining the applicability of Common Article 3 exists; rather, the goal is to interpret Common Article 3 as broadly as possible.[24] Two considerations have proven to be particularly important to courts and tribunals faced with uncertainty about the existence of a non-international armed conflict: "the intensity of the conflict and the organization of the parties to the conflict."[25] These criteria help to "distinguish[] an armed conflict from banditry, unorganized and short-lived insurrections, or terrorist activities, which are not subject to international humanitarian law."[26]

Intensity requires an analysis of the seriousness of the fighting in order to determine whether it has passed from riots and other random acts of violence to engagements more akin to regularized military action. Traditionally, analyzing intensity has encompassed a range of specific factors regarding the actual hostilities. For example, the ICTY has considered factors such as the number, duration, and intensity of individual confrontations; the types of weapons and other military equipment used; the number of persons and types of forces engaged in the fighting; the geographic and temporal distribution of clashes; the territory that has been captured and held; the

[22] Michael N Schmitt (ed), *Tallinn Manual on the International Law Applicable to Cyber Warfare* (New York: Cambridge University Press, 2013) 83. Apropos of the topic of this chapter, the Tallinn Manual notes that the "term 'cyber warfare' is used [therein] in a purely descriptive, non-normative sense." Schmitt, *Tallinn Manual*, 4.

[23] Schmitt, *Tallinn Manual* (n 22) 84.

[24] GC IV Commentary, (n 12) 36 ("Does this mean Article 3 is not applicable in cases where armed strife breaks out in a country, but does not fulfil any of [the suggested criteria]? We do not subscribe to this view. We think, on the contrary, that the Article should be applied as widely as possible.").

[25] *Tadić*, Case No IT-94-1-T, judgment, para 562 (ITCY, May 7, 1997); Sylvain Vite, "Typology of Armed Conflicts in International Law: Legal Concepts and Actual Situations" (2009) 91 *International Review of the Red Cross* 76–7.

[26] *Tadić*, Case No IT-94-1-T, judgment, para 562. Government forces are presumed to be sufficiently organized to be a party to an armed conflict. *Prosecutor v Haradinaj*, Case No. IT-04-84-T, judgment, para 60 (ICTY, April 3, 2008); Vite (n 21) 77.

number of casualties; the extent of material destruction; and the number of civilians fleeing combat zones.[27] The ICTY has also declared that the involvement of the UN Security Council may reflect the intensity of a conflict.[28] The collective nature of the fighting, the state's resort to use of its armed forces, the duration of the conflict, and the frequency of the acts of violence and military operations are all additional factors to take into account as well.

There is little doubt that a cyber-based conflict could, at some point, reach a sufficient level of intensity to satisfy this threshold; however, the evidence or analysis of such intensity could differ from the factual information used in a kinetic scenario. Most or all of these considerations are highly relevant in the cyber context as well (with the exception, perhaps, of the capture of territory), but the analysis will rely overwhelmingly on the effects of attacks rather than the types of operations, the engagement of forces, or the number of persons involved, because those categories of information are extremely difficult to assess in the cyber arena.

In addition to intensity, various international tribunals and other courts have looked to a non-state armed group's level of organization as one way to distinguish armed conflict from unorganized violence and riots. Whether one takes a more formalized approach to the definition of armed conflict, relying heavily on the intensity/organization factors, or a more totality-of-the-circumstances approach,[29] some notion of an opposing party fighting against the state is essential to characterizing a situation as an armed conflict for the application of LOAC. Factors traditionally considered as important in determining whether a group is sufficiently organized to be a party to an armed conflict include a hierarchical structure, territorial control and administration, the ability to recruit and train fighters, the ability to launch operations using military tactics and the ability to enter peace or ceasefire agreements.[30] The International Committee of the Red Cross (ICRC) has also highlighted the non-state actor's authority to launch attacks bringing together different units and the existence or promulgation of internal rules.[31] To the extent that a non-state armed group is engaged in a struggle against government forces in which cyber operations form only one tool in that struggle, the analysis will be similar to that in other situations, such as those highlighted in the ICTY's jurisprudence in which it, for example, examined the nature of the Kosovo Liberation Army in determining whether it constituted an organized armed group such that the violence in Kosovo was an armed conflict.[32]

[27] *Prosecutor v Haradinaj*, Case No IT-04-84-T, judgment, para 49 (ICTY, April 3, 2008); *Prosecutor v Limaj*, Case No IT-03-66-T, judgment, paras 135–43 (ICTY, November 30, 2005); *Prosecutor v Tadić*, judgment, paras 564–5.

[28] *Prosecutor v Haradinaj* (n 23) para 49.

[29] Rigid adherence to specific measures or types of organization have the potential to undermine the effectiveness of LOAC by hindering its application to situations that otherwise seem to obviously fall within the notion of an armed conflict. See Laurie R Blank and Geoffrey S Corn, "Losing the Forest for the Trees: Syria, Law and the Pragmatics of Conflict Recognition" (2012) 46 *Vanderbilt Journal of Transnational Law* 693.

[30] See *Prosecutor v Limaj*, paras 95–109; *Prosecutor v Lukić*, Case No IT- 98-32/1-T, judgment, para 884 (ICTY, Jul 20, 2009); *Prosecutor v Haradinaj*, para 60.

[31] Vite (n 21) 77.

[32] See, for example, *Prosecutor v Limaj*; *Prosecutor v Haradinaj*.

When an organized armed group engages with the government solely in the cyber realm, however, the cyber arena can pose unique challenges for determining the existence of a non-international armed conflict. One could envision a group that is organized with some type of command structure, a decision-making and operational planning process, and the ability to launch operations. In essence, the type of weapon used—cyber—would be the main distinction between this type of group and an organized armed group using kinetic[33] force, and there would be little question that such a group is sufficiently organized to meet the criterion of organization to be a party to an armed conflict. It is much more likely, however, that the non-state actors launching cyber attacks against a government would be independent actors, disparate actors sharing similar goals or even loosely affiliated groups of hackers or other actors. "Autonomous actors who are simply all targeting a State, perhaps in response to a broad call to do so from one or more sources," but without any formal direction or structure, "cannot be deemed to be organized."[34] As the Commentary to the Additional Protocols explains, "individuals operating in isolation" generally do not fit within the understanding of "organized."[35]

The nature of the virtual world, in which members of groups—even ones with a high degree of organization and shared purpose—have no face-to-face contact or connection, compounds the challenges of identifying sufficient organization to meet the definition of armed conflict. For example, during the conflict between Georgia and Russia in the summer of 2008, numerous cyber attacks were launched against Georgia. Most of these attacks were initiated using information from a website that provided cyber tools and lists of Georgian government websites and cyber targets.[36] The attacks were not coordinated with regard to timing, target and effect, or in any other aspect. Based on existing analyses of the *Tadić* definition of armed conflict and the requisite components of the factor of organization, something more than this type of merely collective action would be needed in the solely cyber realm. It has been argued that the determination of whether a group acting for a shared purpose meets the organization criterion should depend on such context-specific factors as the existence of a formal or informal leadership entity directing the group's activities in a general sense, identifying potential targets and maintaining an inventory of effective hacker tools.[37]

[33] Kinetic force refers to the use of actual weapons (guns, munitions, etc) rather than electronic, cyber, electromagnetic, or other tools.

[34] Michael N Schmitt, "Cyber Operations and the Jus in Bello: Key Issues" (2011) 41 *Israel Yearbook on Human Rights* 113, 124–5.

[35] Yves Sandoz, Christophe Swinarski, and Bruno Zimmermann (eds), *Commentary on the Additional Protocols of 8 June 1977 to the Geneva Conventions of 12 August 1949* (Geneva: International Committee of the Red Cross,1987) 512 (hereinafter Protocol Commentary).

[36] Eneken Tikk, Kadri Kaska, and Liis Vihul, *International Cyber Incidents: Legal Considerations* (Cooperative Cyber Defence Centre of Excellence, 2010).

[37] Schmitt (n 28) 125. Although this analysis of factors relevant to organization is presented with regard to the identification of an organized armed group for the purposes of distinguishing members of that group from civilians in the context of a direct participation in hostilities analysis, it is equally useful in the present context of analysing the extent of a group's organization for the purposes of finding an armed conflict that triggers the application of LOAC.

(b) Rhetoric and its consequences

As the above analysis demonstrates, although cyber operations can constitute an armed conflict—either international or non-international—a variety of challenges and uncertainties make it somewhat unlikely that purely cyber activities will trigger LOAC. And yet, pronouncements of "cyberwar" and "cyberwarfare" abound in the news media and commentary, particularly in the context of a threat from China and hackers operating on behalf of or with the tacit approval of the Chinese government.[38] In 2011, then Secretary of Defense Leon Panetta warned that "the next Pearl Harbor could very well be a cyber attack."[39] The immediate consequence of this rhetoric is to conflate cyber security issues with cyberwar; to conflate espionage, crime, and other acts of sabotage or coercion with actual conflict. A number of cyber experts have begun to criticize this conflation from a policy perspective, focusing on the dangers of escalating both threats and responses. Thus:

> [although] cyber crime and cyber espionage are real problems, conflating them under one term limits the possibility for taking the most specific and effective actions in response to each, leading simultaneously to the possibility of miscalculation and overreaction in some cases and a do-nothing, boy-who-cried-wolf response in others.[40]

The consequences extend beyond the borders of the United States and can have significant consequences for diplomacy and international relations as well. The idea of a "cyber arms race" is sobering, indeed.[41] One scholar has taken this analysis a step further and compared so-called "cyberwar" to war as understood by Clausewitz, concluding that cyberwar is not "war" as traditionally understood. Under Clausewitz's definition, war must (1) be a violent act, (2) be instrumental, that is, have a means and an end, and (3) have a political purpose rather than be an isolated act.[42] According to this analysis, "there is no cyber offense that meets all three criteria…and, there are very few cyber attacks in history that meet only one of these criteria."[43]

Beyond the policy criticisms and the concerns about threat escalation and policy confusion, the rhetoric of cyberwar has significant legal consequences as well. The primary

[38] See, for example, Sreeram Chaulia, "Cyber Warfare is the New Threat to the Global Order," *The Nation*, April 11, 2013, at <http://www.nationmultimedia.com/opinion/Cyber-warfare-is-the-new-threat-to-the-global-order-30203813.html>; David E Sanger, David Barboza and Nicole Perlroth, "Chinese Army Unit is Seen as Tied to Hacking Against U.S.," *The New York Times*, February 18, 2013; Richard A Clark and Robert K Knake, *Cyber War* (New York: Ecco, 2010).

[39] Lisa Daniel, "Panetta: Intelligence Community Needs to Predict Uprisings" (American Forces Press Service, February 11, 2011).

[40] Sean Lawson, "Putting the 'War' in 'Cyberwar': Metaphor, Analogy and Cybersecurity Discourse in the United States," 17 *First Monday*, July 2, 2012, at <http://firstmonday.org/ojs/index.php/fm/article/view/3848/3270#author> accessed November 29, 2013.

[41] See Bruce Schneier, "China CyberWar Rhetoric Risks Dangerous Implications," at <http://searchsecurity.techtarget.com/video/Bruce-Schneier-China-cyberwar-rhetoric-risks-dangerous-implications> accessed November 29, 2013.

[42] See Thomas Rid, "Cyber War Will Not Take Place" (2012) 35 *Journal of Strategic Studies* 5, 6, citing Carl von Clausewitz, *On War*. See also Thomas G Mahnken, "Cyber War and Cyber Warfare," in Kristin Lord and Travis Sharp (eds), *America's Cyber Future: Security and Prosperity in the Information Age* (Center for a New American Security, 2011) 53–62.

[43] Rid (n 35) 9.

issue is how the terminology of "war" impacts and changes the overall legal framework. The assertion of the "war" paradigm naturally leads to the assertion of wartime privileges and authorities—appropriate in the context of an actual armed conflict triggering the application of LOAC but potentially highly problematic when the result of rhetorical flourish and policy conflation. Unlike the counterterrorism arena, where rhetoric drove law to the detriment of individual rights, the cyber discourse has, fortunately, not yet produced such results, partly due to extensive prospective efforts to analyze how international law applies to cyber activities, such as the *Tallinn Manual*. Nonetheless, "talking about cyberwar . . . blurs the lines between competition, espionage and military conflict"[44] in a manner than can become problematic. Furthermore, uncertainty about what cyberwarfare actually is translates into uncertainty about the applicable law[45] and, from there, the potential for manipulation of the law. Understanding the consequences of the rhetoric is, therefore, essential to ensuring that cries of "cyberwar!" do not result in diminished rights, unfettered authority, and other unjustified "wartime" excesses.

The use of force is perhaps the starkest example of what an expansive, rhetoric-driven conception of war would mean. Invoking wartime authority is, at base, a decision to harness the authority to use force as a first resort against those identified as the enemy, whether insurgents, the armed forces of another state, or hackers operating as an organized armed group. In contrast, human rights law, which is the dominant legal framework in the absence of a conflict, authorizes the use of force only as a last resort.[46] The former—LOAC—permits targeting of individuals based on their status as members of a hostile force;[47] the latter—human rights law—permits lethal force against individuals only on the basis of their conduct posing a direct threat at that time.[48] LOAC also accepts the incidental loss of civilian lives as collateral damage,

[44] Graham Webster, "Cyber Cold War Rhetoric Haunts the U.S. and China," *Al Jazeera*, December 25, 2011.

[45] See, for example, Cordula Droege, "Get Off Of My Cloud: Cyber Warfare, International Humanitarian Law, and the Protection of Civilians" (2012) 94 *International Review of the Red Cross* 533, 536 ("In the meantime, there is confusion about the applicability of [LOAC] to cyber warfare—which might in fact stem from different understandings of the concept of cyber warfare itself, which range from cyber operations carried out in the context of armed conflicts as understood in [LOAC] to criminal cyber activities of all kinds.").

[46] See Basic Principles on the Use of Force and Firearms by Law Enforcement Officials, Eighth United Nations Congress on the Prevention of Crime and Treatment of Offenders, GA Res 45/166, UN Doc A/CONF.144/28/Rev.1, at 112 (December 18, 1990) (stating that force can only be used "in self-defence or defence of others against the imminent threat of death or serious injury, to prevent the perpetration of a particularly serious crime involving grave threat to life, to arrest a person presenting such a danger and resisting their authority, or to prevent his or her escape, and only when the less extreme means are insufficient to achieve these objectives"); David Kretzmer, "Targeted Killing of Suspected Terrorists: Extra-Judicial Executions or Legitimate Means of Defence?" (2005) 16 *European Journal of International Law* 171, 176 ("Under [international human rights law,] the intentional use of lethal force by state authorities can be justified only in strictly limited conditions. The state is obliged to respect and ensure the rights of every person to life and to due process of law. Any intentional use of lethal force by state authorities that is not justified under the provisions regarding the right to life, will, by definition, be regarded as an 'extra-judicial execution.'").

[47] See Additional Protocol I, (n 6) Article 51 (prohibiting targeting individuals based on their status as civilians).

[48] Geoffrey Corn, "Mixing Apples and Hand Grenades: The Logical Limit of Applying Human Rights Norms to Armed Conflict" (2010) 1 *Journal of International Humanitarian Legal Studies* 76 ("[D]eadly force is presumptively invalid unless and until the state actor determines that a genuine individual necessity to employ force exists.").

within the bounds of the principle of proportionality;[49] human rights law contemplates no such casualties.

These contrasts can literally mean the difference between life and death in many situations. Indeed:

> [i]f it is often permissible to deliberately kill large numbers of humans in times of armed conflict, even though such an act would be considered mass murder in times of peace, then it is essential that politicians and courts be able to distinguish readily between conflict and nonconflict, between war and peace.[50]

At present, this dichotomy in the authority to use force lies at the center of the debates over the use of targeted strikes to combat terrorists.[51] To the extent that the terminology of "war" leads to a willingness to view as war a situation that otherwise does not fit the rubric of armed conflict, this rhetoric can have a profound effect on the protection of individual rights and safety of persons both in the United States and abroad. This is particularly important given the US policy that it can respond in self-defense to purely cyber acts using both cyber and kinetic force, increasing the potential for casualties and injury.[52]

The manner in which rhetoric can expand authority stretches well beyond the use of force, however. LOAC authorizes the detention of enemy personnel or civilians posing an imperative threat to security for the duration of the conflict (or, in the case of civilians, for such time as they continue to pose such a threat)—without charge.[53] This regime is fundamentally different from the parameters for detention during peacetime. The experience with detention in the counterterrorism context over the past twelve years demonstrates, unfortunately, how the rhetoric of "war" can lead to "law of war" detention that does not match the traditional law of war framework for detention.[54] The notion—frequently proclaimed in the first few years after the 9/11 attacks—that the United States is "at war" with terrorists led to a policy of subsuming all counterterror operations and actions within a wartime framework, regardless of whether LOAC actually applied, or how it applied, to a given situation or person. However, the essential prerequisite to the notion of terrorist suspects being held in detention under the law of war is that they are captured in a context that triggers the law of war: an armed conflict. But the rhetoric of "war" superseded any legal

[49] See Additional Protocol I, (n 6) Article 51 (prohibiting attacks expected to cause incidental loss of civilian life if "excessive in relation to the concrete and direct military advantage anticipated"); Additional Protocol I, Article 57 (requiring that parties take precautions to refrain from disproportionate attacks).

[50] Rosa Ehrenreich Brooks, "War Everywhere: Rights, National Security Law, and the Law of Armed Conflict in the Age of Terror" (2004) 153 *University of Pennsylvania Law Review* 675, 702.

[51] See Laurie R Blank, "Targeted Strikes: The Consequences of Blurring the Armed Conflict and Self-Defense Justifications" (2012) 38 *William Mitchell Law Review* 1655.

[52] See US Department of Defense, *Cyberspace Policy Report* (2011) 4, at <http://www.defense.gov/home/features/2011/0411_cyberstrategy/docs/NDAA%20Section%20934%20Report_For%20webpage.pdf>.

[53] See Third Geneva Convention (n 15) Article 4 (describing various prisoner of war categories); Fourth Geneva Convention (n 15) Articles 42, 78 (permitting internment).

[54] See, for example, Laurie R Blank, "A Square Peg in a Round Hole: Stretching Law of War Detention Too Far" (2011) 63 *Rutgers Law Review* 1169.

determination of whether there was actually an armed conflict occurring; as a result, persons captured anywhere in the world were detained in a war paradigm, meaning without charge until the end of hostilities.[55]

No such practical implications have resulted from the widespread use of the term "cyberwar," and perhaps none will. But understanding the potential legal and policy consequences of the rhetorical excess is essential in order to protect against exactly such a development. Imagine, for example, that a state identified certain individuals who were participating in cyber operations against one or more state agencies or entities. Such individuals may well have committed one or more crimes under that state's domestic law and could face prosecution. Translating the rhetoric of "cyberwar" into the belief that the state is actually in a conflict with such persons, however, could lead to the state launching attacks against them or detaining them without charge, actions that certainly do not fit within a non-conflict paradigm. Beyond this expansion of authority, the rhetoric of war can lead to infringement of individual rights based on the claim that "we are at war,"[56] so therefore certain rights must be limited in the name of national security. The government has multiple opportunities to play on fear and uncertainty in such situations, turning cyber security concerns into justifications for any number of restrictive policies. In essence, "[i]ntelligence—cyber espionage, if you will—is not cyberwar. It's just business as usual. But the cyberwar pundits lump everything in the same bucket, pointing the finger at another nation-state and saying we're under attack."[57] The public's fundamental lack of knowledge about cyber security, cyber operations, and other related topics offers fertile ground for rhetoric to morph into policy in this manner.

III. Cyber Attack

The word "attack" figures prominently in the two bodies of law relevant to armed force and military operations: LOAC and jus ad bellum. In both legal regimes, "attack" has a particular meaning that triggers permissive authority to take action, as well as certain obligations regarding how that action is taken. As with cyberwar and the consequences of "war" rhetoric, it is useful to understand how the terminology of "cyber attack" matches—or does not match—with the legal notions of "attack" in the relevant bodies of law. Indeed, rhetoric is not the only source of this confusion: the United States defines "computer network attack" as "actions taken through the use of computer networks to disrupt, deny, degrade, or destroy information resident in

[55] Blank, "A Square Peg in a Round Hole" (n 54).

[56] The relaxation of the *Miranda* protections for suspected terrorists is a prime example: the claim, "We are at war with these people," was all it took to undermine one of the most fundamental constitutional protections of the past half-century. Evan Perez, "Rights are Curtailed for Terror Suspects," *The Wall Street Journal*, March 24, 2011, A1 (describing new rules that relax *Miranda* warning requirements); see also Charlie Savage, "Holder Backing Law to Restrict Miranda Rules," *The New York Times*, May 10, 2010, A1 (discussing US Attorney General Eric Holder's proposal for Congress to loosen the *Miranda* rule for terrorism suspects).

[57] Marcus Ranum, "Cyberwar Rhetoric is Scarier than Threat of Foreign Attack," *US News & World Report*, March 29, 2010.

computers and computer networks, or the computers and networks themselves."[58] That definition of attack, however, could encompass far more activity than that which falls within either the LOAC definition of attack or the jus ad bellum definition of attack.[59] Indeed:

> despite [computer network attack's] practical utility, its use causes measurable disquiet among lawyers, for 'attack' is a legal term of art that has specific meaning in the context of two very different bodies of international law governing State behaviour in times of crisis or conflict.[60]

After a brief analysis of the meaning of "attack" within the jus ad bellum and LOAC, this section will highlight three consequences from rhetoric stretching well beyond legal parameters: the expansion of authority or opportunity to use force in self-defense; the conflation of legal regimes regarding obligations and privileges in both wartime and peacetime; and the incorporation of more individuals and groups within the category of "enemy" during armed conflict.

(a) Two definitions of attack

i. Jus ad bellum: armed attack

The term "armed attack" arises in the jus ad bellum, the law governing the resort to force—that is, when a state may use force within the constraints of the United Nations Charter framework and traditional legal principles. In particular, the UN Charter prohibits the use of force by one state against another: "All members shall refrain in their international relations from the threat or use of force against the territorial integrity or political independence of any State, or in any other manner inconsistent with the Purposes of the United Nations."[61] Jus ad bellum provides for three exceptions to the prohibition on the use of force, only the third of which is relevant to the meaning of "armed attack." First, a state may use force with the consent of the territorial state, such as when a state battling an armed group requests assistance from one or more other states. Second, a state can use force as part of a multinational operation authorized by the Security Council under Chapter VII, as provided in Article 42. Third, a state may use force in accordance with the inherent right of self-defense under Article 51 in response to an armed attack:

> Nothing in the present Charter shall impair the inherent right of individual or collective self-defence if an armed attack occurs against a Member of the United Nations,

[58] See Joint Chiefs of Staff, *Joint Publication 3-13: Information Operations* (2006) II-5, at <http://www.dtic.mil/doctrine/new_pubs/jp3_13.pdf>.

[59] See Paul A Walker, "Rethinking Computer Network 'Attack': Implications for Law and U.S. Doctrine," (2011) 1 *National Security Law Brief* 33, 36 ("Unfortunately, this common use of 'attack' has bled over into legal analysis and military doctrine, specifically into the *Air and Missile Warfare Manual* and the United States doctrine of 'computer network attack'.")

[60] Michael N Schmitt, "'Attack' as a Term of Art in International Law: The Cyber Operations Context" 4th International Conference on Cyber Conflict (2012) 284, at <http://ssrn.com/abstract=2184833>.

[61] UN Charter, Article 2(4).

until the Security Council has taken measures necessary to maintain international peace and security.[62]

Any lawful use of force in self-defense, therefore, depends initially on the existence of an armed attack. Operations, attacks, or acts causing significant death, injury, damage, or destruction by a state or a non-state group[63] will generally be considered to be an armed attack. An armed attack is more severe and significant than a use of force, meaning that a state can be the victim of a use of force without being the victim of an armed attack that triggers the right of self-defense.[64] In assessing whether a particular hostile action directed at a state rises to the level of an armed attack, the ICJ has looked at the scale and effects of the act.[65] For example, if a state deploys its regular armed

[62] UN Charter, Article 51. Any force used in self-defense must comply with the requirements of necessity, proportionality, and immediacy. These obligations form part of customary international law and have been reaffirmed numerous times by the ICJ. See for example, Military and Paramilitary Activities in and Against Nicaragua (*Nicaragua v US*), 1986 ICJ 14, para 176, 194 (June 27) [hereinafter Military and Paramilitary Activities]; Oil Platforms (*Iran v US*), 2003 ICJ 161 paras 43, 73–74, 76 (November 6); Legality of the Threat or Use of Nuclear Weapons, Advisory Opinion, 1996 ICJ 226, para 41 (July 8). The requirement of necessity addresses whether there are adequate non-forceful options to deter or defeat the attack, such as diplomatic avenues, defensive measures to halt any further attacks or reparations for injuries caused. The requirement of proportionality measures the extent of the use of force against the overall military goals, such as fending off an attack or subordinating the enemy. Rather than addressing whether force may be used at all—which is the main focus of the necessity requirement—proportionality looks at how much force may be used, that is, whether the measure of counterforce used is proportionate to the needs and goals of repelling or deterring the original attack.

[63] Nothing in Article 51 specifies that the right of self-defence is only available in response to a threat or use of force by another state. Nonetheless, the precise contours of what type of actor can trigger the right of self-defense remains controversial. The ICJ has held, and some commentators also argue, that only states can be the source of an armed attack—or imminent threat of an armed attack—that can justify the use of force in self-defense. See, for example, Military and Paramilitary Activities, (n 52); Oil Platforms, (n 52); Armed Activities on the Territory of the Congo (*Democratic Republic of Congo v Uganda*), 2005 ICJ 168 (December 19); Legal Consequences of the Construction of a Wall in the Occupied Palestinian Territory, Advisory Opinion, 2004 ICJ 136, 215 (July 9); Antonio Cassese, "The International Community"s "Legal' Response to Terrorism" (1989) 38 *International and Comparative Law Quarterly* 589, 597; Eric Myjer and Nigel White, "The Twin Towers Attack: An Unlimited Right to Self-Defense" (2002) 7 *Journal of Conflict & Security Law* 5, 7 ("Self-defense, traditionally speaking, applies to an armed response to an attack by a state."). However, state practice in the aftermath of the 9/11 attacks provides firm support for the existence of a right of self-defense against non-state actors, even if unrelated to any state. UN Resolution 1368 recognized the right of self-defense in response to the 9/11 attacks and NATO triggered the collective defence provision in Article 5 of the North Atlantic Treaty for the first time. SC Res 1368, para 1, UN Doc S/RES/1368 (12 Sept 2001) (emphasis added); North Atlantic Treaty Article 5, April 4, 1949, 63 Stat 2241, 2244, 34 UNTS 243, 246; Press Release, North Atlantic Treaty Organization, Statement by the North Atlantic Council (September 12, 2001), at <http://www.nato.int/docu/pr/2001/p01-124e.htm>. See, for example, Yoram Dinstein, *War, Aggression and Self-Defence* (Cambridge: Cambridge University Press, 2nd edn, 1994) 214; Christopher Greenwood, "International Law and the Preemptive Use of Force: Afghanistan, al Qaeda, and Iraq" (2003) 4 *San Diego International Law Journal* 7, 17 (discussing the effects of attacks made by non-State actors); Sean D Murphy "The International Legality of US Military Cross-Border Operations from Afghanistan into Pakistan," in Michael N Schmitt (ed), *The War in Afghanistan: A Legal Analysis* (2009) 85 *Internatioanl Law Studies* 109, 126 ("While this area of the law remains somewhat uncertain, the dominant trend in contemporary interstate relations seems to favor the view that States accept or at least tolerate acts of self-defense against a non-State actor.").

[64] Military and Paramilitary Activities (n 52) para 191 (June 27). See also Michael N Schmitt, "Cyber Operations in International Law: The Use of Force, Collective Security, Self-Defense, and Armed Conflict," in *Committee on Deterring Cyberattacks*, National Research Council, Proceedings of a Workshop on Deterring Cyberattacks: Informing Strategies and Developing Options for U.S. Policy (2010) 163, at <http://books.nap.edu/openbook.php?record_id=12997&page=R1>.

[65] Military and Paramilitary Activities (n 52).

forces across a border, that will generally be considered an armed attack, as will a state sending irregular militias or other armed groups to accomplish the same purposes. In contrast, providing weapons or other assistance to rebels or other armed groups across state borders will not reach the threshold of an armed attack.[66]

Although the analysis may seem relatively straightforward in the context of military units, armed bands, and kinetic force, in the cyber realm, identifying and analyzing an armed attack are significantly more challenging. The most common method of analysis with regard to whether cyber actions rise to the level of an armed attack is an effects-based analysis. At present, there is a general consensus that "any use of force that injures or kills persons or damages or destroys property" constitutes an armed attack, including in the cyber arena.[67] Others point to the target of a cyber operation, arguing that any cyber action against critical national infrastructure should qualify as an armed attack,[68] or, alternatively, to an "instrument-based" approach, according to which a cyber operation constitutes an armed attack if "the damage caused by a cyber attack could previously have been achieved only by a kinetic attack."[69] In contrast, economic damage, political embarrassment or coercion, a disruption of communications, and the distribution of propaganda through cyber means do not rise to the level of an armed attack. The *Tallinn Manual* explains that cyber intelligence gathering and theft do not constitute an armed attack, nor would "cyber operations that involve brief or periodic interruption of non-essential cyber services."[70]

ii. LOAC: "attack"

LOAC establishes parameters for the conduct of parties to an armed conflict—how they conduct military operations and how they treat individuals, both fighters and civilians, in their control and in the area of military operations. Additional Protocol I builds on previous treaty law governing the conduct of hostilities—namely the 1899 and 1907 Hague Regulations—and sets forth extensive rules providing "General Protection Against the Effects of Hostilities."[71] These rules are designed to uphold the historic notion that "war is waged between soldiers and that the population should remain outside hostilities[, which] was introduced in the sixteenth century and became established by the eighteenth century."[72] This notion forms the heart of the central LOAC principle of distinction, which mandates that parties to a conflict must distinguish between those who are fighting and those who are not and only target

[66] Based on the *Nicaragua* judgment, arming and training a rebel group in another state would constitute the threat or use of force against that state, but merely funding rebels would not be a use of force. *Nicaragua* judgment, paras 195, 228. See also Michael N Schmitt, "Legitimacy versus Legality Redux: Arming the Syrian Rebels" (2013) 7 *Journal of National Security Law & Policy* 3–5.

[67] *Tallinn Manual* (n 17) 54.

[68] Eric Talbot Jensen, "Computer Attacks on Critical National Infrastructure: A Use of Force Invoking the Right of Self-Defense" (2002) 38 *Stanford Journal of International Law* 207, 221–26.

[69] David E Graham, "Cyber Threats and the Law of War" (2010) 4 *Journal of National Security Law & Policy*,87, 91 (citing Yoram Dinstein, "Computer Network Attacks and Self-Defense," in Michael N Schmitt and Brian T O'Donnell (eds), *Computer Network Attack and International Law*, (1999) 76 *International Law Studies* 99, 103–5.

[70] *Tallinn Manual* (n 18) 55.

[71] This is the title of Part IV, Section I of Additional Protocol I.

[72] Protocol Commentary (n 29) 585.

attacks at the former.[73] Additional Protocol I operationalizes this fundamental principle in multiple ways, most specifically by mandating that (1) "[t]he civilian population and individual civilians shall enjoy general protection against dangers arising from military operations"; and (2) "[t]he civilian population as such, as well as individual civilians, shall not be the object of attack."[74]

For the purposes of this discussion, it is important to note that the definition of attack is only relevant during armed conflict, when LOAC applies, and that LOAC distinguishes between military operations in general and attacks more specifically. The term "military operations" is broader, encompassing "all the movements and activities carried out by armed forces related to hostilities."[75] Not all actions of a military or other party to a conflict will constitute attacks. For example, "non-destructive psychological operations directed at the civilian population, such as dropping leaflets, broadcasting to the enemy population, or even jamming enemy public broadcasts" are all military operations but do not meet the definition of attack under LOAC.[76] There is continued debate whether the obligations of distinction, proportionality, and precautions extend to military operations or only apply to attacks. However:

> it is clear that states did differentiate in Additional Protocol I between the general principles in the respective chapeaux of the rules of distinction and precaution and the specific rules relating to attacks, and that they found it necessary to define attacks specifically in Article 49 of the Protocol.[77]

Cyber or other operations that do not fall within the definition of attack will not be subject to the rules governing attacks under Additional Protocol I and customary international humanitarian law.

The word "attack" triggers certain obligations specific to the launching of attacks—such as distinction, proportionality and precautions, and protections for certain objects[78]—which, therefore, makes the definition of attack relevant for understanding the obligations of each party to a conflict. According to Article 49(1) of Additional Protocol I, "'attacks' means acts of violence against the adversary, whether in offence or in defence."[79] As the Commentary explains, this definition "has a special significance ... [and] is not exactly the same as the usual meaning of the word." Attack in LOAC applies to both offensive and defensive acts and involves violence or "combat action,"[80]

[73] See Additional Protocol I (n 6) Article 48.

[74] Additional Protocol I (n 6) Article 51(1) and (2).

[75] Protocol Commentary (n 29) 617.

[76] Schmitt (n 50) 289.

[77] Droege (n 38) 555. The term "chapeau" refers to introductory text in a treaty or treaty provision that broadly defines its principles, objectives, and background.

[78] Knut Dörmann, "Applicability of the Additional Protocols to Computer Network Attacks," (Paper delivered at the International Expert Conference on Computer Network Attacks and the Applicability of International Humanitarian Law, Stockholm, November 17–19, 2004) 3, at <http://www.icrc.org/eng/assets/files/other/applicabilityofihltocna.pdf> ("The definition of the term 'attack' is of decisive importance for the application of the various rules giving effect to the principle of distinction and for most of the rules providing special protection for certain objects.").

[79] Additional Protocol I (n 6) Article 49(1).

[80] Protocol Commentary (n 29) 603. The Commentary also reinforces that "attack" in LOAC is an entirely distinct notion from "armed attack" in jus ad bellum: "Finally it is appropriate to note that in the sense of the Protocol an attack is unrelated to the concept of aggression or the first use of armed force; it refers simply to the use of armed force to carry out a military operation at the beginning or during the

as the Commentary notes. The prohibition on attacks against civilians in Article 51 of Additional Protocol I demonstrates that the concept of attack includes violent acts against enemy personnel or objects as well as violent acts against civilians. Furthermore, the "term 'acts of violence' denotes physical force. Thus, the concept of 'attacks' does not include dissemination of propaganda, embargoes, or other non-physical means of psychological or economic warfare."[81] The criterion of violence is not limited to the method of the attack, however, but applies more generally to the consequences of the attack. For this reason, the use of biological, chemical, and radiological weapons constitute attacks even though they do not use physical force.[82] This focus on consequences can be seen as well in the language used throughout Additional Protocol I that focuses on the effects of attacks on the civilian population and on civilian objects, on dangers to civilians and other harms that are identified by the consequence of an act rather than the nature of the act itself.[83]

Cyber operations pose a particular challenge for the definition of attack because cyber acts do not involve violence directly. Such operations may, of course, result in violent effects, but the act itself is not violent in the way that a kinetic operation involves violence. The *Tallinn Manual* defines an attack in the cyber realm as "a cyber operation, whether offensive or defensive, that is reasonably expected to cause injury or death to persons or damage or destruction to objects."[84] The consequences of an act thus ultimately determine, in the consensus of the *Tallinn Manual* experts, whether a cyber operation constitutes an attack as defined in Article 49 of Additional Protocol I. Acts that cause consequential damage (i.e., not *de minimis* harm) to individuals or objects, or interfere in the functionality of an object, will meet the threshold for an attack.[85] Some scholars also include acts that result in neutralization of objects, based on the definition of military objective, which talks of the destruction or neutralization of an object that offers direct military advantage.[86] The ICRC characterizes cyber attack in the LOAC paradigm as "cyber operations by means of viruses, worms, etc., that result in physical damage to persons, or damage to objects that goes beyond the computer program or data attacked."[87]

Although experts continue to disagree on the margins of the definition of cyber attack under LOAC, certain cyber operations fall squarely within—or outside—the definition. A cyber operation that destroys an object and causes damage to persons or objects is an attack. A denial of service attack that interferes with propaganda, or blocks a television broadcast or website would not be an attack—cyber operations that cause inconvenience are generally not considered to be attacks. Similarly, espionage or embargoes accomplished through cyber means are not attacks under the meaning in LOAC.[88] Nonetheless, finding the line between cyber operations that are an attack

course of an armed conflict. Questions relating to the responsibility for unleashing the conflict are of a completely different nature." Protocol Commentary (n 29) 603.

[81] Michael Bothe et al, *New Rules for Victims of Armed Conflicts* (Leiden: Martinus Nijhoff, 1982) 289.

[82] See *Prosecutor v Tadić*, Decision on the Defence Motion for Interlocutory Appeal, (ICTY, October 2, 1995) paras 120 and 124; Droege (n 37) 557.

[83] See Schmitt (n 50) 291; Bothe et al (n 71) 325; Droege (n 38) 557.

[84] *Tallinn Manual* (n 18) 106. [85] *Tallinn Manual* (n 18) 106–10.

[86] Dörmann (n 68) 4. [87] ICRC (n 1) 37.

[88] See Oona A Hathaway, Rebecca Crootof, Philip Levitz, Haley Nix, Aileen Nowlan, William Perdue, and Julia Spiegel, "The Law of Cyber-Attack" (2012) 100 *California Law Review* 817, 829.

and those that are below the threshold of an attack can be difficult, especially given the extent to which both the military and the civilian population rely on computers, the internet and communications systems.

(b) Rhetoric and its consequences

i. Expanded authority to use force in self-defense

Under international law, an armed attack triggers the victim state's right to use force in self-defense. Acts falling short of the threshold of "armed attack" may justify the use of countermeasures or other responses short of the use of force, but do not provide justification for using force in self-defense. As a result, the characterization of the initial act is essential to the subsequent rights and obligations of the parties involved. Indeed, "determining that a cyberattack is an act of war would be more than just a conclusion; it is a decision that could initiate a war of choice."[89] If the expansive rhetoric of cyber attack were to translate into an expansive reading of the legal architecture for the use of force and the right of self-defense, the potential harm is significant and dangerous.

Cyber attack is commonly used to describe nearly any cyber act that causes or potentially causes a negative effect, infiltrates a network or server, manipulates a computer network or system, steals information, or many other possible actions. Most of these acts fall far short of any notion of use of force, let alone the more heightened threshold of armed attack. For example, many experts use the term "cyber exploitation" to refer to a large subset of cyber operations, namely "deliberate activities designed to penetrate computer systems or networks used by an adversary, for the purposes of obtaining information resident on or transiting through these systems or networks."[90] Cyber exploitation is akin to espionage and, if carried out successfully, leaves no trace and the victim unaware of the operation or its effects. Due to the attack-centered rhetoric of the cyber discourse, however, the press regularly refers to cyber exploitations, cyber espionage, and related activities as "cyber attacks."[91] A likely—and unfortunate—effect of this expanded rhetoric will be to ease the threshold for characterizing an armed attack in the cyber arena, which correspondingly weakens the prohibition against the use of force. Indeed, for this very reason, many experts now urge that "because so many different kinds of cyberattack are possible, the term 'cyberattack' should be understood as a statement about a methodology for action—and that alone—rather than as a statement about the scale of the action's effect."[92] The prohibition on the use of force in the UN Charter is a central component of the UN framework and forms a bulwark

[89] Martin C Libicki, "Don't Buy the Cyberhype: How to Prevent Cyberwars From Becoming Real Ones," *Foreign Affairs*, August 14, 2013, at <http://www.foreignaffairs.com/articles/139819/martin-c-libicki/dont-buy-the-cyberhype>

[90] Herbert Lin, "Cyber Conflict and International Humanitarian Law" (2012) 94 *International Review of the Red Cross* 515, 519.

[91] Lin, "Cyber Conflict and International Humanitarian Law" (n 90).

[92] National Research Council (NRC), William Owens, Kenneth Dam, Herbert Lin (eds), *Technology, Policy, Law, and Ethics Regarding U.S. Acquisition and Use of Cyberattack Capabilities* (National Academies Press, Washington, DC, 2009) 11, at <http://www.nap.edu/catalog.php?record_id=12651>.

of the international structure for security and peaceful settlement of disputes. As the preamble to the UN Charter states, among the UN's (and thus the international community's) primary goals are "to save succeeding generations from the scourge of war" and "to ensure, by the acceptance of principles and the institution of methods, that armed force shall not be used, save in the common interest."[93] No less, the Charter's first stated purpose is to maintain international peace and security, partly through the suppression of the use of force and the promotion of peaceful settlement of disputes.[94] This emphasis leaves little room for doubt that any lawful use of force must meet a high threshold to ensure the continued respect for and fulfillment of the UN's primary goals.

Cyber rhetoric that ultimately equates cyber attack as colloquially understood with armed attack as legally defined in the jus ad bellum poses a significant danger to this established framework. "Nearly all of the adversarial actions known to have been taken in cyberspace against the United States or any other nation, including both cyber attack and cyber exploitation, have fallen short of any plausible threshold for defining them as 'armed conflict', 'use of force', or even 'armed attack'."[95] Headlines, media reports, and public discourse still cry "cyber attack" regularly in response to almost any type of cyber operation, regardless of type, severity, victim, or effect.[96] Just as rhetoric of war leads to an acceptance of policies that limit rights in the name of national security and wartime necessity, so a rhetoric of attack will enable responses unjustified by the actual facts of the cyber operation, but seemingly justified by the overheated rhetoric. The result will only be to weaken "the ban on force [that] forms part of the deep structure of the Charter and one of the core motivating premises of the United Nations."[97]

ii. Conflation of legal regimes, rights and obligations

Using the term "cyber attack" to describe acts that fall short of or outside the definition of attack in LOAC—whether by outside parties against entities in the United States or by US actors against persons or entities abroad—risks conflating the authorities and obligations inherent in the different legal regimes governing peacetime and wartime conduct. Uncertainty about whether certain acts occur within armed conflict or not

[93] UN Charter, preamble.

[94] UN Charter preamble, Article 1(1) ("The Purposes of the United Nations are: 1. To maintain international peace and security, and to that end: to take effective collective measures for the prevention and removal of threats to the peace, and for the suppression of acts of aggression or other breaches of the peace, and to bring about by peaceful means, and in conformity with the principles of justice and international law, adjustment or settlement of international disputes or situations which might lead to a breach of the peace...").

[95] Lin (n 79) 530.

[96] See, for example, Alan Wong, "Cyber Attack on Hong Kong Vote Was Among Largest Ever, Security Chief Says," *The New York Times*, June 21, 2014, at http://sinosphere.blogs.nytimes.com/2014/06/21/cyber attack-on-hong-kong-vote-was-among-largest-ever-security-chief-says/; Stuart Corner, "Australian Companies Hit by Increasing Number of Cyber Attacks," *Sydney Morning Herald*, June 30, 2014.

[97] David Kaye, "Harold Koh's Case for Humanitarian Intervention," *JustSecurity.org*, October 7, 2013, at <http://justsecurity.org/2013/10/07/kaye-kohs-case/>.

leads to a variety of analytical and implementation challenges with regard to LOAC, human rights law, jus ad bellum, and other relevant legal regimes. At the most fundamental level, to the extent that the rhetoric of cyber attack creates the impression—or worse, the understanding—that the United States is involved in an armed conflict, the conflation of two vastly different legal regimes regarding the use of force is highly problematic. During armed conflict, LOAC authorizes the use of force as a first resort against those identified as the enemy, whether insurgents, terrorists, or the armed forces of another state. In contrast, human rights law, which is the dominant legal framework in the absence of armed conflict, authorizes the use of force only as a last resort.[98] The failure to have clear lines between the two regimes, and clear distinctions between their relevant parameters, can have dramatic and deadly effects.

Under LOAC, an individual can be attacked solely based on his or her identification as a member of a hostile force, regardless of that person's conduct at the time of the attack.[99] Although LOAC flatly prohibits denial of quarter or attacks on individuals who are *hors de combat* due to sickness, wounds or detention, there is no obligation to seek to capture before killing.[100] In addition, LOAC's principle of proportionality focuses on the innocent casualties of attacks, mandating that parties refrain from launching attacks where the expected civilian casualties will be excessive in light of the anticipated military advantage to be gained.[101] This rule of proportionality is unconcerned with the type or degree of force used against the target or object of the attack (protections provided for in other LOAC rules[102]) but rather seeks to mitigate the suffering of persons not participating in hostilities, the unintended victims of the use of force.

Human rights law takes a wholly opposite approach, prohibiting the use of force except when absolutely necessary based on an individual's conduct posing a direct threat at that time. If non-forceful measures can foil the attack or threat of attack without the use of deadly force, then the state may not use force in self-defense.[103] As a result, state forces have an obligation to attempt to capture before any use of force: the supremacy of the right to life means that "even the most dangerous individual must be captured, rather than killed, so long as it is practically feasible to do so, bearing in mind all of the circumstances."[104] Finally, proportionality in human rights law refers

[98] See Corn (n 41) 74–5.

[99] See Additional Protocol I, art 51 (prohibiting attacks on civilians); Geoffrey S Corn, "Back to the Future: De Facto Hostilities, Transnational Terrorism, and the Purpose of the Law of Armed Conflict" (2009) 30 *University of Pennsylvania Journal of International Law* 1345, 1352–3.

[100] See Geoffrey S Corn, Laurie R Blank, Christopher Jenks, and Eric Talbot Jensen, "Belligerent Targeting and the Invalidity of a Least Harmful Means Rule" (2013) 89 *International Law Studies* 536–626; Blank (n 44) 1683–5.

[101] Additional Protocol I (n 6) Articles 51(5)(b), 57(2)(a)(iii), and 57(2)(b).

[102] For example, LOAC prohibits the use of means or methods of warfare that cause unnecessary suffering, a rule specifically designed to protect those who are fighting. See Additional Protocol I (n 5) Article 35(2).

[103] See *McCann v United Kingdom*, 21 European Court of Human Rights 97 (1995) (noting that lethal force is disproportionate whenever non-lethal alternatives are available).

[104] Marko Milanovic, "When to Kill and When to Capture?," *European Journal of International Law*: Talk (May 6, 2011) at <http://www.ejiltalk.org/when-to-kill-and-when-to-capture/>; see also Jan Romer, "Killing in a Gray Area Between Humanitarian Law and Human Rights: How Can the National Police of Columbia Overcome the Uncertainty of Which Branch of International Law to Apply?" (Berlin: Springer-Verlag, 2010) 116; Michael N Schmitt, "The Interpretive Guidance on the Notion of

to the measure of force directed at the intended target of the attack. Force can only be used "when strictly necessary and to the extent required for the performance of [law enforcement officials'"] duty.'[105] In the law enforcement paradigm, therefore, "the principle of proportionality operates to protect the object of state violence by allowing only that amount of force necessary to subdue a hostile actor."[106]

The expansive rhetoric of "cyber attack" threatens this bright line differentiation between armed conflict and peacetime because the commentary and discourse suggest acts that are taken within an armed conflict and LOAC even if the facts of a given operation or situation do not constitute an armed conflict or qualify as an attack within the definition of LOAC. Cyber rhetoric's danger lies primarily in the transposition of LOAC principles and authorities to a non-conflict situation. If a cyber act is believed to be an "attack" as understood in LOAC, then those engaging in that act need to comply with LOAC's principles of distinction, proportionality and precautions. On first glance, this seems to be highly protective: LOAC mandates that parties launching attacks distinguish between those who are fighting and those who are not, only launch attacks that meet the principle of proportionality, and take a series of precautions to protect individual civilians and the civilian population from the effects of the attack.[107] In the context of an armed conflict, these principles are indeed highly protective and help ensure one of LOAC's core purposes: protecting civilians from the hazards of war and mitigating suffering during wartime.

Analyzed in a non-conflict situation, however, LOAC's core principles are not necessarily the most protective. As explained already, LOAC authorizes the use of force as a first resort, classifies certain persons and groups as enemy personnel or forces who can be targeted with lethal force at all times, and identifies objects that qualify as military objectives that can be attacked and destroyed. Human rights law—also called the law enforcement paradigm—contemplates none of these methods or authorities. Human rights law does not classify persons or objects as legitimate targets, does not authorize the use of force against designated targets, and does not include any notion of incidental casualties or collateral damage. Rather, human rights law prohibits the use of force except as a last resort and requires that the use of force be proportional to the aim to be achieved—law enforcement authorities can use no more force than is absolutely necessary to effectuate an arrest, defend themselves, or defend others from attack.[108]

Direct Participation in Hostilities: A Critical Analysis" (2010) 1 *Harvard National Security Journal* 5, 42 ("A requirement does exist in human rights law to capture rather than kill when possible. It applies primarily during peacetime as well as in certain circumstances when occupying forces are acting to maintain order.")

[105] Code of Conduct for Law Enforcement Officials, GA Res 34/169, Article 3 (December 17, 1979).

[106] Jimmy Gurulé and Geoffrey S Corn, *Principles of Counter-Terrorism Law* (St Paul: West Academic Publishing, 2011) 80.

[107] See Additional Protocol I, art 51 (prohibiting attacks on civilians); Article 51(5)(b) (prohibiting disproportionate attacks); Article 57(2) (setting forth precautions attacking parties must take when launching attacks).

[108] International Covenant on Civil and Political Rights, Article 6, GA Res 2200A (XXI), UN GAOR, Supp No 16, UN Doc. A/6316 (December 16, 1966) (hereinafter ICCPR); Convention for the Protection of Human Rights and Fundamental Freedoms Article 2, *opened for signature* 4 Nov 1950, 213 UNTS 222 (entered into force September 3, 1953); Basic Principles on the Use of Force and Firearms by Law Enforcement Officials, Eighth United Nations Congress on the Prevention of Crime and Treatment of Offenders, GA Res 45/166, UN Doc A/CONF.144/28/Rev.1, at 112 (December 18, 1990).

Applying LOAC's principles to acts that do not fall within LOAC's paradigm—either because there is no armed conflict or because the relevant acts are not "attacks" as understood in LOAC—can, therefore, actually produce a less protective environment. This same issue arises in the context of the debate over US targeted strikes outside of Afghanistan. The on-going discourse about whether such strikes occur within an armed conflict—either a global transnational conflict or one or more localized conflicts in Yemen or other locations—is outside the scope of this chapter. However, the debate highlights an important aspect of the conflation between LOAC and jus ad bellum in the application of international law to the use of force in the current complex environment. The United States has regularly asserted that the right of self-defense is one justification for such strikes, based on the imminent threat posed by the particular terrorist operative targeted.[109] It then presents a rule of law paradigm for such strikes in which LOAC's principles of military necessity, humanity, distinction, and proportionality govern all strikes, even those taking place outside of an armed conflict and solely in the context of self-defense.[110]

However, the right to use force in self-defense, outside of armed conflict, is not governed by LOAC, but rather by the jus ad bellum, which requires that any such use of force be necessary, proportionate, and in response to an armed attack or imminent threat of such attack.[111] Those obligations only govern the right to resort to force, not the parameters for the use of force itself. In the absence of an armed conflict, international human rights law and the principles governing the use of force in law enforcement are the relevant paradigm. Here the use of lethal force is—appropriately—tightly prescribed and extraordinarily restricted.

> Under [international human rights law,] the intentional use of lethal force by state authorities can be justified only in strictly limited conditions. The state is obliged to respect and ensure the rights of every person to life and to due process of law. Any intentional use of lethal force by state authorities that is not justified under the provisions regarding the right to life, will, by definition, be regarded as an "extra-judicial execution."[112]

[109] See, for example, Harold Hongju Koh, Legal Adviser, US Department of State, Keynote Address at the Annual Meeting of the American Society of International Law: The Obama Administration and International Law (March 25, 2010).

[110] See, for example, Remarks of CIA General Counsel Stephen W Preston at Harvard Law School, April 10, 2012, at <https://www.cia.gov/news-information/speeches-testimony/2012-speeches-testimony/cia-general-counsel-harvard.html>; Attorney General Eric Holder Speaks at Northwestern University School of Law, March 5, 2012, at <http://www.justice.gov/iso/opa/ag/speeches/2012/ag-speech-1203051.html>. See also Kenneth Anderson and Benjamin Wittes, *Speaking the Law: The Obama Administration's Addresses on National Security Law* (Lawfare blog, 2013) 26. (This assertion can be interpreted "as a broadening of the commitment of the US government to adhere to law-of-war principles even in circumstances in which the United States uses force outside of formal armed conflicts but where it does not acknowledge the applicability of human rights law.")

[111] Military and Paramilitary Activities (n 52).

[112] Kretzmer (n 39) 176; see also Basic Principles on the Use of Force and Firearms by Law Enforcement Officials (n 39) 112 (stating that force can only be used "in self-defence or defence of others against the imminent threat of death or serious injury, to prevent the perpetration of a particularly serious crime involving grave threat to life, to arrest a person presenting such a danger and resisting their authority, or to prevent his or her escape, and only when the less extreme means are insufficient to achieve these objectives").

The use of lethal force against individuals or groups that pose a threat outside of armed conflict can, therefore, only be used when absolutely necessary to protect potential victims of any attacks. Using LOAC principles instead means that individuals may be targeted on the basis of status or membership in a hostile group—certainly less protective than the human rights mandate of absolutely necessary based on conduct. Using LOAC principles also means that innocent bystanders may be killed or injured as "collateral damage" that is not unlawful under LOAC, a paradigm that makes sense in the context of armed conflict but produces a significant weakening of the right to life in peacetime. Beyond the use of force, using LOAC principles means that individuals posing a threat, or believed to be part of a hostile group, can be detained without charge. When the use of the word "attack" can change the relevant legal paradigm (inappropriately, but changed nonetheless), rhetorical choice and terminology matter greatly.

The combination of expansive cyber rhetoric and the difficulty of identifying different types of cyber operations can thus easily lead to the application of LOAC to acts that fall far short of any definition of attack. "Terms like 'cyber attacks' or 'cyber terrorism' may evoke methods of warfare, but the operations they refer to are not necessarily conducted in an armed conflict."[113] However, because "the distinction between cyberattack and cyber exploitation may be very hard to draw from a technical standpoint, and may lie primarily in the intent of the user,"[114] the rhetoric can cause cyber exploitation to be swept into the category of cyber attack. Even a cursory look at major policy and legal articles on cyber attack evince this danger. For example, a major National Research Council report on cyber capabilities includes a section on "Cyberattack and Domestic Law Enforcement" that focuses on how "cyberattack may be relevant to at least two other constituencies—the domestic law enforcement community and the private sector."[115] It is unlikely that cyber attack in that context means "attack" in the LOAC understanding of the term, but the opportunity for conflation is ripe. Similarly, more than one law journal article highlights the need to distinguish between "the different types of computer network operations, as a cyber-attack, cyber-crime, or cyber-espionage... in analyzing an appropriate legal response," but nonetheless uses "cyber-attack... broadly to describe [computer network operations] that include attacks which fall under the law of war and cyber-crime attacks."[116] Definitions are essential to the effective application of the law in any legal regime; rhetoric that blurs, conflates, or erases clear definitional lines can thus undermine the law's ability to carry out its core goals.

iii. Incorporating more individuals and groups into the notion of the enemy

A final way in which expansive cyber rhetoric regarding the term "cyber attack" can raise significant dangers is to create the impression—and thus potentially the

[113] Droege (n 38) 542 (further adding that "cyber operations can be and are in fact used in crimes committed in everyday situations that have nothing to do with war.").

[114] National Research Council (n 81) 22.

[115] National Research Council (n 81) 200.

[116] See, for example, David Weissbrodt, "Cyber-Conflict, Cyber-Crime, and Cyber-Espionage" (2013) 22 *Minnesota Journal of International Law* 347, 349, 354.

reality—that more individuals and groups are part of an "enemy" against whom we are fighting. Individuals who engage in criminal activity, espionage, terrorism, or other similar acts—whether in the "real world" or the cyber realm—are committing crimes and should face the full force of the state's law enforcement powers. Rhetoric of "war" and "attack," however, suggests that such individuals are engaged in a war and, therefore, must be treated through a war paradigm rather than the law enforcement system. As was unfortunately evident throughout the first decade after the 9/11 attacks, introducing a war paradigm to terrorism and other criminal acts raises a host of questions and problems regarding individual rights and the expansion of executive authority. The rhetoric of "war" broadened existing paradigms to encompass more individuals, more conduct, and more geographic space.

The debate over what to call terrorist operatives is a useful example. Before the 9/11 attacks, terrorists were nearly uniformly considered criminals and were pursued, arrested, prosecuted, and punished within the criminal justice system.[117] Individuals that the United States fought against in armed conflict, such as in Iraq in the first Gulf War, were considered to be enemy fighters—usually prisoners of war if they met the relevant criteria—and treated in accordance with the LOAC.[118] By all measures, the post-9/11 paradigm has involved a merging—in one form or another—of these two frameworks, as numerous scholars have analyzed and critiqued over the past decade.[119] Although LOAC includes significant protections for persons detained in the course of armed conflict, it nonetheless provides the authority to detain individuals, whether prisoners of war, fighters, or civilians, without charge until the end of hostilities. This authority is fundamentally different from that found in the law enforcement paradigm, where detention without charge is highly unusual and requires special legislative authority, such as in the case of registered sex offenders or other special categories.

The rhetoric of cyber attack and cyberwar leads naturally to the conclusion that those persons or groups who are launching cyber attacks or engaged in cyberwarfare are enemies in such a cyberwar and, therefore, should all be treated as enemy personnel in wartime. Indeed, "it is unsurprising... that with a wide variety of 'hostile or malicious action in cyberspace' conflated under the term cyberwar—including crime, espionage, protest, and activism—policy-makers and political leaders have looked primarily to war-related historical analogies and metaphors to understand their understanding of responses to cyber security challenges."[120] Before long, a lone hacker becomes a member of a virtual armed group, potentially subject to targeting

[117] See, for example, *United States v Yousef*, 327 F.3d 56, 77, 172 (2d Cir 2003) (upholding the sentences of individuals accused of terrorist activities); *United States v Rahman*, 189 F.3d 88, 103 (2d Cir. 1999) (affirming convictions of ten individuals accused of terrorist activities).

[118] US Department of Defense, *Conduct of the Persian Gulf War: Final Report to Congress* (1992) app. L (stating that the most important requirements for enemy POW operations are defined by the four Geneva Conventions).

[119] See, for example, Robert Chesney and Jack Goldsmith, "Terrorism and the Convergence of Criminal and Military Detention Models" (2008) 60 *Stanford Law Review* 1079, 1109 ("The government has responded to these pressures by incorporating into the military detention model many of the procedural constraints associated with the criminal justice system.").

[120] Lawson (n 33).

or detention as a member of a hostile force in an armed conflict. Identifying organized armed groups in the cyber arena for the purposes of conflict recognition or classification of persons is extraordinarily difficult; characterizing direct participation in hostilities for the purposes of targeting in an armed conflict is equally challenging.[121] Heightened rhetoric about cyber attack that turns cyber operations, cyber crime or cyber exploitation into cyber attacks in the midst of a cyberwar lowers the barriers to participation in conflict and undermines protections for persons in both peacetime and wartime.

IV. Conclusion

Cyber activities and cyberspace are extraordinarily difficult to understand, given the complexities of the computer networks, capabilities, persons and groups involved, and real and potential consequences of various acts, whether offensive or defensive. In a wide range of situations, law often plays an important role in defining the content and categories of specific behaviors, the authorities and obligations of different actors and groups, and the rights of persons affected by the law and the acts it seeks to regulate. Policy-makers and other actors in a particular paradigm will often eschew the careful, even compulsive, definitional specifics of the law, however, in order to maximize flexibility of action and reaction, especially in a rapidly changing environment like cyberspace.

Rhetoric that blurs definitional categories, conflates legal regimes, or inappropriately triggers legal authorities or obligations can, therefore, have substantial consequences for rights and protections. The overly expansive use of the terms "cyberwar" and "cyber attack" for actions and situations that do not fit within the specific legal definitions of the terms "war" or "attack" in LOAC or the jus ad bellum pose several significant risks for both the protection of individual rights and the stability of the international order. First, rhetoric that suggests a wartime footing facilitates the use of force in the first resort; the characterization of individuals who engage in threatening cyber activity as "the enemy," with all the attendant consequences to such a characterization; and the use of law of war-based detention without the requisite foundational basis. Second, the indiscriminate use of the term "attack" for acts falling far outside the jus ad bellum or LOAC definitions of attack can weaken the prohibition on the use of force in the international system and introduce the more permissive LOAC rules for use of force and treatment of persons when the situation does not actually justify such a framework—with the result of grave consequences for individual rights. The runaway rhetoric of cyberwar and cyber attack has the potential to lead to these results unless careful attention is paid to ensure that rhetoric does not become law.

[121] See, for example, Schmitt (n 28) 97–102; Laurie R Blank, "International Law and Cyber Threats from Non-State Actors" (2012) 89 *International Law Studies* 406.

6

The Rise of Non-State Actors in Cyberwarfare

Nicolò Bussolati

I. Introduction

Even though the academic debate on the use of code weapons in warfare has dramatically increased over the past ten years, no state has ever officially deployed cyber weapons during an armed conflict. Cyber war appears more as a fearful, probable future than a present reality. Alarmingly though, cyber weapons are increasingly demonstrating their potential to be destructive and profitably employed in warfare.[1] In point of fact, one of the most remarkable elements of past cyber events is the substantial involvement of non-state actors. In 2007, following the relocation of a Soviet era monument, Estonian government websites and banks were struck with distributed denial of service attacks (DDoS).[2] Among the consequences of the attack, the emergency communication lines were rendered unavailable, although for a short period of time. One year later, during the 2008 South Ossetia war, Georgian digital systems suffered a similar attack.[3] Analysis of these incidents indicated the likely involvement of hacker[4] groups.

[1] This chapter will not deal with an extensive analysis of cyber events that took place until today and their possible categorization as armed attacks. However, it may be important to point out that the increase of political and military attention (and investments) on cyberspace and digital devices, and the ever-increasing reliance of state critical infrastructure on new technologies—thus their augmenting vulnerability—are, by themselves, sufficient reasons to analyze how the legal framework regulating armed conflict may react to the solicitations of new means and methods of war that are gaining momentum, approaching on the horizon.

[2] See, inter alia, Rain Ottis, "Analysis of the 2007 Cyber Attacks Against Estonia from the Information Warfare Perspective," in Dan Remenyi (ed), *Proceedings of the 7th European Conference on Information Warfare and Security* (Plymouth: Academic Publishing Limited, 2008) 163–8. A denial of service (DoS) attack is a type of cyber attack aimed at making a digital machine or network resource unavailable, usually by saturating the system with external communication requests. A *distributed* denial of service (DDoS) attack is a type of DoS attack involving the use of multiple compromised systems—usually through a botnet—in conducting the attack. See S V Raghavan and Ed Dawson (eds), *An Investigation into the Detection and Mitigation of Denial of Service (DoS) Attacks: Critical Information Infrastructure Protection* (London: Springer Science & Business Media, 2011).

[3] See, inter alia, Eneken Tikk, Kadri Kaska, Kristel Rünnimeri, Mari Kert, Anna-Maria Talihärm, and Liis Vihul, "Cyber Attacks against Georgia: Legal Lessons Identified" (NATO Cooperative Cyber Defence Centre of Excellence 2008) at <http://www.carlisle.army.mil/DIME/documents/Georgia%201%200.pdf> accessed July 7, 2014.

[4] For the sake of simplicity, in this chapter the term "hacker" will be mainly used in its "pejorative" meaning of a person who uses computer technology to cause damage, gain unauthorized access to data, and so on—although this use is occasionally contested on the basis of a semantic precision in distinguishing between different types of approaches to digital technology within the broad "hacker" umbrella term (mainly in terms of ethical principles and legal/illegal purposes),

With the outbreak of international terrorism, non-state actors became a major threat to internal and external state security. Their expanded influence on international politics induced a reformulation of the criminal and public international law structures, which necessarily touched the state-centric architecture of ius ad bellum and in bello.[5] New technologies profoundly reshaped the boundaries of this phenomenon.

The edification of a "cyberspace" undermined the foundations of the traditional Westphalian system.[6] Surely, this *alter*-space, which responds to different rules from the physical world, deeply eroded the importance of geopolitical barriers. Its transcendence gave any conduct performed therein a strong transnational character. It nullified geographical distances, warping space and time: everything and everyone on the web may only be at a "click" distance away. As a point of fact, cyberspace—intended as a virtual space that covers and permeates the physical world, created by a web of connecting servers and machines—is based on a metaphysical architecture weaved on completely new concepts of time and space which hardly fit the established models of criminal and international law.

This virtual space became a fertile place for the proliferation of various malicious groups. Some found on the web a place to gather and express their political, ideological, or criminal interests. Some, already existing in the physical world, translated their activities online. Cyberspace became their virtual headquarters and a means of propagation of their actions. These groups equipped themselves with digital weapons—cheap, powerful, and easy to use, to obtain, or to "manufacture." Indeed, new technologies strongly enhanced their relevance, granting them a central position in the international system.

This chapter focuses on the rise of non-state actors in cyber warfare and its impact on international law. The first part considers how digital technologies stimulated an increasing role for non-state actors in the international system and accelerated the demise of the state as primary actor of international law. Moreover, it conducts a taxonomical analysis of non-state actors operating in cyberspace, highlighting their present and future role in cyber warfare. Finally, it examines how they relate to the states, and how the peculiarities of cyberspace affect their structures and modus operandi. The second part evaluates, in light of these morphological and operative features, the challenges to international law posed by the non-state actors' involvement in this new paradigm of warfare. In the first place, it considers how their participation in cyber war may be covered by the traditional corpus of norms regulating armed conflicts. Moreover, it analyzes issues related to the attribution of the act, and the ability of the state to respond to cyber threats deriving from non-state actors under the law of self-defense.

[5] International Law and International Humanitarian Law (IHL) started to take into consideration non-state actors—namely groups engaging in wars of liberation—during the 1950s–1980s process of decolonization. See M Cherif Bassiouni, "The New Wars and the Crisis of Compliance with the Law of Armed Conflict by Non-State Actors" (2008) 98 *The Journal of Criminal Law and Criminology* 711, 734. See also Georges Abi-Saab, "Wars of national liberation," in *Collected Courses of the Hague Academy of International Law in the Geneva Conventions and Protocols* (Leiden: Martinus Nijhoff, 1979).

[6] See David Betz and Tim Stevens, *Cyberspace and the State: Toward a Strategy for Cyber-power* (London: Routledge, 2011) 55–74.

II. The Rise of Hacker Groups as Central Actors in the International Cyber System

Digital systems and computers are a production of the human hand and mind. As a fair reflection of their faultiness, they are necessarily affected by some imperfections. Their lack of intelligence, their inanimate stupidity, makes them sufficiently easy to "deceive" or exploit for illegal purposes. Man takes advantage of these flaws for his personal interest, whether economical, ideological, political, or military.

The computer was the first manmade apparatus able to receive instructions (programs) and to process them. At first, computers were capable of relatively simple calculations,[7] then, in an exponentially increasing rate of technologization,[8] they became faster, cheaper, and capable of executing a higher number of instructions. Rapidly, the new technology proliferated, expanding its diffusion, finding a massive, irreplaceable use in society, particularly in industry, communications, state infrastructure (finance, politics, health-care, etc.), and in war.[9] In the 1970s, the "personalization" of the computer opened the use and programming of digital technologies—previously limited to "insiders"[10]—to the wider public. Laymen learnt to program the computer and use it for various purposes, being legal or illegal. In the 1990s, the connectivity revolution created a virtual web linking together the digital systems. This interconnection system boosted the virulent diffusion and dangerousness of cyber attacks, at the same time strongly augmenting their transnational character. Attractive targets at a "click distance"—private companies, state infrastructure, military, and government nerve centers are now exposed to hackers, criminal organizations, terrorists, and foreign authorities. Increasing cases of attacks on public computerized systems and political espionage—such as the attack on US military computers by German hackers exploiting ARPANET and MILNET,[11] and the selling of data to the KGB[12]—clearly evidenced the vulnerability of the newborn information society.[13]

On the other hand, the technological revolution produced extremely interesting sociological phenomena. The virtual space created by the web of interconnected machines became a container where public, private, and civil society started to transfer

[7] For instance, ENIAC (Electronic Numerical Integrator and Computer), the first electronic general-purpose computer, was designed in the mid forties to calculate artillery firing tables.

[8] The so-called "accelerating change." See Ray Kurzweil, *The Age of Spiritual Machines* (London: Viking, 1999).

[9] See, inter alia, Robert McGinn, *Science, Technology, and Society* (London: Prentice Hall, 1991).

[10] Mainly scientists and information technology students.

[11] A section of ARPANET used for unclassified United States Department of Defense traffic. In 1983 ARPANET and MILNET (then NIPRNET, now Sensitive but Unclassified IP Data) were split. For the history of MILNET and ARPANET (and its replacement by the National Science Foundation Network (NSFNET)) see, inter alia: Subrata Goswami, *Internet Protocols: Advances, Technologies and Applications* (London: Springer Science & Business Media, 2003) 7–10; Janet Abbate, *Inventing the Internet* (Cambridge, MA: MIT Press, 2000), 81ff.

[12] See David Curry, *UNIX System Security: A Guide for Users and System Administrators* (Boston, MA: Addison Wesley, 1992); Robert J McCarthy, "3 West German 'Hackers' Held as KGB Spies," *The Washington Post*, March 3, 1989.

[13] See Ulrich Sieber, "Legal Aspects of Computer-related Crime in the Information Society" (Comcrime Study, 1998) 2–3, 19. See also, generally, on the technological development and its social effects: Frank Webster, *Theories of Information Society* (London: Routledge, 3rd edn, 2006).

power, data, and knowledge. It created new cultural, political, and economic opportunities. Many human actions in the physical world—in either individual or social/political group capacities—migrated online: commerce, cultural and political debates, knowledge sharing, and relations with physical and juridical persons including administrations, governments, or political parties.[14] New virtual *agorai* (fora, social networks, electronic petition platforms, etc.) connected people from all over the world. Groups formed around these virtual places. Bodies already existing in the physical world moved their affairs online.

Among other activities, crime and political/ideological strife migrated to cyberspace. Exploiting the peculiar compression of time and distances, the opacity of the web—which hinders attribution of acts performed therein—and the lack of regulations, individuals and groups started to perform online illegal acts against private citizens, corporations, states, and international institutions, their main motivations being economic, political, or ideological.[15] Increasingly, in the last decade, war is also moving towards the cyber realm:[16] state armies, but also violent non-state actors, are thither extending their scope of action. Recent cyber events linked to military operations, such as the Georgian cyber attacks[17] or "Operation Israel" conducted by the hacker group Anonymous,[18] highlight

[14] For instance, the s.c. "electronic democracy" (see OECD, "Promise and Problems of E-Democracy: Challenges of Online Citizen Engagement" (2003)) at <http://www.oecd.org/governance/public-innovation/35176328.pdf> accessed July 7, 2014; Chung-pin Lee, Kaiju Chang and Frances Stokes Berry, "Testing the Development and Diffusion of E-Government and E-Democracy: A Global Perspective" (2011) 71 *Public Administration Review* 444.

[15] Notwithstanding their distinctive transnationality, these conducts were usually dealt with through domestic criminal law (on domestic and regional responses to cybercrime, see UNODC, "Comprehensive Study on Cybercrime" (2013); ITU, "Understanding Cybercrime: Phenomena, Challenges and Legal Response" (2012)). Still, it is important to point out that, within the cybercrime *genus*, acts of a sufficient gravity aimed at spreading fear among the population or coercing governments or international organizations and driven by political, ideological, or religious motivations may be considered—at least theoretically—as terrorist crimes. On international terrorism and problems of definition, see, inter alia, Ben Saul, *Defining Terrorism in International Law* (Oxford: Oxford University Press, 2006); Kai Ambos, "Judicial Creativity at the Special Tribunal for Lebanon: Is There a Crime of Terrorism under International Law?" (2011) 24 *Leiden Journal of International Law* 655.

[16] The use of digital technologies in warfare may present numerous advantages compared to the traditional physical combat means. Firstly, their use may result in fewer human casualties. Certainly, some types of attack may have a wider "circular error probability" (the ballistic measure of a weapon system's precision), since they bear the risk of producing extremely spatially diffused collateral damages—for instance striking computer systems with similar IP addresses. However, a cyberweapon may strike the designated target without the risk of collateral killings normally associated with the impact or explosion of a physical weapon, and without involving "boots on the ground." From a strategic point of view, technology and obfuscation techniques complicate the attribution of the attack. Furthermore, cyberweapons are relatively cheap and easy to construct, since they do not require complex industrial processes or expensive and rare materials as components. Storage of such technology is also easier and cheaper than with conventional weapons. While kinetic weapons are produced through industrial apparatus, technologies and knowledge necessary for constructing (and operating) digital weapons are easily available to non-state groups and individuals.

[17] See Ariel Cohen and Robert E Hamilton, *The Russian Military and the Georgia War: Lessons and Implications* (Strategic Studies Institute 2011); John Markoff, "Before the Gunfire, Cyberattacks" New York Times, 12 August 2008; "RBN (Russian Business Network) now nationalized, invades Georgia Cyber Space," *RBNExploit Blog*, September 8, 2008, at <http://rbnexploit.blogspot.it/2008/08/rbn-georgia-cyberwarfare.html> accessed July 7, 2014.

[18] See Kelly Jackson Higgins, "Israel Draws Ire Of Anonymous," *Dark Reading*, November 19, 2012, at <http://www.darkreading.com/attacks-breaches/israel-draws-ire-of-anonymous/240142355?cid=fbook_DarkReading&wc=4> accessed July 7, 2014; Matthew Schwartz, "Anonymous Launches OpIsrael DDoS

the central role acquired by non-state actors in such context, clearly demonstrating their ability and attitude to participate in cyberwarfare either alongside or against state actors.

The peculiar character of the virtual world—the altered nature of time and space in which it operates, its accessibility and geographical diffusion, the availability of technology and knowledge, and the absence of superordinate structures—offers the basis for a new, flattened and expanded international political structure, whose internal relations between asymmetric actors appear levelled. May this phenomenon be read as the definitive passage from a state-centric political system to post-international politics?[19] Surely, in the cyber age, van Creveld's assertion of the final decline of Clausewitz's trinitarian model of war (people, army, and government) appears sounder than ever.[20]

III. An Essential Classification of Non-State Actors Operating in Cyberspace

Non-state actors active in cyberspace having the potential to employ digital force or, to various degrees, to be involved in cyber military operations may substantially differ according to size, internal structure, motivational grounds, and relation with the state. Their size may vary from simple (even "unicellular") organisms to large transnational groups. Their organizational structure may be informal, lacking a chain of command, or complex, formal, and stably hierarchical. They may be driven by economic, political, ideological, or religious motivations. Usually, such organisms do not pursue purely military goals, such as power-outcome, typical of state actors or traditional non-state groups that engage in kinetic warfare. Further, they may be directed or stimulated by states or be fiercely opposed to any connection with state political entities.

(a) Individual hackers

In cyberspace, the individual represents an important carrier of knowledge and technical skills (which are often located outside official educational structures). Further, new technologies have strongly augmented the capabilities of the individual beyond one's physical potentials, enhancing the capacity and role as an autonomous actor in cyberwar.

Individual hackers may serve as a valuable asset in the hands of the state, which can benefit from their technical skills. IT experts and hackers may be formally or informally

Attacks After Internet Threat," *InformationWeek*, November 15, 2012, at <http://www.informationweek.com/security/attacks/anonymous-launches-opisrael-ddos-attacks/240142149> accessed July 7, 2014; "Anonymous Leaks Personal Information of 5,000 Israeli Officials," *RT*, November 18, 2012, at <https://rt.com/news/anonymous-israel-officials-leaked-002> accessed July 7, 2014. See also § III (d).

[19] See James N Rosenau, "Global Changes and Theoretical Challenges: Toward a Postinternational Politics for the 1990s," in Ernst-Otto Czempiel and James N Rosenau (eds), *Global Changes and Theoretical Challenges. Approaches to World Politics for the 1990s* (Lanham, MD: Lexington Books, 1989).

[20] See Martin van Creveld, *The Transformation of War* (New York, NY: Free Press, 1991). See also Umberto Gori, "Cyberspazio e relazioni internazionali: implicazioni geopolitiche e geostrategiche," in Umberto Gori and Serena Lisi (eds), *Information Warfare 2012: Armi cibernetiche e processo decisionale* (Milan: FrancoAngeli, 2013).

employed in electronic warfare army units, such as the Israeli Defence Forces Unit 8200 or the Chinese People's Liberation Army (PLA) Unit 61398.[21] Moreover, state authorities may hire private hackers for specific operations or buy valuable information from them—in particular, software vulnerabilities, through which a digital system may be penetrated. As a matter of fact, vulnerabilities possess extreme economic and military importance: the so-called "bug hunters"[22] may sell them directly to states or to private companies—such as iDefense, Revuln, Tipping Point, and Netragard—which may re-sell them to interested buyers.[23] For instance, the famous malware Stuxnet was based on the exploitation of several vulnerabilities, likely bought from bug hunters in the digital black market (darknet).[24] Further, hackers may be employed in cyber espionage: one of the first cases relates to German hackers—some of them affiliated to the famous hacker group Chaos Computer Club—who in the 1980s sold information stolen from US military computers to the KGB.[25]

The increased capability of a single individual to independently produce and use malevolent software, and the loose structure of many hacker groups (in particular, politically or ideologically driven groups), elect the individual hacker as the essential monad of the hacking panorama. Singular hackers unaffiliated to any group may independently launch attacks, as well as extemporarily join groups and take part in concerted cyber operations. Even if equipped with very basic expertise and technological means,[26] they may participate in operations through the use of automated attack tools available on the web or to which they may be directed to through hacker fora. For instance, during the 2008 Russia–Georgia war—the first cyber attack launched alongside a kinetic military attack—fora such as *stopgeorgia.ru* and *xaker.ru* were providing users (including unaffiliated hackers) with attack toolkits, information, and vetted target lists of Georgian objectives.[27]

Finally, even ordinary citizens may participate in conflicts: unconsciously, as "zombified" computers (a "zombie" is a computer infected with a bot, which is controlled by the operator of the botnet and used for an attack—with consequent

[21] See Jeffrey Carr, *Inside Cyber Warfare* (Sebastopol, CA: O'Reilly Media, 2nd edn, 2012) 217ff. For instance, according to computer analysts, Tan Dailin (aka Wicked Rose), leader of the Chinese Network Crack Program Hacker group, regularly works for the Chinese army. See David Harley, *AVIEN Malware Defense Guide for the Enterprise* (Amsterdam: Elsevier, 2011) 200–201. See also, with regards to the allegations about the use of *Anonymous* hackers by the United States government: "Hacker Accuses US Government of Tricking Anonymous into Attacking Foreign Targets," *RT*, August 23, 2013, at <http://rt.com/usa/anonymous-sabu-hammond-sentence-910> accessed July 7, 2014).

[22] Hackers that search for vulnerabilities.

[23] See Paul N Stockton, Michele Golabek-Goldman, "Curbing the Market for Cyber Weapons" (2013) 32 *Yale Law & Policy Review* 101, 109ff.

[24] See Ralph Langner, *To Kill a Centrifuge: A Technical Analysis of What Stuxnet's Creators Tried to Achieve* (Arlington, VA: The Lagner Group, 2013) at <http://www.langner.com/en/wp-content/uploads/2013/11/To-kill-a-centrifuge.pdf> accessed July 7, 2014; Chris Strohm and Jordan Robertson, "Military Bid for Next Stuxnet Confronts Hacker Resistance," *Bloomberg*, August 2, 2013, at <http://www.bloomberg.com/news/2013-08-02/military-bid-for-next-stuxnet-confronts-hacker-resistance.html> accessed July 7, 2014. See also Michael Reilly, "How Long Before All-Out Cyberwar?" (2008) 2644 *New Scientist* 23, 24–5.

[25] See n 12.

[26] See Rain Ottis, "From Pitchforks to Laptops: volunteers in cyber conflicts," in Christian Czosseck and Karlis Podins (eds), *Conference on Cyber Conflict: Proceedings* (CCD COE Publications 2010).

[27] See Carr (n 21) 15.

problems of attribution),[28] or passively, providing their devices to the army as operational platforms.[29]

(b) Criminal organizations

Due to the lucrative potential of cybercrime, criminal organizations have flourished in cyberspace.[30] Criminal consortia with a presence or ramifications online are among the non-state actors operating in cyberspace that present a higher degree of structural formality and aptitude to interconnect with the state. Their digital manifestations likely reflect the structured, hierarchical organization typical of the crime syndicates.[31] The exclusively financial interest that fuels the organization[32] is the focal point that permits its usability by the state, expressed through an economic relationship. Indeed, such organizations may prove to be ruthless enough and possess the necessary technology, knowledge, and structure to be effective in a cyber scenario. For instance, the Russian Business Network, a criminal organization which administers various illegal activities related to computer crimes—such as child pornography, phishing, spam, and malware distribution[33]—appears, according to some commentators, to have contributed (in all likelihood, under state stimulus) to the cyber attacks conducted in 2008 during the Russia–Georgia war[34] and against Kyrgyzstan in 2009.[35]

(c) Cyber mercenaries

Species of the criminal organization genus, cyber "mercenary" groups are composed of highly skilled hackers specialized in sophisticated cyber attacks. They may sell their skills to the public or private sector, which hires them to conduct precise and specific attacks. Their motivation is exclusively economic and the relationship with the contractor, although not permanent, is expressed in a specific agreement.

[28] See Johan Sigholm, "Non-State Actors in Cyberspace Operations" (2013) 4 *Journal of Military Studies* 1, 13.

[29] See Scott Applegate, "Cybermilitias and political hackers: Use of irregular forces in cyberwarfare" (2011) 9 *IEEE Security and Privacy* 16, 17; Sigholm (n 28) 13.

[30] For an analysis of traditional organized crime groups having online ramifications and organized cybercrime groups that operate exclusively online, see K K Raymond Choo, "Organized Crime Groups in the Cyberspace: a Typology" (2008) 11 *Trends in Organized Crime* 270.

[31] The power structure of a criminal organization appears hierarchical, where the specialization of group roles and a stable membership reduce the permeability in the entering or exiting from the structure. See David Luban, Julie R O'Sullivan, and David P Stewart, *International and Transnational Criminal Law* (New York, NY: Aspen, 2010) 505.

[32] Although some criminal organizations may have political interests, usually the political outcome is aimed at strengthening the solidity of the organization and widening its power with the aim of a greater financial gain. See Luban et al (n 31) 505: "Traditionally, an 'organized crime' group is motivated by money or power, not by ideology; the nonideological nature of such group means that any political activity (generally corrupt) is an instrument for achieving their criminal aims or shielding them from law enforcement, not a goal in itself."

[33] See Thomas J Holt and Bernadette H Schell, *Hackers and Hacking: A Reference Handbook* (Santa Barbara, CA: ABC-CLIO, 2013) 223–224.

[34] See n 17.

[35] See Danny Bradbury, "The Fog of Cyberwar" *The Guardian*, February 5, 2009, at <http://www.theguardian.com/technology/2009/feb/05/kyrgyzstan-cyberattack-internet-access> accessed July 7, 2014.

The Kaspersky lab, in collaboration with KISA (Korea Internet Security Agency) and Interpol, recently published an interesting analysis of a cyber mercenary group, active in Japan and South Korea since 2011. According to the research, the group already targeted—inter alia—governmental institutions, military contractors, communication operators and industrial/high-tech companies.[36]

(d) Hacktivists

Digital activist (or "hacktivist," a portmanteau of hacker and activist) groups are independent, politically or ideologically driven hacker groups. They may range from local units composed of no more than a dozen persons to large transnational organisms with several satellite sub-groups.[37] Their internal structure, essentially informal, is shaped by the virtual composition and social life of the group. Operations are usually planned and organized on digital platforms, such as fora or IRC channels. The platform represents the (virtual) place where members meet, discuss, and share knowledge and information.[38] There, directions, information, and toolkits for specific cyber operations are distributed. Given their activist nature, their actions are mainly driven by ideological and political motivations. Their theoretical foundations are grounded in the cyber countercultures linked with the libertarian conception of cyberspace as a space free from the political influence of the state.[39] They stand as a consequence of the virtualization and internationalization of the urban square—understood as an *agora* for exchanging and developing political and cultural idea(l)s—and of political and economic power.[40] Digital acts of protest and civil disobedience began to go along with physical street protests, the latter losing its main aim of message diffusion (now easily reachable through the web, bypassing the filter of the media) and hindering or stopping the political and economical flux (now moved from the physical streets to the virtual arteries of the web).

Political activist groups started to perform acts of electronic protest at the end of the 1980s.[41] At the edge of the new millennium the numbers of attacks radically

[36] See Kaspersky Lab Global Research and Analysis Team, "The 'Icefog' Apt: A Tale Of Cloak And Three Daggers" (Kaspersky Lab Zao, 2013) at <http://kasperskycontenthub.com/wp-content/uploads/sites/43/vlpdfs/icefog.pdf> accessed July 7, 2014: "In the future, we predict the number of small, focused APT-to-hire groups to grow, specializing in hit-and-run operations, a kind of 'cyber mercenaries' of the modern world."

[37] Such as Anonymous or Lulzsec, which possess a global extension. Parmy Olson, *We Are Anonymous: Inside the Hacker World of Lulzsec, Anonymous and the Global Cyber Insurgency*, (New York, NY: Little, Brown and Co., 2012).

[38] See Dorothy E Denning, "Cyber Conflict as an Emergent Social Phenomenon," in Thomas J Holt and Bernadette Hlubik Schell (eds), *Corporate Hacking and Technology-Driven Crime: Social Dynamics and Implications* (Hershey, PA: IGI Global, 2010) 172.

[39] See John P Barlow, "A Declaration of the Independence of Cyberspace," in Peter Ludlow (ed), *Crypto Anarchy, Cyberstates, and Pirate Utopias* (Cambridge, MA: MIT Press 2001) 27: "Governments of the Industrial World, you weary giants of flesh and steel, I come from Cyberspace, the new home of Mind. On behalf of the future, I ask you of the past to leave us alone. You are not welcome among us. You have no sovereignty where we gather… "

[40] See Critical Art Ensemble, *Electronic Civil Disobedience and Other Unpopular Ideas* (New York, NY: Autonomedia, 1996).

[41] See John Arquilla and David Ronfeldt, *Networks and Netwars: The Future of Terror, Crime, and Militancy* (Santa Monica, CA: Rand Corporation, 2001) 278–280 (discussing 1989 NASA Galileo Probe attack).

increased.[42] Hacktivist groups are gradually extending their reach on every aspect of domestic and international politics that—according to their own view—expresses examples of injustice. This has let, and expectably will increasingly lead to, substantial participation of hacktivist groups in armed conflicts. In 2012, as a response to the Israeli military operation in Gaza Pillar of Defence, the hacktivist community Anonymous conducted a DDoS attack against several Israeli websites and posted online names, identification numbers, and personal emails of 5,000 Israeli Defence Force officials.[43] In March 2014, it launched two operations (OpRussia and OpUkraine) striking Russian cyberspace with digital attacks in reaction to the Russian maneuvers in Crimea.[44]

Thus far, digital activists have mainly performed low-level demonstrative actions, primarily operating web defacements[45] and denial of service attacks.[46] Still, in a prospective view, these types of attacks are progressively acquiring dangerousness as the technology evolves (e.g., new species of DDoS attacks,[47] SQL injections,[48] or "recombinant DNA" malwares[49]), becomes cheaper and more easily available[50] and the state infrastructure reliance on technology grows,[51] also geographically.

(e) Patriotic hackers

Patriotic hacker groups present strong similarities—principally structural—to hacktivist groups. They are permanently or extemporaneously created (usually in the aftermath of precise events)[52] hacker groups, exclusively driven by a patriotic devotion to

[42] See, for example, Olson (n 37) Timeline. [43] See n 18.

[44] ...and, on the opposite side, a pro-Russian hacker group named *Cyber Berkut* repeatedly attacked NATO and Ukrainian websites. See Jeffrey Carr, "Rival Hackers Fighting Proxy War Over Crimea," CNN.com, March 25, 2014, at <http://edition.cnn.com/2014/03/25/opinion/crimea-cyber-war/> accessed July 4, 2014.

[45] Cyber attacks on a website aimed at changing its visual appearance.

[46] See Jose Nazario, "Politically Motivated Denial of Service Attacks," in Christian Czosseck and Kenneth Geers (eds), *The Virtual Battlefield: Perspectives on Cyber Warfare* (Amsterdam: IOS Press, 2009).

[47] See Alex Pinto, "How Anonymous Attacks—and How to Protect Your Site from the New Breed of #DDoS," *Yottaa*, August 22, 2012, at <http://www.yottaa.com/blog/bid/215371/How-Anonymous-Attacks-And-How-To-Protect-Your-Site-from-the-New-Breed-of-DDoS> accessed July 7, 2014.

[48] See Justin Clarke, *SQL Injection Attacks and Defense* (Amsterdam: Elsevier, 2012).

[49] See Jon Brodkin, "Government-Sponsored Cyberattacks on the Rise, McAfee Says," *Network World*, November 29, 2007, at <http://www.networkworld.com/news/2007/112907-government-cyberattacks.html> accessed July 7, 2014.

[50] See, for example, Pauline Jelinek, "Iraq Insurgents Hack U.S. Drones For Under $26," *Huffington Post*, December 17, 2009, at <http://www.huffingtonpost.com/2009/12/17/iraq-insurgents-hack-us-drones_n_395337.html> accessed July 7, 2014.

[51] For instance, several American critical infrastructures are controlled by a Supervisory Control and Data Acquisition (SCADA) system, a computer-control system that manages and controls physical processes. See Stuart A Boyer, *Scada: Supervisory Control And Data Acquisition* (International Society of Automation, 2010). Metropolises are becoming smart technological cities where digital services—such as Wi-Fi hotspots, energy smart grids, surveillance systems, and intelligence transportation systems—communicate and interact between them and with interconnected citizens. See, for example, Symantec, *Executive Report: Smart Cities. Transformational "Smart Cities": Cyber Security And Resilience* (2013) at <http://eu-smartcities.eu/sites/all/files/blog/files/Transformational%20Smart%20Cities%20-%20Symantec%20Executive%20Report.pdf> accessed July 7, 2014.

[52] See Denning (n 38) 178–80.

defend the interests of their country. Such a strong ideological linkage with the state may translate to a relationship of control by state authorities, and usually carves their social form within national borders. Furthermore, when the patriotic motivations are linked to geopolitical claims based on religion, such hacker collectives may have connections with terrorist groups.[53]

Their actions are triggered by events that, according to their sensitivity, may damage the interests of, or pose a threat to, their country. For instance, cyber attacks against Estonian government websites and banks, in 2007[54] (which, inter alia, interfered with the telephone access to emergency services)[55] were triggered by the Estonian government's decision to relocate a monument of the Soviet Era. The involvement of the Russian government was never proved; however, the attack was in all likelihood conducted by patriotic groups of hackers. The assistant of a Russian State Duma deputy and prominent member of the Nashi—a Russian youth political organization with digital protractions, strongly linked with the Russian government[56]—openly admitted to having participated in the attack.[57] A year later, following the enactment of a Lithuanian law banning the display of Soviet emblems, groups of patriotic hackers attacked Lithuanian websites and diffused online a "Hackers United against External Threats to Russia" manifesto.[58] Russian hackers may have been implicated in numerous other cyber incidents.[59] Active patriotic hacker groups may also be observed in other countries.[60] Among them, Chinese patriotic hacker groups, such as the Red Hacker Alliance,[61] are particularly active since the late 1990s.[62]

IV. Participation in Conflict

As individual and group hackers take part in armed conflict, their classification according to the categories of ius in bello is of pivotal importance. Within the persons actively or passively involved in an armed conflict, IHL recognizes a fundamental dichotomy between fighters and civilians. Participation through the use of (kinetic or digital) weapons in an armed conflict brings about a series of legal consequences. In international armed conflicts, individuals having official combat tasks in the armed forces of a belligerent party are considered lawful combatants.[63] As

[53] Denning (n 38) 171–80. [54] See n 2.

[55] See "Marching off to Cyberwar," *The Economist*, December 4, 2008.

[56] See Carr (n 21) 115.

[57] See Carr (n 21) 3; Noah Shachtman, "Kremlin Kids: We Launched the Estonian Cyber War," *Wired*, November 3, 2009, at <http://www.wired.com/2009/03/pro-kremlin-gro/> accessed July 7, 2014.

[58] See Brian Krebs, "Lithuania Weathers Cyber Attack, Braces for Round 2," *The Washington Post*, July 3, 2008, at <http://voices.washingtonpost.com/securityfix/2008/07/lithuania_weathers_cyber_attac_1.html> accessed July 7, 2014.

[59] See Athina Karatzogianni, "Blame it on the Russians: Tracking the Portrayal of Russians During Cyber Conflict Incidents," (2010) 4 *Digital Icons: Studies in Russian, Eurasian and Central European New Media* 128. See also n 46.

[60] See Denning (n 38) 179ff.

[61] See Scott Henderson, "Evolution of the Red Hacker Alliance," *The Dark Visitor*, November 17, 2007, at <http://www.thedarkvisitor.com/2007/11/evolution-of-the-red-hacker-alliance> accessed July 7, 2014.

[62] See Scott Henderson, *The Dark Visitor: Inside the World of Chinese Hackers* (Raleigh, NC: Lulu, 2007).

[63] See Protocol Additional to the Geneva Conventions of August 12, 1949, and relating to the Protection of Victims of International Armed Conflicts (Protocol I), June 8, 1977, Article 43(2).

a consequence, they can be lawfully targeted, and they enjoy combatant immunity and prisoner of war status.[64] Combatant status is also afforded to members of groups assimilated to official military forces of the state. According to Article 4A(2) of the Third Geneva Convention—which reflects customary international law—in international armed conflicts "members of militias or volunteer corps, including those of organized resistance movements, belonging to a Party to the conflict" are to be considered combatants, provided the fulfilment of four conditions: (a) that of being commanded by a person responsible for his (or her) subordinates; (b) that of having a fixed distinctive sign recognizable at a distance;[65] (c) that of carrying arms openly; (d) that of conducting their operations in accordance with the laws and customs of war.[66]

When outside the sphere of a combat role in the armed forces of a belligerent party or in a group assimilated to the regular forces, the law of armed conflict affords civilians protection against the dangers and the consequences of war. However, if civilians take a direct part in hostilities—even through the use of digital weapons—they lose their non-combatant immunity and may be lawfully targeted.[67] The ICRC issued an "Interpretative Guidance" to "provide recommendations concerning the interpretation"[68] of the conventional and customary rules[69] regulating direct participation in conflict.[70] It indicates three cumulative basic requirements, namely: a sufficient threshold of harm, a direct link between the act and the harm, and a nexus with the Parties to the conflict through either opposing or supporting them. When these requirements are met, individuals—and

[64] See Knut Ipsen, "Combatants and non-combatants," in Dieter Fleck (ed), *The Handbook of International Humanitarian Law* (Oxford: Oxford University Press, 2nd edn, 2008).

[65] The requirements to display a fixed distinctive sign recognizable at a distance—usually satisfied in physical warfare with the use of uniforms—and to carry weapons openly, are founded on the necessity of the opposition forces to distinguish between civilians and combatants. For cyber warriors, the obligation to wear uniforms may be useful in distinguishing combatants during a physical attack on their physical headquarters. See Michael N Schmitt (ed), *Tallinn Manual on the International Law Applicable to Cyber Warfare* (New York, NY: Cambridge University Press, 2013) 86–7. On the other hand, a digital counterpart of the uniform—for example, in forms of recognizable military IP—might also cover the requirement of "carrying arms openly" (which is clearly untranslatable in the cyber warfare context) and reduce the risk of attacks against civilian digital systems.

[66] Convention (III) relative to the Treatment of Prisoners of War, Geneva (Third Geneva Convention), August 12, 1949, Article 4A(2).

[67] On direct participation in cyber conflicts, see Jody M Prescott, "Direct Participation in Cyber Hostilities: Terms of Reference for Like-Minded States?," in Christian Czosseck, Rain Ottis and Katharina Ziolkowski (eds), *4th International Conference on Cyber Conflict* (CCD COE Publications 2012); Emily Crawford, "Virtual Battlegrounds: Direct Participation in Cyber Warfare" (2012) Sydney Law School Research Paper 12/10, at <http://papers.ssrn.com/sol3/papers.cfm?abstract_id=2001794> accessed July 7, 2014.

[68] Nils Melzer, *Interpretive Guidance on the Notion of Direct Participation in Hostilities under International Humanitarian Law* (ICRC, 2009) 9.

[69] See Protocol I (n 63) Article 51(3); Protocol Additional to the Geneva Conventions of August 12, 1949, and relating to the Protection of Victims of Non-International Armed Conflicts (Protocol II), June 8, 1977, Article 13(3). As for the customary nature of the rule, *see, inter alia*: Israeli Supreme Court, *The Public Committee Against Torture in Israel v The Government of Israel*, (2006) HCJ 769/02, paras 29–30.

[70] See Interpretive Guidance on the Notion of Direct Participation in Hostilities under International Humanitarian Law, Adopted by the Assembly of the International Committee of the Red Cross on February 26, 2009.

among them, hackers—are considered to be directly participating in the conflict. As a result, they do not enjoy the protection accorded to civilians against attacks. However, direct participation does not afford combatant or prisoner of war status.

According to Article 47(1) of Additional Protocol I, mercenaries do not enjoy combatant immunity or prisoner of war status.[71] Such a category is defined by Article 47(2) according to six cumulative criteria: being recruited (locally or abroad) to fight in an armed conflict; taking direct participation in the conflict; being essentially moved by economic motivations (private gain); not being a national of a party to the conflict nor resident in its territory; not being on official duty in the armed forces of a belligerent party; not being sent by a state which is not a party to the conflict on official duty as a member of its armed forces. Thus, hackers having a nationality or residential tie with the territory of the embroiled state—such as, for instance, patriotic hackers—are excluded from this category.

Of particular interest is the application to non-state groups operating in the cyberspace of the concept—although quite marginal in the current IHL system—of *levée en masse*. IHL affords combatant status to persons participating in a spontaneous civilian resistance to an invader. Article 4A(6) of the Third Geneva Convention[72] (and Article 2 Hague Regulations),[73] which reflects customary law,[74] provides that "inhabitants of a non-occupied territory, who on the approach of the enemy spontaneously take up arms to resist the invading forces without having had time to form themselves into regular armed units" enjoy combatant status. Theoretically, the provision may cover hackers that, autonomously and detached from any state connection, conduct acts of cyber defense against an invading power. Indeed, the *levée en masse* participants, as the term suggests, must be a considerable segment of the population able to carry arms. Considering the necessary basic knowledge to participate in cyber attacks, it may be disputed that in case of cyberwarfare such a requirement must be referred only to the cyber-capable population.[75] In any case, small groups of hackers are to be excluded from the scope of the provision. Moreover, the territorial requirement precludes any form of "digital resistance" encountering a wider ideological mobilization and thus encompassing a transnational character. Although archaic and hardly applicable, this self-help option against foreign invasion through civil resistance seems to appropriately sketch out the traits of a political and ideological extemporaneous resistance characterized by a lack of proper organizational structure of some hacktivist and patriotic hacker groups.

[71] See Protocol I (n 63) Article 47.

[72] See Third Geneva Convention (n 66) Article 4A(6).

[73] See Regulations concerning the Laws and Customs of War on Land. The Hague, October 18, 1907, Article 2.

[74] See the ICRC study on customary international humanitarian law: a contribution to the understanding and respect for the rule of law in armed conflict, Commentary on Rule 106.

[75] See *Tallinn Manual* (n 65) 89. However, the diffusion of computers and portable devices and the considerable growth of computer literacy are, at least in the richest countries, pushing towards an overlapping of the group (all citizens able to carry arms) with the subgroup (cyber-capable citizens).

V. IHL and Organizational Structure of the Group

The internal social structure of most hacker groups is constructed around a virtual place. There, the members of the group gather, communicate, share knowledge and information, and take decisions. The "virtuality" of the group's social relations, which influences its internal structure, may produce peculiar frictions with the rules governing armed conflicts, calibrated on the organizational architecture of traditional military groups. Notably, the organizational structure of a group involved in an armed conflict is a central element of the IHL system. A stable internal structure may be necessary to provide discipline and compliance with the rules regulating armed conflicts. As pointed out, within international armed conflicts this is directly related to combatant status. According to Article 43(I) of Additional Protocol I, the armed forces of a party to a conflict "shall be subject to an internal disciplinary system which, 'inter alia,' shall enforce compliance with the rules of international law applicable in armed conflict."[76] Furthermore, Article 4A(2) of the Third Geneva Convention states that "members of militias or volunteers corps, including those of organized resistance movements, belonging to a Party to the conflict" enjoy combatant status provided that—among other requirements—they are commanded by a person responsible for his or her subordinates and that they conduct their operations in accordance with the laws and customs of war.[77]

Additionally, the organizational structure has implications when categorizing the conflict as either international (i.e., use of armed forces by a state against another state)[78] or non-international (i.e., use of armed force by a state against a non-state group, or between such groups).[79] In the former case, a connection with a belligerent party is a necessary element for IHL to cover the conduct of a non-state actor: this issue will be analyzed infra. In the latter case, the law of armed conflicts finds application only where hostilities reach a certain level of intensity and the armed group(s) involved possess a minimum degree of structural organization.[80]

The issue of whether a digital group can satisfy the necessary internal structure requirements to meet the threshold of these organizational criteria is not self-evident.[81] A fundamental marker in the analysis appears to be the ability to plan and carry out concerted military operations, acting in a coordinated fashion through a hierarchical command structure able to make the group act as a unit and speak with

[76] Protocol I (n 63) Article 43(I). [77] See Third Geneva Convention (n 66) Article 4(2).

[78] See Christopher Greenwood, "Scope of Application of Humanitarian Law," in Dieter Fleck (ed), *The Handbook of International Humanitarian Law* (Oxford: Oxford University Press, 2nd edn, 2008) 46–9.

[79] See International Criminal Tribunal for the former Yugoslavia (ICTY), Appeals Chamber, *Prosecutor v Dusko Tadic a/k/a "Dule,"* Decision on the Defence Motion for Interlocutory Appeal on Jurisdiction, IT-94-1, October 2, 1995, para 70: "an armed conflict exists whenever there is a resort to armed force between States or protracted armed violence between governmental authorities and organized armed groups or between such groups within a State."

[80] See, ex plurimis, ICTY, Trial Chamber, *The Prosecutor v Fatmir Limaj, Haradin Bala and Isak Musliu*, judgment, IT-03-66-T, 30 November 2005, paras 88ff.

[81] See *Tallinn Manual* (n 65) para 79.

one voice.[82] Article 1 of Additional Protocol II, applicable to non-international armed conflicts, conjoins the internal command structure of the group to its ability to "carry out sustained and concerted military operations" and to implement the provisions of the Protocol.[83] Common Article 3 of the Geneva Conventions lacks such a provision. Indeed, compliance with IHL is another central element of the issue: logically, it is imperative to provide for the applicability of a normative system only to subjects that are, at least theoretically, able to comply with it.

IHL is aimed at "humanizing the conflict" by limiting the harmful effects of war for the civilian population through the protection of specific categories of persons and property and the prohibition of certain means and methods of warfare. In the aftermath of World War II, as non-state groups began to substantially participate in conflicts—in particular, in post-colonial national liberation wars and regime revolutions—problems regarding the compliance with the rules regulating the conduct of hostilities by those actors arose.[84] Indeed, the loosened internal hierarchical command and control structure, the lack of training, and the asymmetry in their fight against a state—often balanced with resorting to unlawful means and methods of warfare—led to novel problems of non-compliance with IHL.[85] In digital warfare, however, such problems appear exacerbated.

First of all, the peculiar characteristics of the environment in which "cyber" non-state actors operate tend to impair observance of the rules regulating armed conflict. The "virtuality" of the act may hinder hackers with no military training from understanding the belligerent context. With no real/conventional weapons in their hands, they perform actions associated more with programming or playing online videogames. Nor do they visually perceive, on a screen or in the flesh, the physical consequences of their acts. The moral restraints are thus reduced. Besides, no efficient deterrence is provided by international or domestic criminal law. Hackers are protected by the turbidity of the web (and, more specifically, by the use of obfuscating techniques and anonymizing tools),[86] and, possibly, by the state itself, when interconnected with it.[87] Statistical data on the prosecution of hackers belonging to groups responsible for past cyber incidents confirm this fact: in the aftermath of the Estonian

[82] See ICTY, Trial Chamber, *Prosecutor v Boskoski and Tarculovski*, IT-04-82-T, July 10, 2008, para 250ff; *Limaj* (n 80) para 129.

[83] See Protocol II (n 69) Article 1; ICRC Commentary on the Additional Protocols of June 8, 1977 to the Geneva Conventions of August 12, 1949, para 4470.

[84] See Bassiouni (n 5). [85] Bassiouni (n 5).

[86] See Peter Loshin, *Practical Anonymity: Hiding in Plain Sight Online* (Waltham, MA: Syngress, 2013). Interestingly, Snowden's leaks have recently revealed the development of anti-obfuscation (onion routing, i.e., TOR) techniques by the NSA. See "Peeling Back the Layers of Tor with EgotisticalGiraffe'—Read the Document,"(*The Guardian*, October 4, 2013, at <http://www.theguardian.com/world/interactive/2013/oct/04/egotistical-giraffe-nsa-tor-document> accessed July 7, 2014.

[87] See Carr (n 21) 29. Per contra, states have a positive duty to take legislative and prosecution steps in order to repress violations of IHL through the enactment of "any legislation necessary to provide effective penal sanctions" for grave breaches of the Geneva Conventions and to search and trial those responsible for such breaches. Further, they have to "take measures necessary for the suppression of all acts contrary to the provisions of the present Convention" (see Convention (IV) relative to the Protection of Civilian Persons in Time of War, Geneva, August 12, 1949, Article 146).

DDoS attacks, which was treated under domestic criminal law,[88] investigations ended with the sole fining of an ethnic Russian living in Estonia.[89]

Hacktivist groups may indeed cultivate idealistic values that may include humanitarian sensibilities. However, as the motivation of the act shifts from economic to political, ideological, or religious, the personal shade of view of the individual member, which is acting outside the behavioural influence and control (physical—mimic—sight) of the "leaders," may substantially differentiate from the common denominator propelling the group. Moreover, it is important to consider that hacker groups—contrarily to traditional non-state actors in kinetic conflicts—are not seeking legitimacy or political recognition as an overall goal, which may function as a positive factor to induce compliance with IHL.

As a point of fact, however, the main risk of IHL violations may directly follow the lack of a hierarchical organizational structure that ensures a sufficient disciplinary system. The structural scheme of non-state actors operating in cyberspace appears inherently different from the physical military group command and control structure.[90] Among the various non-state groups operating in cyberspace, however, the internal structures may substantially vary.

Criminal organizations and cyber mercenaries are likely to possess the highest degree of internal hierarchy, stability, and formalization of structure, which may concretely translate to the ability to plan and carry out concerted military operations, and act in a coordinated fashion. Indeed, the very efficiency of the organization depends on such capability. Conversely, patriotic hacker groups and hacktivists usually possess a more informal structure.

Small groups of hackers—due to their limited size—may enjoy a more accentuated "organisational stability and effectiveness"[91] which, given their capability to engage in military operations,[92] may likely permit the group to move in a uniform direction and, at least theoretically, assure compliance with IHL.[93]

Larger organisms, on the other hand, may often lack a proper pyramidal hierarchy. Most of such groups (even the least formal: liquid and amorphous organisms permeable to the constant entrance and leaving of new "members") present a typical

[88] Interestingly, on that occasion the Estonian Minister of Foreign Affairs stated that: "(a)t present, NATO does not define cyber-attacks as a clear military action. This means that the provisions of Article V of the North Atlantic Treaty, or, in other words collective self-defence, will not automatically be extended to the attacked country." See Ian Traynor, "Russia Accused of Unleashing Cyberwar to Disable Estonia. Parliament, Ministries, Banks, Media Targeted. Nato Experts Sent in to Strengthen Defences," *The Guardian*, May 17, 2007, at <http://www.theguardian.com/world/2007/may/17/topstories3.russia> accessed July 7, 2014.

[89] See Thomas Claburn, "Estonian Hacker Fined for Cyber Attack," *InformationWeek*, January 25, 2008, at <http://www.informationweek.com/estonian-hacker-fined-for-cyber-attack/205918839> accessed July 7, 2014.

[90] The first factor that may vitiate the control of the superior over its subordinates and the establishment of a stable internal discipline is the lack of a physical contact between members. See *Tallinn Manual* (n 65) para 79. As a matter of fact, in cyberspace even the psychological effect of the physical presence of the superior is absent.

[91] See *Limaj* (n 80) para 129.

[92] It is worthy to note that military cyber operations require substantially fewer men than traditional war operations.

[93] Compare *Tallinn Manual* (n 65) para 79.

double-layered structure, which is induced by the very nature of the digital platform around which they gather: administrators—which supervise and manage the platform—and affiliated users. With regards to the latter category, the degree of affiliation with the group may vary, from a stable association to an extemporaneous participation of sympathizing or curious lone hackers in the attack. Also, the affiliated user's technical abilities may vary. Frequently, its function is relegated to mere executor of given instructions for large-scale coordinated attacks such as DDoS—often automated—for which basic knowledge and technical skills may suffice. The core directing group, on the other hand, may be composed of recognized and stable leaders, which maintain the power of command over the affiliated,[94] or be moulded in a more dynamic and meritocratic way around members according to personal charisma or experience.[95] During the Georgian cyber attacks, for instance, one of the main digital fora from which the attacks were orchestrated (*Stopgeorgia.ru*) was structured around a core group of leaders—which provided directions, toolkits, and target lists—and those carrying out the actual attacks.[96] However, even though the higher level of administrators or coordinators—likely, hackers with superior technical skills—may supply tools and provide directions for the attacks, the decision on the concrete actions is often taken on the basis of impromptu discussions among members or individual considerations, which may be influenced by different experiences, technical knowledge or even ideological shades. Additionally, in the largest collectives such as Anonymous (which, however, may be intended more as a flag under whose name the attacks are carried than a stable group) assorted hackers may team up for attacks having very wide and vague objectives, such as the general disruption of a country's digital infrastructures.

Such an organizational structure may indeed translate into a critical lack of any internal disciplinary systems—whose main component is the command and control power of the superior over its subordinates[97]—thereby affecting one of the most effective systems to ensure respect for IHL. As a matter of fact, it provides no differentiation of functions according to rank.[98] Further, it lacks a military vertical command and control structure or a system of direct hierarchical superiors responsible for preventing or punishing crimes perpetrated by their subordinates. Often, the control of the higher hierarchical level of the group is limited to the power of banning

[94] For example, the hacktivist group Lulz Security (LulzSec) had six core members. See Olson (n 37): "...the group would later operate keeping strategic decisions to the core six but working with a second tier of trusted supporters to help them carry out attacks."

[95] See, inter alia, Thomas J Holt, "The Attack Dynamics of Political and Religiously Motivated Hackers," in *Proceedings of the Cyber Infrastructure Protection Conference* (CUNY, 2009) 173: "(...) For example, one site established its leadership and attack command structure based on individual performance in a hacking challenge set up through their website. Individuals must progress through 13 missions, and their performance establishes how they will participate in the larger group."

[96] See Carr (n 21) 16.

[97] See Protocol I (n 63), Articles 86 and 87.

[98] Although different functions and specific skill-sets may be offered by the various members (such as finding the vulnerabilities, providing directions and toolkits, assembly armies of botnets), these are not rationalized in a pyramidal hierarchic chain of command differentiated according to function and through which orders may flow. See Aunshul Rege-Patwardhan, "Cybercrime against Critical Infrastructures: a Study of Online Criminal Organization and Techniques" (2009) 22 *Criminal Justice Studies: A Critical Journal of Crime, Law and Society* 261, 266–7).

a member from a platform, or simply based on charismatic influence and respect. Members are not steered by superiors with military training, which hold a concrete and material power of control over their subordinates. Singular actions by individual users are impossible to prevent. Concretely, many hacker groups may consist of a tangle of individual hackers (ranging from highly experienced hackers to simple "farmers with laptops"[99]) who know each other only by their nickname, whose actions are driven by differing shades of motivations—such as money, entertainment, ego, cause, entrance to a social group, or status[100]—cultural background, personal views, or obsessions on the issue at stake. The bearing of such factors may unpredictably direct the attack outside the expected objectives.

The lack of a tight internal organizational structure, coupled with the peculiar problems of virtual command and control, may complicate the concrete implementation of the law of armed conflicts. Moreover, it may exclude hacker groups from the scope of the norms regulating combatant status in international conflict. Thus, hacker groups lacking a sufficiently structured internal organization, even when participating in a conflict under the direction or stimulation by a state, are to be classified as civilians directly participating in the conflict. Further, a structural deficit may exclude non-state actors from being party to a non-international conflict.

VI. Attributing Cyber Conducts to the State

States may hold substantial interests in directing or stimulating non-state actors, covertly using them as irregular forces in cyberwarfare, and benefitting from the skills of computer-savvy, talented hackers in order to prepare or conduct cyber attacks (or defenses). Indeed, hackers offer reasonably cheap knowledge, skills, and technology. In some cases, participation in cyber attacks requires very little experience, knowledge, and technological equipment, often limited to a computer and some basic skills such as the ability to use web browsers.[101] There, non-state actors represent a valuable quick-to-mobilize, cheap army, to which the attack may be "crowdsourced." A famous Chinese martial adage states: "kill with a borrowed knife." Not only the state takes advantage of high numbers of technologically equipped persons to attack an enemy; they may avoid the political, legal, and military burden of attribution exploiting the difficulty in tracing back the attack from the non-state actors to the state and to prove any substantial involvement. In point of fact, the use of obfuscation techniques or botnets may substantially impede the traceability of the attack.[102] As a result, legal and political consequences may be avoided.

Per contra, hacker groups are not always easy-to-use tools by the state. The main drawbacks relate to the difficulties in controlling or stopping the attack.[103] As pointed out, the action may exceed the predetermined limits and deviate to non-agreed targets, with the consequent possible risk of IHL violations. Nevertheless, the lack of

[99] See Ottis (n 26).

[100] See Max Kilger, Ofir Arkin and Jeff Stutzman, "Profiling," in The Honeynet Project (ed), *Know Your Enemy: Learning about Security Threats* (Boston, MA: Addison Wesley Professional, 2nd edn, 2004).

[101] For example, some types of DoS attacks. See Ottis (n 26).

[102] See n 86.

[103] See Applegate (n 29) 20.

necessary adherence to military protocols or humanitarian law requirements may be even considered by the state an inherent value of non-state actors, who may move on the virtual battlefield free from the "burdens" of IHL rules and limitations.

Assessing the existence and the degree of connection between a non-state actor and a state may have important consequences as to the nature of the attack, the character of the conflict, the political and legal responsibility for the act and a possible military response in self-defense.

According to international law, conduct by state organs (de jure and de facto) and agents directed or controlled by the state are attributed to such a state.

Firstly, conducts brought about by a de jure organ of a state considered as such by its internal law are attributed to the state.[104] For instance, units of the official electronic warfare military are considered as states' de jure organs. Moreover, if a factual relationship of complete dependence and control exists, the entity is de facto equated to an organ of the state: its acts are thus attributed to the state. The necessary threshold of control identified in the international case law amounts to a "strict control" of a "great degree,"[105] expression of a "complete dependence" of the entity from the state.[106] Any "margin of independence" precludes such a relationship.[107] The high threshold of control tends to concretely exclude hacker groups not officially or covertly included in the armed forces of a state and possessing a sufficient hierarchical structure on which strict control is exerted.

According to Article 8 of the Draft Articles on Responsibility of States for Internationally Wrongful Acts—which reflects customary international law[108]—"[t]he conduct of a person or group of persons shall be considered an act of a State under international law if the person or group of persons is in fact acting on the instructions of, or under the direction or control of, that State in carrying out the conduct."[109] The persons or the group are thus located outside of the state structure, and act on the basis of a factual relationship of agency. The instruction and direction link must relate to specific tasks or operations in which the influence of the state on the group is exerted.[110] Conducts associated with the fulfilment of such operations are thus attributed to the state. The criterion of control, on the other hand, may relate to a more diffuse relation between the state and the group. The non-state actors, although not completely dependent on the state, are controlled by it. ICJ case-law on the issue introduced a rather stringent criterion of "effective control" of the state over non-state

[104] See ICJ, *Case Concerning Application of the Convention on the Prevention and Punishment of the Crime of Genocide (Bosnia-Herzegovina v Yugoslavia)*, July 11, 1996, paras 385–95; International Law Commission, Draft Articles on Responsibility of States for Internationally Wrongful Acts, November 2001, Article 4.

[105] See *Bosnian Genocide* case (n 104) paras 391–3.

[106] *Bosnian Genocide* (n 104) paras 391–4.

[107] *Bosnian Genocide* (n 104) paras 391–4.

[108] *Bosnian Genocide* (n 104) para 406. Compare Antonio Cassese, "The Nicaragua and Tadić Tests Revisited in Light of the ICJ Judgment on Genocide in Bosnia" (2007) 18 *European Journal of International Law* 649, 651.

[109] See Articles on State Responsibility (n 104) Article 8.

[110] See *Bosnian Genocide* case (n 104) para 400; ICJ, *Case Concerning United States Diplomatic and Consular Staff in Tehran (United States of America v Iran)*; May 12, 1981, para 58; Articles on State Responsibility (n 104), Commentary on Article 8, points 2, 3.

actors.[111] According to such criterion, the control exerted by the state remains articulated on the specific operations. In the *Tadic* case, the ICTY diverged from this criterion, criticizing the high threshold of the "effective control" test enunciated by the ICJ as not "consonant with the logic of the law of State responsibility."[112] In relation to attribution to the state of acts of "organised and hierarchically structured" groups, the Tribunal indicated a far less stringent criterion of "overall control."[113]

The flexible test enounced by the ICTY[114] differentiates between individuals and unorganized groups on the one hand, and organized and hierarchically structured groups on the other hand. In the latter case, the agency relationship will be articulated through a general control of the state over the group, unrelated to specific acts. The linkage may concretely be a connection with the highest hierarchical levels of the organization in terms of coordination and help in the general planning. Per se, financing or equipping the group (for instance, with hardware or software) does not satisfy the necessary threshold for attribution.[115] The "overall control" criterion was strongly criticized by the ICJ, as stretching "too far, almost to breaking point, the connection which must exist between the conduct of a State's organs and its international responsibility."[116]

Attribution to the state of conduct brought about by individuals and groups lacking a sufficiently structured internal organization, on the other hand, follow an "effective control" test (according to both the ICTY and ICJ case-law). Specific instructions concerning the particular conducts must be shown. According to the ICTY, it is necessary that "the State exercised some measure of authority over those individuals but also that it issued specific instructions to them concerning the performance of the acts at issue, or that it ex post facto publicly endorsed those acts."[117] In this case, considering the typical two-tiered structure exhibited by several hacker groups,[118] the state will likely have to establish a tight connection with the higher level of that structure (the "administrative" level), or directly form part of this level with its agents. According to the ICJ, the test will be satisfied when the state "directed or enforced the perpetration of the acts."[119] Therefore, the state must have issued specific directions about the conducts, which filter through the administrative level of the group to all the members. For instance, it may provide specific toolkits, target lists or indicate particular objectives. Direct enforcement of specific operations may be intended as concretely

[111] See ICJ, *Case Concerning Military and Paramilitary Activities In and Against Nicaragua (Nicaragua v United States of America)*, Merits, June 27, 1986, para 115.

[112] ICTY, Appeal Chamber, *Prosecutor v Dusko Tadic (Appeal Judgement)*, IT-94-1-A, July 15, 1999, 116.

[113] *Tadic appeal* (n 112) 131.

[114] On case law and practice supporting the ICTY position, see, generally, Cassese (n 108).

[115] See *Tadic appeal* (n 112) para 175.

[116] See *Bosnian Genocide* case (n 104) para 406. The Court, contrary to the ICTY view expressed in Tadic (n 112) paras 103–5, points out how that the "overall test" applied by the ICTY in determining whether an armed conflict is international, is inapplicable on the different issue of state responsibility. According to the Court, "It should (...) be observed that logic does not require the same test to be adopted in resolving the two issues, which are very different in nature: the degree and nature of a State's involvement in an armed conflict on another State's territory which is required for the conflict to be characterized as international, can very well, and without logical inconsistency, differ from the degree and nature of involvement required to give rise to that State's responsibility for a specific act committed in the course of the conflict." *Bosnian Genocide* case (n 104) para 405.

[117] *Tadic appeal* (n 112) para 118.

[118] See § V.

[119] *Nicaragua* case (n 111) para 115.

leading the group through its execution. Given the difficulty to directly control members' individual actions, any actual enforcement of conduct will be hard to carry out.

Concretely, attributing cyber conduct of non-state actors to the state may prove to be particularly difficult. The first set of problems relates to technical issues. The web's "opacity" and the volatility of digital evidences may hinder the ability to prove the linkage with the state, or even to determine the origins of the attack itself. Indeed, since any instructions, directions, or proof of control by the state—where they exist—are likely to be mostly in digital format, they might be easily and quickly erased or altered. Technical problems of traceability[120]—possibly exacerbated by the transnational character of the attack, which may originate from different jurisdictions,[121] and the use of obfuscation methods—may combine with issues of interstate cooperation. In assessing the source of the attack, the attacked state may need assistance from countries in which the attack originated, passed through, or in which the physical devices exploited to prepare or perpetrate the attack are located. Often, they might even need assistance from the state that may actually have sponsored the attack. In the aftermath of the 2007 attacks on its infrastructure, Estonia submitted several requests to help track down the origin of the attack to Russia, which rejected them notwithstanding the obligations provided by a bilateral legal assistance treaty.[122]

As cyber attacks propagate in cyberspace and materialize in the target state with extreme speed, rapidity in determining the origin and the nature of the attack may be essential. Attributing the act of a non-state group to a state is necessary for assessing its political and legal consequences. Among them, is the activation of the right to use force to suppress the cyber threat under the law of self-defense.

VII. Self-Defense against Cyber Attacks by Non-State Actors

According to both conventional[123] and customary law,[124] states enjoy the "inherent right"[125] to individually or collectively resort to force in self-defense against an unlawful armed attack—in exception to the prohibition of threat or use of force set out by Article 2.4 of the UN Statute. The main conditions of such a right are the gravity of the armed attack suffered, and necessity, proportionality, and immediacy of the response.[126]

[120] See, for example, Daniel Ramsbrock and Xinyuan Wang, "The Botnet Problem," in John R Vacca (ed) *Computer and Information Security Handbook* (Waltham, MA: Morgan Kaufmann, 2012); Guillaume Lovet, "Fighting Cybercrime: Technical, juridical and ethical challenges," in *Proceedings of the Virus Bulletin Conference* (2009) 65ff.

[121] Given the transnational character of several hacker groups and the use of attack methods that involve botnets.

[122] See Duncan B Hollis, "Why States Need an International Law for Information Operations" (2007) 11 *Lewis & Clark Law Review* 1023, 1026.

[123] See United Nations, Charter of the United Nations, October 24, 1945, Article 51.

[124] In the *Nicaragua* case (n 111), the ICJ clarified that the conventional and customary sources do not overlap and present slightly different content. According to the Court "it cannot be held (...) that Article 51 is a provision which 'subsumes and supervenes' customary international law"; "customary international law continues to exist alongside treaty law." *Nicaragua* case (n 111) 176.

[125] See UN Charter, Article 51.

[126] See, inter alia, ICJ, *Case Concerning Oil Platforms (Islamic Republic of Iran v United States of America*, November 6, 2003, para 52; Yoram Dinstein, *War Aggression and Self Defence* (Cambridge: Cambridge University Press, 2011) 230–4, 212.

The right to respond in self-defense against attacks possessing a sufficient degree of intensity that originates from a state's regular armed forces, or from armed groups under its instruction, direction, or control, is undisputed.[127] However, in case of attacks originating from a non-state actor with connivance with a state, evolving state practice, developed in particular with regards to international terrorism, seems to have pushed the scope of the right beyond the limits of the traditional attribution criteria. The issue, indeed quite controversial, has to be read in the light of the increasing threat posed by non-state armed groups, and of the international political consensus gathered around the fight against international terrorism since the exacerbation of the phenomenon in the 1990s. As non-state actors acquire the stable capability to pose a threat to the state, menacing its self-preservation, the area where the state cannot lawfully respond to suppress the deadly threat will naturally tend to be reduced to the minimum.[128]

In 2001, following the (in)famous twin towers attacks, the US militarily reacted—on the basis of the right of self-defense—against al-Qaeda, from whom the attack directly originated, and the Taliban regime,[129] accused of having tolerated and harbored the terrorist activities of the group.[130] The UN SC Resolutions 1368 (2001) and 1373 (2001) endorsed the US military operation Enduring Freedom, acknowledging the threat to international peace and security generated by the terrorist attacks and recognizing the right of the United States to act in self-defense as consistent with Article 51 of the UN Charter. It is evident from the defensive force exerted against the group *and* the state that tolerated and harbored it, that the acts of the non-state actor were considered directly linked to the state entity. Such a broadened approach to the law of self-defense marks a sharp shift from the traditional attribution criteria, stretching the necessary link between the state and the non-state group far beyond the "overall control" criterion set up by the ICTY. What appears now is a further loosening of the attribution criteria to encompass mere assistance, exhibited when allowing the use of the state's territory as a base for the operation[131] and tolerating and harboring the activities of the non-state actor. The criterion falls completely outside of the scope of specific instructions, directions, or control by the state. Nor might it be attributed to a hypothesis of "indirect aggression" of the state pursuant to Article 3(g) of UNGA Resolution 3314 (XXIX) that defines aggression:[132] on the point, the ICJ excluded that

[127] See Institut de Droit International, Session De Santiago—2007, 10a Resolution, October 27, 2007, *Present Problems Of The Use Of Armed Force In International Law: A. Self-Defence*, para 10.

[128] See *Nicaragua* case (n 111) Judge Jennings Dissenting Opinion: "it seems dangerous to define unnecessarily strictly conditions for lawful self-defence, so as to leave a large area where both a forcible response to force is forbidden, and yet the United Nations employment of force, which was intended to fill that gap, is absent." Generally, on extraterritorial interventions against individuals and non-state actors see Noam Lubell, *Extraterritorial Use of Force Against Non-State Actors* (Oxford: Oxford University Press, 2010).

[129] Thus, at the moment—notwithstanding the scarce international recognition—the de facto government of (the Islamic Emirate of) Afghanistan. See Yoram Dinstein, "Terrorism and Afghanistan," in Michael N Schmitt (ed) *The War in Afghanistan: A Legal Analysis* (*US Naval War College International Law Studies*, Vol. 85, 2009) 50–1.

[130] See US Presidential Address to the Nation (September 11, 2001); *Authorization for Use of Military Force*, Pub L No 107-40, 115 Stat 224 (2001).

[131] See Letter from the Permanent Representative of the United States of America to the United Nations addressed to the President of the Security Council, October 7, 2001.

[132] See UN General Assembly Resolution 3314, A/RES/3314, December 14, 1974, Article 3: "Any of the following acts, regardless of a declaration of war, shall, subject to and in accordance with the provisions

"the concept of 'armed attack' includes (...) also assistance to rebels in the form of the provision of weapons or logistical or other support."[133] According to this construction of the right of self-defense, the link with the state legitimizing the use of force is detached from a role of command and direction of the group. Instead, it lowers the criterion to simply aiding and abetting the non-state actor.

A slightly different issue pertains to armed attacks conducted by non-state groups without any involvement or connivance by the state. State practice—supported by several commentators[134]—upholds the legitimacy of the use of force in self-defense directly against non-state groups where the state from which territory the attack is conducted is "unwilling or unable" to suppress the threat represented by the non-state actor.[135] This theory, whose roots may be traced back to the famous *Caroline* case,[136] stood as a legal basis for, inter alia, the 2002 Russian attack in Georgia against the Chechen rebels and the 2006 Israeli self-defense actions against Hezbollah.[137]

In *Democratic Republic of Congo v Uganda*,[138] the ICJ failed to cast light on the issue of whether international law provides for a right of self-defense against attacks originating from non-state actors. However, in the *Construction of a Wall* case, it explicitly recognized the sole existence of "an inherent right of self-defence in the case of an armed attack by one State against another State."[139] The traditional, restrictive approach, based on the reading of Article 51 of the UN Statute in the light of the systematic categories of the Charter—which holds that armed attacks in the meaning of such Article are acts of states only—is supported by the Amendments to the Rome Statute of the International Criminal Court on the Crime of Aggression, which still retains a necessary state element.[140]

The legal grounds of such a broadened construction to the law of self-defense reside, on the one hand, in the right of self-preservation of the state and possibly in the fulfilment of an *erga omnes* obligation to act in defense of values represented by the fight against international terrorism.[141] To this regard, it is important to point out that cyber

of article 2, qualify as an act of aggression: (...) (g) The sending by or on behalf of a State of armed bands, groups, irregulars or mercenaries, which carry out acts of armed force against another State of such gravity as to amount to the acts listed above, or its substantial involvement therein."

[133] See *Nicaragua* case (n 111) para 195.

[134] See, inter alia, Dinstein (n 129), 272ff; Thomas M Franck, "Terrorism and the Right of Self-Defense" (2001) 95 *American Journal of International Law* 839, 840; Ruth Wedgwood, "Responding to Terrorism: The Strikes against Bin Laden" (1999) 24 *Yale Journal of International Law* 559, 566.

[135] See Ashley Deeks, "'Unwilling or Unable': Toward a Normative Framework for Extra-Territorial Self-Defense" (2012) 52 *Virginia Journal of International Law* 483.

[136] See Christopher Greenwood, "The Caroline," in *Max Planck Encyclopedia of Public International Law* (Oxford: Oxford University Press, 2009) at <http://opil.ouplaw.com/view/10.1093/law:epil/9780199231690/law-9780199231690-e261?rskey=2cHAqm&result=1&q=&prd=EPIL> accessed July 7, 2014.

[137] See Deeks (n 135). On the state practice, see also Raphaël Van Steenberghe, "Self-Defence in Response to Attacks by Non-State Actors in the Light of Recent State Practice: A Step Forward?" (2010) 23 *Leiden Journal of International Law* 183.

[138] See ICJ, *Case Concerning Armed Activities on the Territory of the Congo (Democratic Republic of the Congo v Uganda)*, December 19, 2005.

[139] See ICJ, *Legal Consequences of the Construction of a Wall in the Occupied Palestinian Territory*, July 9, 2004, para 139.

[140] See UN General Assembly, Rome Statute of the International Criminal Court (last amended 2010), July 17, 1998, Article 8bis.

[141] See Massimo Condinanzi, "L'uso della forza e il sistema di sicurezza collettiva delle Nazioni Unite," in Sergio M Carbone, Riccardo Luzzato and Alberto Santa Maria, *Istituzioni di Diritto Internazionale*

attacks by hacker groups could likely fall under the scope of the definition of international terrorism.[142] On the other hand, they are based on the state's obligation not to knowingly allow its territory to be used for acts contrary to the rights of other states.[143] Since hacker groups may possibly use servers, or be composed by members, located in various nations, the spatial extent of this obligation connected with such groups may touch multiple jurisdictions. In the dualistic nature of the new technologies, cyber conduct echoes on both a digital and a physical place. The scope of the toleration and harboring criterion may extend to cover the offering of a "virtual shelter" to the group, such as the use of servers located within the territory of the state. States will, thus, have to apply a tight control over their information and communication systems. Such an increased duty may tend to entail the enactment of an efficient domestic criminal law aimed at successfully criminalizing illegal cyber conducts, and implementation of appropriate technology improving the law enforcement authorities' ability to detect and suppress the threat. Moreover, the use of obfuscation techniques and disguising technology—such as botnets—may further complicate territorial problems in responding to attacks, generating the risk of attacking an "innocent" digital system or, at least, inducing the state to undergo long and difficult investigations to trace back the origin of the attack.

Given the transnational character of cyberspace, individual members of a non-state group that participates in the conflict may be physically located outside the belligerent territory. According to IHL, combatants, civilians directly participating in the conflict and military objectives[144] can be lawfully targeted. However, this may conflict with the neutrality of the state on whose territory the attacker or the digital system exploited are located, which is inviolable.[145] While a neutral power is not responsible for the fact of its citizens crossing the frontier independently to offer their services to one of the belligerents,[146] the duty to prevent and punish activities against a state that are committed within its territory relies on the neutral state.[147]

Those who advocate a broader construction for the use of self-defense may perceive requiring the state to powerlessly suffer attacks and to renounce preserving its security as unjust, especially in light of the acquired potentialities of non-state actors to

(Turin: Giappichelli, 2011) 371; Christian J Tams, "The Use of Force against Terrorists" (2009) 20 *European Journal of International Law* 359, 395.

[142] See n 15.

[143] See ICJ, *Corfu Channel Case (United Kingdom v Albania)*, Merits, April 9, 1949, 22. See also Tal Becker, *Terrorism and the State: Rethinking the Rules of State Responsibility* (Portland, OR: Hart, 2006); Antonio Cassese, "Terrorism is Also Disrupting Some Crucial Categories of International Law" (2001) 12 *European Journal of International Law* 993, 997.

[144] See, inter alia, Protocol I (n 63), Article 52(2): "(i)n so far as objects are concerned, military objectives are limited to those objects which by their nature, location, purpose or use make an effective contribution to military action and whose total or partial destruction, capture or neutralization, in the circumstances ruling at the time, offers a definite military advantage."

[145] See Convention (V) respecting the Rights and Duties of Neutral Powers and Persons in Case of War on Land, The Hague, October 18, 1907, Article 1.

[146] Convention (V) (n 145) Article 6.

[147] Convention (V) (n 145) Article 5. See also n 143. Further specific duties of prevention come from the Convention against recruitment, use, financing and training of mercenaries, which at Article 6 set forth a specific duty for the territorial state to prevent the commission or preparation of the offences envisaged in the Convention.

conduct disruptive attacks through digital means. Indeed, the traditional concepts which permeate the system of peace and security maintenance and regulation of the use of force expressed in the UN Charter—calibrated on state entities—may appear unable to accommodate the fight against the ever more frequent threats coming from non-state actors. In a prospective view, cyber threats originating from non-state actors may exacerbate such friction. Shall these extensive approaches indicate a possible future evolution of the premises on which the law of self-defense is grounded?[148] The answer is beyond the scope of this work. Indeed, in the event of cyber attacks by non-state groups, these expanded modulations of the law of self-defense may overtake the peculiar problems related to attributing the act to the state according to the traditional criteria of instruction, direction, and control. In all likelihood, the state's right to act in self-defense will expand to sufficiently cover urgent situations that leaves "no moment for deliberation"[149]—including for assessing state responsibility. Given the technical peculiarities of the cyber events, the urgency and difficulty of attribution may be intensified in case of cyber attacks. Such a loosened approach will widen the scope of use of active defenses[150] and reduce the time of response—fairly limited, given the peculiar temporal characteristic of cyber attacks—strongly augmenting the chances to prevent further attacks and damages. Besides, it will likely have a deterrent effect on the use of hacker groups by the state. On the other hand, however, it may create the risk to stretch the scope of the right of self-defense to cover impulsive reactions lacking a sufficient analysis of the origin of the attack, the connivance of the state, or its concrete inability or unwillingness to suppress the threat itself.

VIII. Conclusion

This chapter pointed out how the digitalization of warfare has augmented the importance of non-state actors in the twenty-first century digital conflict, both as autonomous actors and instruments in the hands of the state. Furthermore, it evidenced how such a process challenged international law, burdening the existing norms regulating armed conflicts and generating several regulatory issues. The rise of non-state actors in cyberwarfare perpetuated the process of erosion of the role of the state as primary actor in the international scenario, following the path set in motion by international terrorism. However, as pointed out in the preceding sections, the digitalization of war offers completely new boundaries to the phenomenon: it strongly enhances the role and the capacity of non-state actors involved, and offers them structural and operative characteristics which deeply challenge the traditional corpus of norms regulating conflicts. At the same time, it complicates the attribution of the act and the ability of the state to respond to the threat under the law of self-defense. These regulatory challenges must be resolved, possibly through a reinterpretation of the relevant norms of

[148] See Van Steenberghe (n 137) 191ff.

[149] US Secretary of State Daniel Webster, note of April 24, 1841, British–American Diplomacy on the Caroline incident, at < http://avalon.law.yale.edu/19th_century/br-1842d.asp> accessed July 7, 2014.

[150] See Matthew J Sklerov, "Solving the Dilemma of State Responses to Cyberattacks: A Justification for the Use of Active Defenses against States Who Neglect Their Duty to Prevent" (2009) 201 *Military Law Review* 1.

international law in light of the peculiarities of cyberwarfare: the *Tallinn Manual on the International Law Applicable to Cyber Warfare*[151] is a shining example to be followed.[152] As the fight against international terrorism has taught us, the inefficiency of the corpus of norms regulating non-state actors' participation in cyberwarfare may generate an inequality between the state and the non-state actor which forcibly leads the state to respond outside of the tracks of law. Unilateral responses aimed at filling the existing gaps should be avoided.

A final consideration, however, relates to the unique characteristics of "fluidity" and variability offered to this process by the pivotal role of new technology. As technology evolves, non-state actors may present completely new morphological aspects, and their modus operandi may substantially change. Attribution of cyber conducts may be complicated by the creation of new anonymity techniques, or be facilitated by the development of new traceback methods. New technologies in the hands of non-state actors may increase their dangerousness, while, on the other hand, new cyber defenses at the state's disposal may decrease the gravity of the phenomenon. Finally, cyberspace itself may change its form and structure with the introduction of new technology or through a political reorganization, finding new modulations in the oscillating struggle between its pure internationalization and a subdivision into separated zones:[153] such modifications will surely reverberate on the actors operating in it.

[151] See n 65.

[152] See Michael N Schmitt, "Cyber Operations in International Law: The Use of Force, Collective Security, Self-Defense, and Armed Conflict," in *Proceedings of a Workshop on Deterring Cyberattacks: Informing Strategies and Developing Options for US Policy* (Washington, DC: National Academies Press, 2010): "Far from being counter-legal, this process of reinterpretation is natural; understandings of international legal norms inevitably evolve in response to new threats to the global order."

[153] The s.c. Internet "balkanization" (using a geopolitical term which describes a process of fragmentation of a political entity into smaller units). See Jonah Force Hill, "Internet Fragmentation: Highlighting the Major Technical, Governance and Diplomatic Challenges for US Policy Makers" (2012) Harvard Belfer Center for Science and International Affairs Working Paper, at <http://belfercenter.ksg.harvard.edu/files/internet_fragmentation_jonah_hill.pdf> accessed July 4, 2014. *See also* Charlotte Alfred, "Web At 25: Will Balkanization Kill The Global Internet?," *The Huffington Post*, March 19, 2014, at http://www.huffingtonpost.com/2014/03/19/web-balkanization-national-internet_n_4964240.html> accessed July 4, 2014.

PART III

CYBERSECURITY AND INTERNATIONAL HUMANITARIAN LAW: THE ETHICS OF HACKING AND SPYING

7

Re-Thinking the Boundaries of Law in Cyberspace

A Duty to Hack?

Duncan B Hollis

I. Introduction

Warfare and boundaries have a symbiotic relationship. Whether as its cause or effect, nation-states historically used war to delineate the borders that divided them.[1] Laws and borders have a similar relationship. Sometimes laws are the product of borders; today, national boundaries constitute states that regulate relations within or among their respective territories. But borders may also be the product of law; laws often draw lines between permitted and prohibited conduct and bound off required acts from permissible ones.[2]

Thus, it is not surprising to see the logic of boundaries prevalent in discussions of international law in cyberspace. To date, that discussion has largely centered on the role of boundaries in generating legal authority. Some have characterized the internet as a distinct, self-governing "space," entitling it to autonomy or non-state governance. In contrast, others insist cyberspace is just a technological medium that states can govern by reference to national boundaries. Over time, conventional wisdom has oscillated from one pole to the other, with the second now holding pride of place. Today, states regularly claim territorial borders as the basis for prescribing law in cyberspace and increasingly assert a right to decide "who decides" on its governance.

However, boundaries play an equally significant—if less acknowledged—conceptual role as the product of legal line-drawing in cyberspace. For international law, with few exceptions, these boundaries are not tailor-made for cyberspace.[3] Rather, they derive from outside sources—"law-by-analogy"—invoking lines drawn for other times and contexts.[4] Significant uncertainty remains on exactly what behavior these legal boundaries prohibit/permit/require.[5] Nonetheless, few dispute that international law operates in cyberspace by reference to its pre-existing regimes and their borders.

[1] Today, of course, the UN Charter prohibits the use of force except in self-defense or pursuant to Chapter VII authorization. See UN Charter, Articles 2(4), 40–2, 51.

[2] See, for example, M Render, "Introduction," in *The Meador Lectures on Boundaries* (Tuscaloosa: University of Alabama, School of Law, 2013) 8.

[3] To date, the Cybercrime Convention is the only treaty specifically tailored to the cyber context. Council of Europe, Convention on Cybercrime (Budapest, November 23, 2001) CETS No 185 ("Cybercrime Convention").

[4] For a paradigmatic example, see MN Schmitt (ed), *Tallinn Manual on the International Law Applicable to Cyber Warfare* (New York: Cambridge University Press, 2013) 5–6.

[5] For example, the most notorious cyber operation to date, Stuxnet, reportedly infected SCADA systems globally, but disrupted one set in particular—those at Iran's Natanz uranium enrichment

For example, international law is widely assumed to govern "cyberwar" by analogies that delimit the boundaries of the jus ad bellum (the set of laws regulating *when* force can be used) and the jus in bello, also known as international humanitarian law or "IHL" (the set of laws regulating *how* states may use force).[6]

This chapter proposes re-thinking prevailing approaches to legal boundaries in cyberspace generally and for state cyber operations in particular.[7] To be clear, I do not dispute that boundaries *can* dictate authority in cyberspace. Nor do I question if international law currently regulates cyberspace by analogy—I think it clearly does.[8] My critique is a normative one—asking if the current emphasis on drawing law from boundaries and boundaries from law is a sufficient or effective way to regulate cyberspace and its conflicts.

Current thinking on cyberspace's boundaries may be questioned on both theoretical and functional grounds. In terms of *theory*, cyberspace lacks a uniform rationale for *why* it needs boundaries. Three decades into the internet's existence, what cyberspace "is" remains highly contested, making agreement on its boundaries nearly impossible. Defining what cyber operations "are" (i.e., acts of war, criminal behavior, or something else?) has generated equally intractable questions as to *which* sets of international law rules apply and how they do so.

However, even where theoretical disputes do not occupy the foreground, the *functions* boundaries serve are open to challenge as a matter of: (1) *accuracy*, (2) *effectiveness*, and (3) *completeness*.[9] Relying on law-by-analogy to delineate specific IHL thresholds, for example, can generate *inaccurate* lines if they do not accommodate non-analogous characteristics of cyberspace and cyber operations. In other cases, the question is not *where* law's boundary lines go, but *whether* boundaries can work at all. A boundary's *effectiveness* depends on certain criteria—a degree of determinacy, impermeability, and constraint—that can be difficult to guarantee given cyberspace's prevailing conditions (e.g., anonymity).[10] Furthermore, the constraining logic

facility—causing significant damage to approximately 1,000 centrifuges. D Albright et al, *Did Stuxnet Take out 1000 Centrifuges at the Natanz Enrichment Plant?* (ISIS, 2010). States have remained largely silent on Stuxnet's international legality, while many of the world's most prominent experts cannot agree if it constituted an "armed attack" (entitling Iran to respond in self-defense) or triggered IHL. See, for example, *Tallinn Manual* (n 4) 58, 83–4.

[6] *Tallinn Manual* (n 4) 13. As a concept, cyberwar remains under-theorized. Debate continues on whether conventional operations must accompany cyber operations or if a cyberwar could occur exclusively in cyberspace. See, for example, T Rid, "Cyber War Won't Take Place" (2012) 35 *Journal of Strategic Studies* 5.

[7] See n 118 for a discussion of the various forms of cyber operations that may occur.

[8] There is, for example, near universal acceptance of IHL's application to cyberspace. *Tallinn Manual* (n 4) Rule 20. The previous hold-out—China—now appears to accept this position. See US-China Economic and Security Review Commission, *2013 Annual Report to Congress* (2013) 249; L Zhang, "A Chinese Perspective on Cyber War" (2012) 94 *International Review of the Red Cross* 801, 804.

[9] A law's functions may, of course, be contingent on, or derivative of, a satisfactory theory for why the law exists. But, theoretical differences do not have to forestall agreement on such functions entirely. Sometimes, consensus is possible even without agreement on *why* this is the case. As Cass Sunstein noted, in a world of diverse interests "incompletely theorized agreements" should be expected if not encouraged. CR Sunstein, "Incompletely Theorized Agreements in Constitutional Law" (2007) 74 *Social Research* 1 (noting consensus on valuing religious liberty without agreement on why it deserves value: "Some people may stress what they see as the need for social peace; others may think that religious liberty reflects a principle of equality and a recognition of human dignity; others may invoke utilitarian considerations; still others may think that religious liberty is itself a theological command").

[10] See nn 100–2 and accompanying text.

of borders may generate an *incomplete* regulatory response to cyber operations, not just because of inaccurate or ineffective line-drawing, but also for failing to emphasize law's other capacities (e.g., empowering behavior). Such prevailing difficulties on *whether, where*, and *why* borders are needed in cyberspace suggests the time is ripe for re-appraising the current landscape.

What form should such a reappraisal take when it comes to international law and cyberwar? One could, of course, begin at the macro-level. Resolving which boundaries delineate authority for rule making in cyberspace could go a long way to reducing the status quo's theoretical and functional challenges. Similar benefits might follow adjustments to—or reconciliation among—the boundaries of competing international legal regimes (e.g., IHL, criminal law, espionage). The problem with such efforts, however, lies in their viability. The very size and complexity of the issues involved casts doubt—at least at present—on any reasonable prospects for success.

Given such challenges, this chapter adopts a more modest approach. I propose beginning to re-think how legal boundaries operate in cyberspace by focusing first on discrete areas within specific international law regimes, offering IHL's rules on weaponry and the conduct of hostilities as a case study. By focusing on boundaries *within* discrete legal regimes, some of the more intractable theoretical problems may be avoided (or at least minimized). Hopefully, over time, success in re-designing law's boundaries in particular cyberspace contexts like IHL may build momentum for tackling the more fundamental questions of governance and relationships among legal regimes.

The legal boundaries IHL imposes are particularly well suited for reassessment. Even as the precise contours of IHL's rules are questioned, there is little dispute as to their fundamental purpose, namely balancing military necessity with humanitarian principles.[11] That purpose offers a base line for assessing the (in)accuracy of IHL line-drawing done by analogy. Similarly, several unique attributes of cyberspace and cyber operations challenge the capacity of IHL-by-analogy to effectively or adequately regulate state cyber operations. To correct for such deficiencies, I propose IHL adopt a new, tailor-made rule for state cyber operations—what I call the "Duty to Hack."

The Duty to Hack would require states to use cyber operations in their military operations whenever they are the least harmful means available for achieving military objectives. Simply put, if a state can achieve the same military objective by bombing a power station or using a cyber operation to disable it temporarily, the Duty to Hack requires that state to pursue the latter course. Although novel in several key respects, I submit the Duty to Hack would better account for IHL's fundamental principles and cyberspace's unique attributes than existing efforts to foist legal boundaries upon state cyber operations by analogy.

My argument proceeds as follows. Section II examines and critiques invocations of borders to generate legal authority in cyberspace. Section III undertakes a similar effort with respect to borders drawn between and within legal regimes in international law. Section IV turns to a case study of IHL's boundaries for the means and

[11] See, generally, M Schmitt, "Military Necessity and Humanity in International Humanitarian Law: Preserving the Delicate Balance" (2010) 50 *The Virginia Journal of International Law* 795.

methods of warfare. It proposes IHL adopt the Duty to Hack in lieu of its current emphasis on analogies, reviewing both the benefits and costs of doing so. The chapter concludes with a call for more attention to relationships in cyberspace among various categories of law and their boundaries.

II. Law as the Product of Boundaries: Cyberspace Governance

(a) Bordering off cyberspace and its governance

In 1996, John Perry Barlow famously offered a "Declaration of Independence of Cyberspace." His manifesto proclaimed cyberspace "immune" from the territorial sovereignty of nation-states.[12] Despite its fanciful rhetoric, Barlow's Declaration continues to serve as a rallying cry for visions of cyberspace as a unique "space" warranting its own form of governance. For some, the vision is self-governance. The most famous proponents of this idea, David Post and David Johnson, argued in explicitly bordered terms, describing "the rise of an electronic medium that disregards geographical boundaries" that "cannot be governed satisfactorily by any current territorially based sovereign."[13] Their solution? Accept a "legally significant border between Cyberspace and the 'real world'," with cyberspace's rules coming from the generative community that uses it rather than externally imposed directives from nation-states.[14]

Johnson and Post's self-governing vision was greeted with skepticism, but not so the idea that cyberspace is a distinct "place" with the potential for autonomy.[15] Larry Lessig adopted the premise of cyberspace as "place" in (famously) declaring "Code is Law."[16] In lieu of users, Lessig promoted the capacity of the internet's architects (i.e., technical institutions like the Internet Engineering Task Force (IETF) and the Internet Corporation for Assigned Names and Numbers (ICANN)) to govern cyberspace by the operations they permit (or prohibit) in the code they write and the equipment they build.[17] This view has gained renewed salience following recent disclosures of widespread US surveillance and monitoring via its National Security Agency.[18] Various internet institutions, including IETF and ICANN, have signaled an intention

[12] JP Barlow, "A Declaration of Independence of Cyberspace" (February 8, 1996), at <http://homes.eff.org/~barlow/Declaration-Final.html>.

[13] DR Johnson and DG Post, "Law and Borders—The Rise of Law in Cyberspace" (1996) 48 *Stanford Law Review* 1367, 1375–6.

[14] DR Johnson and DG Post, "And How Shall the Net Be Governed?: A Meditation on the Relative Virtues of Decentralized, Emergent Law," in B Kahin and JH Keller (eds), *Coordinating the Internet* (Cambridge, MA: MIT Press, 1997).

[15] D Hunter, "Cyberspace as Place, and the Tragedy of the Digital Anticommons" (2003) 91 *California Law Review* 439, 443 ("cognitive science investigations provide ample evidence that, purely as a descriptive observation, we do think of cyberspace as a place").

[16] L Lessig, *Code and Other Laws of Cyberspace* (London: Basic Books, 1999) 190; see also J Kang, "Cyber-Race" (2000) 113 *Harvard Law Review* 1130, 1186–7.

[17] Lessig (n 16) 206–7.

[18] See, for example, A Leonard, "Edward Snowden's Unintended Internet Revolution," *Salon*, October 15, 2013, at <http://www.salon.com/2013/10/15/edward_snowdens_unintended_internet_revolution/>.

to seek more autonomy from nation-states—especially the United States—in governing how cyberspace operates.[19]

In addition to self- or engineered-governance, international regulation offers a third vision for treating cyberspace as its own, novel space for governance. Efforts to empower the International Telecommunications Union (ITU) to take on such a governance role failed in 2012.[20] Yet, the ITU's claims remain, as do other proposals for an international organization, to "run" the internet.[21] Thus, the idea of cyberspace as analogous to—but distinct from—"real" space with its own borders persists, even as various candidates (e.g., users, technical institutions, international organizations) vie for the job of governing it.[22]

(b) Governing cyberspace via territorial boundaries

A competing—and now prevailing—view treats boundaries very differently. It accepts the boundaries of nation-states—rather than cyberspace—as *the* appropriate method for distributing legal authority to govern the internet and information technology. Some, like Jack Goldsmith and Tim Wu, do so by rejecting the idea that cyberspace is "space" at all.[23] They insist the internet is just a technology that operates via hardware located in discrete physical locations much like other communication methods (e.g., radio, television). To the extent those physical locations lie within the territorial borders of a given state, that state has the legal right to exercise control.[24] Where those territorial rights overlap, existing domestic and international law rules can resolve conflicts or regulate state behavior.[25]

Even some of those who accept cyberspace as a distinct space admit that autonomous governance is not automatically required. On the contrary, just as nation-states once colonized Africa, Asia, and the Americas, they may divide and sub-divide cyberspace. As Dan Hunter presciently argued in 2003, the spatial metaphor of cyberspace actually facilitates its commodification through the extension of property rights.[26]

[19] See, for example, Montevideo Statement on the Future of Internet Cooperation (October 7, 2013), at <http://www.icann.org/en/news/announcements/announcement-07oct13-en.htm>; S Yegulalp, "'Core Internet Institutions' Snub US Government," *InfoWorld*, October 11, 2013. The United States appears willing to support such a move, albeit under defined conditions. C Timberg, "US to Relinquish Remaining Control over the Internet," *The Washington Post*, March 14, 2014. Others fear internet institutions simply pushing away from the United States towards different state powers (e.g., Brazil, China, Russia). Z Keck, "Has Snowden Killed Internet Freedom?" *The Diplomat*, July 7, 2013.

[20] See, for example, Council on Foreign Relations, *Task Force Report: Defending an Open, Global, Secure, and Resilient Internet* (2013) 14–15 (describing ITU's 2012 World Conference on International Telecommunications).

[21] See, for example, D Ignatius, "After Snowden a Diminished Internet," *The Washington Post*, February 5, 2014; Council of Foreign Relations (n 20) 15; J Scott, "Chatham House founds Global Commission on Internet Governance," *ComputerWeekly.com*, January 22, 2014.

[22] Sometimes, the autonomy position characterizes cyberspace as having "no borders" but inevitably does so to claim states cannot govern it. I thus view the "no borders" position as a variant on drawing borders around cyberspace to differentiate it from other "real" space.

[23] JL Goldsmith and T Wu, *Who Controls the Internet? Illusions of a Borderless World* (Oxford: Oxford University Press, 2008); see also OS Kerr, "The Problem of Perspective in Internet Law" (2003) 91 *Georgetown Law Journal* 359–61.

[24] Goldsmith and Wu (n 23) 181. [25] Goldsmith and Wu (n 23) 148–9, 173.

[26] Hunter (n 15) 518–19; see also M Lemley, "Place and Cyberspace" (2003) 91 *California Law Review* 521, 523 (challenging inevitability of cyberspace privatization).

And, since property rights derive from sovereign states, national boundaries become definitive markers for lines of authority there.

Today, states regularly assert legal authority over actors and activities in cyberspace based on their borders; they can regulate based on the physical location of the network(s) or server(s) employed, or the physical location where the effects of such activity occur.[27] In other words, states claim to govern cyberspace using boundaries the same way they use them to govern other places (e.g., physical territory) or mediums (e.g., telecommunications).

Of course, using national boundaries to gain authority does not dictate *how* states should do so. The United States has (largely) employed its boundaries—within which much of the original internet's infrastructure exists—to promote a global and open internet where multiple stakeholders participate in governance issues with minimal content restrictions. Other states employ territorial boundaries quite differently. The so-called "Great Firewall" of China involves technological tools that constrain communications and technology within China while attempting to wall off those residing in China's territory from access to certain external sites and communications.[28] Efforts to impose national boundaries within cyberspace, however, are no longer the sole province of authoritarian governments.[29] Brazil has signaled an intention to localize e-mail and other communications within its territory to ward off foreign surveillance.[30] For many, such moves signal a "Balkanization," where the global internet becomes disaggregated into various networks distributed along national lines, producing less communication and cooperation and more competition and conflict.[31]

(c) The limits of boundaries in cyberspace governance

Today, the precise boundaries for internet governance are hotly contested in a bipolar discourse. Boundary claims inevitably end up as a version of one boundary project or the other—that is, "autonomous" boundaries *for* cyberspace or "national" boundaries *over* cyberspace. This "either-or" approach makes reconciliation unlikely in the absence of a unifying theory that might reconcile *why* boundaries should divide authority with respect to cyberspace. But even if one theory were to prevail (by consensus or conquest), it is not clear that it could accurately describe how cyberspace actually works,

[27] See, for example, Computer Fraud and Abuse Act, 18 U.S.C.A. §1030 (regulating "protected computers" defined in territorial terms to include computers "used in or affecting interstate or foreign commerce or communication, including a computer located outside the United States that is used in a manner that affects interstate or foreign commerce or communication of the United States"); see also The Computer Misuse Act 1990 (UK) §§5(2)(b) and (3)(b).

[28] See, for example, L Eko et al, "Google This: The Great Firewall of China, the It Wheel of India, Google Inc., and Internet Regulation" (2011) 15 *Journal of Internet Law* 3, 5; Jyh-An Lee and Ching-Yi Liu, "Forbidden City Enclosed by the Great Firewall: The Law and Power of Internet Filtering in China" (2012) 13 *Minnesota Journal of Law, Science & Technology* 125, 133.

[29] Cuba and Iran have sought to impose similar technological borders. See, for example, F Fassihi, "Iran Mounts New Web Crackdown," *The Wall Street Journal*, January 6, 2012.

[30] See L Chao and P Trevisani, "Brazil Legislators Bear Down on Internet Bill," *The Wall Street Journal*, November 13, 2013; J Watts, "Brazil to Legislate on Online Civil Rights Following Snowden Revelations," *The Guardian*, November 1, 2013.

[31] See S Meinrath, "We Can't Let the Internet Become Balkanized," *Slate*, October 14, 2013.

suggesting that such boundaries would not actually be effective. Boundaries, moreover, are not the sole source of legal authority; a more complete menu would include shared governance and regulation of "hybrid" spaces. Such challenges suggest a need to focus less on what cyberspace "is" and more on delineating boundaries for cyber behavior (e.g., rules for cyber operations).

i. The absence of a unifying theory

Unlike natural boundaries for a mountain range or a watershed, legal boundaries are social constructs.[32] There is nothing inherent or inevitable about a decision to govern using territorial boundaries—or any boundaries for that matter—as a basis for governance.[33] Such decisions require external justification. Unfortunately, cyberspace's two boundary rationales exist in contradistinction. There is presently little chance of a consensus position; indeed, further boundary discourse is only likely to inflame tensions over the distribution of legal authority in cyberspace.

The primary—and default—justification for governing authority in international law today remains "territorial sovereignty."[34] International law apportions "territory" to individual states who assume primary authority to govern there.[35] Although land is the main type of "territory," international law attaches the label to other resources as well, including man-made infrastructure lying on a state's land territory, the air space above it, mineral and oil resources below the surface, and twelve miles of the adjacent sea and seabed.[36]

Although states constitute themselves around territorial sovereignty, international law does not insist it serve as the *exclusive* basis for governance. Some resources are treated as categorically distinct from "territory" and its apportionment among states. For hundreds of years the high seas have qualified as res communis, a resource belonging to everyone, subject to appropriation by no one. As the "common heritage of mankind," the same premise extends (albeit with greater controversy) to the international seabed, the Moon, and outer space.[37] Other resources—such as the ozone layer—receive no such label, but international regulation effectively accords them

[32] Render (n 2) 3.

[33] RT Ford, "Law and Borders," in *The Meador Lectures on Boundaries* (Tuscaloosa: University of Alabama, School of Law, 2013) 55.

[34] Territorial sovereignty is distinct from other conceptions of sovereignty focused on identifying a single source of political authority within a state or membership in the international community. See, for example, SD Krasner, *Sovereignty: Organized Hypocrisy* (Princeton: Princeton University Press, 1999) 9–25; A Chayes and AH Chayes, *The New Sovereignty* (Boston: Harvard University Press, 1995) 27.

[35] See, for example, *Island of Palmas Arbitration (Netherlands v United States)* (1928) II RIAA 829, 839. State authority is exclusive, barring some agreement or peremptory international law rule.

[36] See UN Convention on the Law of the Sea, December 10, 1982, 1833 UNTS 3, Article 3 ("UNCLOS"). Where territorial sovereignty is unclaimed or contested, international law provides—and regulates among—various methods to assign sovereignty: occupation, cession, accretion and prescription. Earlier accepted methods are now prohibited (conquest) or impracticable (discovery). R Jennings and A Watts (eds), 1 *Oppenheim's International Law* (Oxford: Oxford University Press, 9th ed 1992) 678–9.

[37] UNCLOS (n 36) Article 137 (re seabed); Treaty on Principles Governing the Activities of States in the Exploration and Use of Outer Space, including the Moon and Other Celestial Bodies, January 27, 1967, 18 UST 2410, Article 2 ("Outer space, including the moon and other celestial bodies, is not subject to national appropriation by claim of sovereignty...").

such status.[38] In all these cases, the res communis concept justifies some form of collective governance autonomous from the dominant territorial sovereignty model.[39]

To justify the boundaries of authority in cyberspace, therefore, requires a theory of what kind of resource cyberspace "is." But that theory is as contested as the boundaries themselves. For some, cyberspace is inherently unapportionable; a res communis requiring distinct boundaries accompanied by collective governance of one form or another.[40] For others, cyberspace can (and should) be subject to territorial sovereignty, either because cyberspace is "just a network" indistinguishable from other man-made infrastructure, or—even if cyberspace is a distinct "place"—because states may use territorial sovereignty to claim and apportion it as "new" territory.[41] These theories are mutually exclusive; by definition, res communis is not subject to territorial sovereignty and vice versa.

Resolving how boundaries delineate cyberspace governance appears to require choosing one theory of cyberspace's identity, and dispensing with the other. At present, however, neither camp appears willing to concede, meaning competing boundary claims are likely to persist for some time. *Any* attempt to construct cyberspace governance or norms by reference to *its* boundaries (or those of nation states) will proxy for more fundamental and unresolved theoretical questions about what cyberspace "is." As such, instead of helping devise a governance structure for cyberspace, employing the boundaries rubric might actually forestall the effort.

ii. The inaccuracies of both boundary claims

Whatever theoretical debates they engender, using either set of boundaries to delineate cyberspace governance is descriptively inaccurate. The competing theories assume a binary choice that may operate as a disabling assumption. In other words, *both* sides in the debate over what cyberspace "is" may be wrong. As Julie Cohen's work emphasizes, cyberspace is "neither separate from real space nor simply a continuation of it," but rather constitutes an "extension and evolution" of the ways individuals (including those representing states) reason spatially.[42] Thus, attempts to draw a line around cyberspace does too much to divorce it from real space just as simply extending the boundaries of real space to cyberspace does too little to accommodate its unique operating features.

As a practical matter, moreover, neither boundary claim adequately describes the existing governance structure. Proponents of national boundaries emphasize the

[38] Like other res communis areas, the ozone layer is subject to collective governance. See Montreal Protocol on Substances that Deplete the Ozone Layer, September 16, 1987, 1522 UNTS 3, Article 2 et seq.

[39] There are various reasons for *why* a resource receives the *res communis* label; it may be to (1) preserve equitable use of the resource by multiple states; (2) ensure the resource's *sustainability*; or (3) accommodate a *functional necessity* where individual state governance is impossible or unworkable. See DB Hollis, "Stewardship versus Sovereignty? International Law and the Apportionment of Cyberspace," *Cyber Dialogue 2012*, University of Toronto, 2012.

[40] See, for example, Hollis, "Stewardship versus Sovereignty? (n 39) 7; WH von Heinegg, *Legal Implications of Territorial Sovereignty in Cyberspace* (NATO, CCD COE, 2012) 9.

[41] Goldsmith and Wu (n 23); J Cohen, "Cyberspace as/and Space" (2007) 107 *Columbia Law Review* 210, 226.

[42] See Cohen (n 41) 213.

reality of state governance in cyberspace. Yet, it would be a gross mischaracterization to catalog cyberspace governance by listing only states and the products of their authority—domestic laws and international treaties. Other actors play significant governing roles. Non-governmental organizations like IETF and ICANN have *key* roles in cyberspace governance via the protocols and systems they oversee.[43] International organizations like the ITU and the United Nations (UN) have also asserted *some* governance authority, even if their full extent is not yet fully determined.[44]

Emphasizing laws derived from territorial sovereignty also overestimates the extent to which formal, legal frameworks govern cyberspace. Cyberspace operates according to a much larger array of "norms," including—but not limited to—domestic and international laws. The most basic governing features of cyberspace—TCP/IP (transmission control protocol/internet protocol) and DNS (domain name system)—are products of a technical process of consultation and review that has no legal foundation.[45] To simply treat cyberspace governance as an unexceptional extension of national borders thus misses much, if not most, of the actual distribution of authority over cyberspace.

At the same time, it would be a mistake to conclude cyberspace has its own borders because of the predominance of the so-called "multi-stakeholder governance" model. Cyberspace governance may include non-state actors and norms other than law, but states and law still play significant roles. For example, ICANN is a US-incorporated non-profit organization. Thus, it owes its existence to—and can be regulated by—US laws, whether or not it has a future contractual relationship with the federal government.[46]

iii. The (in)effectiveness of existing boundary claims

Even if theory and practice aligned to define one type of boundary for cyberspace governance, there are good reasons to question its effectiveness. Cyberspace is a vast, open, global "network of networks" that operates on (at least) four levels: (1) *hardware* such as servers and routers, which allow the network to operate; (2) *communications*—for example, TCP/IP, which regulates communication among networks; (3) *applications*—the

[43] See, for example, "What's the Effect of ICANN's Role and Work on the Internet?" at <https://www.icann.org/en/about/participate/effect>; K Das, "Internet Engineering Task Force (IETF)" (2008) at <http://ipv6.com/articles/organizations/IETF-History-IPv6.htm>.

[44] See, for example, "ITU Global Cybersecurity Agenda (GCA) High-Level Experts Group (HLEG) Global Strategic Report" (2008) at <http://www.itu.int/osg/csd/cybersecurity/gca/docs/global_strategic_report.pdf>.

[45] Nor are law and technology the only sources of norms for governing cyberspace. Socially accepted norms of particular communities of users, engineers, states, and other actors may provide another useful way to decide "who decides" with respect to cyberspace. See, generally, M Finnemore, "Cultivating International Cyber Norms," in K Lord and T Sharp (eds), *America's Cyber Future—Security and Prosperity in the Information Age* (Center for New American Security, 2011).

[46] In addition to being incorporated in California, ICANN has had a contractual relationship with the US Department of Commerce since its inception, a relationship the United States is now willing to reconsider. See Timberg (n 19). Whatever the future of this US relationship, ICANN has increasingly accommodated more state participation via a Government Advisory Committee (GAC). GAC Operating Principles, Article 1, principle 1; J Weinberg, "Governments, Privatization, and 'Privatization': ICANN and the GAC" (2011) 18 *Michigan Telecommunications & Technology Law Review* 189, 191–207.

content of those communications; and (4) *users*—those who operate in cyberspace.[47] A single type of border or boundary line for regulating all four levels is either impracticable or impossible. A state may invoke territorial sovereignty to regulate hardware and users based on where they reside, but no state can regulate the communications layer alone; by definition it operates on a global scale.[48] Similarly, a state may control applications coded or used within its territory, but extraterritorial applications are unavoidable, including those specifically designed to pierce state boundaries.[49]

Adopting "autonomous" borders for cyberspace would be equally ineffective. A central repository can regulate communications and applications (e.g., existing internet institutions or an international regulatory body). But when it comes to controlling when, where, and how specific technology does (or does not) get used in the real world, the reality of territorial sovereignty inevitably comes into play. And states are very unlikely to cede control over physical activities within their borders. Individual or collective state involvement is inevitable whenever there is a need to identify or punish bad actors, whether individuals, organizations, or states themselves.

Part of the effectiveness problem for cyberspace may reside in the boundary concept itself. By definition, a boundary involves a degree of definiteness (or certainty) as to where its line lies accompanied by a degree of impermeability between the actors, activities, or authorities the line separates. In cyberspace, however, boundary lines are far from definite. As the foregoing discussion reveals, their location(s) are open to challenge. Even mapping cyberspace has proven extraordinarily difficult given rapidly evolving technology. That technology, moreover, makes impermeability a difficult prospect. China shows some electronic barriers are feasible, but other examples (e.g., the Arab Spring) provide cautionary notes on the ability of states to completely wall off or otherwise control what happens via cyberspace in their territory.[50] Nor can internet institutions wall off cyberspace from state interference, especially given recent revelations about how widely governments are surveilling and collecting intelligence in cyberspace. In short, cyberspace governance via boundary lines appears unworkable, at least if the effort focuses only on choosing one type of boundary over another (i.e., national versus autonomous boundaries).

iv. The incompleteness of a boundary-based governance model

Fortunately, international law does not require an "all-or-nothing" choice between res communis or territorial sovereignty any more than it mandates governance by line-drawing generally. International law has recognized an array of middle-ground

[47] N Choucri and D Clark, "Who Controls Cyberspace," *Bulletin of Atomic Scientists*, September 1, 2013.

[48] The root servers that certify addresses for internet traffic, for example, are distributed globally with redundant systems. See <http://www.root-servers.org/>.

[49] The most prominent example may be Tor, a software program designed to protect users from "traffic analysis." See "Tor Overview," at <https://www.torproject.org/about/overview.html.en>.

[50] During the Arab Spring, certain governments (e.g., Egypt) temporarily denied domestic users internet access, but sophisticated users were still able to circumvent those controls while international condemnation was widespread. See Freedom House, "Egypt" *Freedom on the Net 2012* (2012) 4, 6, available at <https://www.freedomhouse.org/sites/default/files/Egypt%202012.pdf>.

or "hybrid" models where in lieu of definite, impermeable boundaries, we see shared governance structures.

In the oceanic context, for example, cost-effective technologies now allow individual states to control or exploit more of the oceans and their seabeds, removing one of the primary rationales for regarding it as res communis.[51] As such, portions of the high seas have been re-categorized to allow for *some* state appropriations without subjecting them to territorial sovereignty per se. A *contiguous zone* now extends twelve additional miles beyond a state's territorial sea where the state can extend some (but not all) of its laws.[52] An *exclusive economic zone* (EEZ) runs two hundred miles from a state's shores over which a state has "sovereign rights," but not full sovereignty; states must still allow other states free transit through the EEZ and conserve its resources.[53]

Thus, governance of the oceans involves a spectrum of authority; territorial sovereignty over the territorial sea, limited sovereignty over the contiguous zone, sovereign rights in the EEZ, and res communis for the high seas. Cyberspace governance might follow a similar spectrum with certain aspects within state sovereignty, others with autonomous governance, and a range of intermediate, shared set of authorities in between.

In other instances, de facto governance does not match the de jure governance model. For example, even as states insist on territorial sovereignty over those parts of the radio-frequency spectrum in their airspace, they agreed to delegate that sovereignty to the ITU.[54] This model may go a fair way to describing the current roles of major internet institutions (although, as NGOs, they ostensibly are not agents of nation states as is the ITU).

In a few cases, like Antarctica, boundary disputes have ended in stalemates. Prior to 1959, seven states claimed sovereignty over various (and often overlapping) portions of that continent.[55] Rather than resolving these claims, however, states figuratively froze them via the Antarctic Treaty.[56] By its terms, the Treaty does not undermine or endorse any of the pre-existing claims. Instead, it sets Antarctica aside as a "scientific preserve," prohibiting military activities and giving *all* states certain freedom to engage in research there, with accompanying requirements to conserve its resources

[51] At first glance, the ocean environment is too dissimilar from cyberspace to warrant comparison. Ocean governance is a function of geography and distance, criteria that information technology largely circumvents given its capacity for global, instantaneous communication. On the other hand, ocean and cyberspace governance both share the problems of multiple users and capacity—both are public goods that no single state could regulate even if it wanted to.

[52] UNCLOS (n 36) Article 33 (authorizing state regulation of "customs, fiscal, immigration, or sanitary" matters).

[53] UNCLOS (n 36) Articles 56(1)(a) and 58 (authorizing EEZ state to explore and exploit natural resources within the EEZ); UNCLOS (n 36) Article 76(1) (conveying similar rights to the continental shelf even if beyond 200 nautical miles).

[54] Constitution and Convention of the International Telecommunication Union, 12 December 1992, 1825 UNTS 3, Article 1 (emphasizing "sovereign right of each State to regulate its telecommunication[s]" while giving ITU responsibility for managing that spectrum and setting technology standards for using it).

[55] P Beck, "Who Owns Antarctica? Governing and Managing the Last Continent" (1994) 1 *Boundary and Territory Briefing* 6–7.

[56] Antarctica Treaty, December 1, 1959, 402 UNTS 71, Article 4.

in doing so.[57] To the extent cyberspace faces a similar stalemate, the Antarctica model offers a means for moving forward.

These examples—oceans, telecommunications, and Antarctica—involved international law. Governance, however, is also possible via non-legal institutions or norms. Prominent examples include the Kimberley Process governing trade in conflict ("blood") diamonds mined in a war zone, and the Basel Committee on Banking Supervision.[58] In citing these examples, I do not mean to favor any one in particular. I only seek to unsettle the "all-or-nothing" choice suggested by existing boundary projects. Cyberspace's scale and complexity suggest a need for shared governance, where various actors accept authority for some purposes and not for others.[59]

Of course, simply because cyberspace needs better governance does not mean a new system will emerge. On the contrary, the significant theoretical and practical problems already described suggest a stalemate at best. As such, it may make sense to sidestep the governance question and focus on a different type of boundary—drawn by the law itself. Instead of getting caught up in deciding what kind of space cyberspace "is," why not first decide on appropriate standards of behavior? Norms can emerge when all stakeholders—states, internet institutions, international organizations (IOs), individuals—agree that certain behavior is out of bounds even absent agreement on who governs.[60] An agreement to prohibit cyber attacks on hospital networks, for example, need not require an overarching decision on the nature of cyberspace or who should govern it. If there is agreement that hospital cyber attacks are beyond the pale, everyone can take appropriate action to avoid or mitigate that behavior. Thus, when it comes to re-thinking the boundaries of law and cyberspace, the project is not simply a question of deciding the boundaries that produce law, but also the various ways law can produce boundaries.

III. Boundaries as the Product of Law: Cyberspace's Legal Regime(s) and Their Rules

Employing the language of boundaries to delineate lines of authority in cyberspace is a relatively familiar exercise given its use to separate authority elsewhere in international and domestic law. But the logic of boundaries applies equally *within* the law. A legal system may draw borders to delineate the application (or not) of a particular set of rules. Common law systems routinely bound off "criminal" from "civil" law or procedure. International law similarly sub-divides into different regimes—for example, trade, environment, human rights—each of which purports to regulate a defined

[57] Antarctica Treaty (n 56) Articles 1–3.

[58] See, generally, Finnemore (n 45); Joost Pauwelyn, "Informal International Lawmaking: Framing the Concept and Research Questions," in Joost Pauwelyn et al (eds), *Informal International Lawmaking* (Oxford: Oxford University Press, 2012) 3; KW Abbott, "Enriching Rational Choice Institutionalism for the Study of International Law" (2008) 2008 *University of Illinois Law Review* 5, 27.

[59] See Hollis (n 39) 8–9.

[60] This would be via an incompletely theorized agreement, where actors agree on a particular cyberspace norm without agreeing on why it should be the norm. See Sunstein (n 9).

(conceptual) space.[61] Within each regime, legal rules draw lines as well. Domestically, laws designate boundaries separating permissible from prohibited behavior or differentiating categories or consequences. International law does the same for states (and to a lesser extent IOs and individuals). Sometimes these borders purport to be impermeable *lex specialis*; in other cases, more than one set of laws or legal regimes may purport to regulate the same matter at the same time.[62]

How does law draw boundaries in cyberspace? Two methods predominate: (1) *tailor-made law* created specifically for the cyber context; or (2) *law-by-analogy* where pre-existing fields of law and their contents are mapped in cyberspace via analogical reasoning.[63] Over the years, states have employed both approaches in domestic law.[64] International law, however, has relied almost entirely on law-by-analogy to regulate cyberspace, particularly when it comes to issues involving the use of force and IHL. This has ensured cyberspace is not some law-free zone. But relying on boundaries drawn by analogy comes at a cost. Existing international legal regimes in cyberspace for cyberwar exhibit some of the same problems—in terms of both *theory* and *function*—encountered in the governance context.

(a) The law's lines in cyberspace: tailor-made or law-by-analogy?

Legal systems regularly draw lines tailor-made for specific areas (e.g., oceans), activities (trade), or sources (treaties) of law.[65] Even within an existing area such as criminal law, a new rule may emerge tailor-made for specific actors, behavior, or victims. Tailor-made fields and the rules they contain function based on pedigree; assuming they are the product of a valid law-making process, tailor-made law governs directly, subject, of course, to standard issues of interpretation and application.[66]

Since its inception, the internet has generated calls for its own field of law—cyberlaw—and laws within that field tailor-made for the technology involved. As a matter of domestic law, these calls have had a modicum of success. Today, cyberlaw is its own field. And many tailor-made rules now accommodate the fact that

[61] Usually, international lawyers disaggregate international law into sub-fields along subject-matter lines, although this is not the only way to divide it. See DB Hollis, *Contemporary Issues in the Law of Treaties* (2007) 101 *American Journal of International Law* 695 (book review).

[62] For example, although controversial, the ICJ has suggested that attacks by non-state actors can only trigger criminal law and *not* rules governing the use of force. See, for example, *Legal Consequences of the Construction of a Wall in the Occupied Palestinian Territory (Advisory Opinion)* (July 9, 2004) 43 *International Legal Materials* 1009, 1050; *accord Armed Activities on the Territory of the Congo (Democratic Republic of Congo v Uganda)* (December 19, 2005) [2005] ICJ Rep 116, [146]–[147]. But there are other situations in which international human rights law is alleged to continue to apply in concert with international humanitarian law. See nn 97–8 and accompanying text.

[63] See, for example, CR Sunstein, "On Analogical Reasoning" (1993) 106 *Harvard Law Review* 741.

[64] See, for example, Computer Fraud and Abuse Act, 18 U.S.C. §1030 (cyber specific criminal law). For an example extending existing US law by analogy see *United States v Forrester*, 512 F.2d 500 (9th Cir. 2008); OS Kerr, "Applying the Fourth Amendment to the Internet: A General Approach" (2010) 62 *Stanford Law Review* 1005, 1008.

[65] See, for example, UNCLOS (n 36) (oceans law); Marrakesh Agreement Establishing the World Trade Organization, April 15, 1994, 1867 UNTS 187 (trade); Vienna Convention on the Law of Treaties, May 23, 1969, 1155 UNTS 331 (treaties).

[66] Interpretation is quickly becoming its own field of international law. See A Bianchi et al (eds), *Interpretation in International Law* (forthcoming; Oxford: Oxford University Press, 2015).

computers and networks are both new tools for, and targets of, various forms of activity (e.g., criminal, economic, political).[67] Thus, states have enacted "cybercrime" laws alongside procedural and evidentiary rules drafted specifically for the cyber context.[68]

In the international law context, tailor-made laws for cyberspace are notably absent. International law does not (yet) have a distinct field for cyberspace. Nor are there cyber-specific rules within existing fields of international law. The Cybercrime Convention represents one notable exception.[69] In it, states agreed to define specific activities as cybercrimes and to take domestic and international steps to prosecute such behavior. The Convention provides boundaries around what constitutes cybercrime, limiting their reach in the process. In particular, negotiators were careful to draw a line excluding government sponsored cyber operations from the cybercrime laws required by the treaty.[70] Thus, for Convention parties at least, there is a boundary line between cybercrime and state-sponsored cyber operations such as cyber espionage.[71]

In many other areas of cyberspace, however, law's boundaries continue to be borrowed by analogy from other contexts. In some cases, this may be done for theoretical reasons as in Judge Easterbrook's famous critique of the "law of the horse."[72] In other instances, law-by-analogy applies to avoid gaps that would otherwise exist in the absence of tailor-made law. In either case, the analogy adds an additional step to the law's application beyond verifying its pedigree. Analogical reasoning requires that

[67] Crimes using computers include cyber theft and cyber extortion, while crimes against computers may occur via viruses, worms, Trojan horse programs, and other malware along with directed denial of service (DDOS) attacks that disrupt communications with one or more websites.

[68] For example, Computer Fraud and Abuse Act, 18 U.S.C. §1030; Electronic Communications Privacy Act, 18 U.S.C. §§2510 et seq; Federal Wiretap Act, 18 U.S.C. §2510; Communications Assistance for Law Enforcement Act, 47 U.S.C. §§1001 et seq; Foreign Intelligence Surveillance Act (FISA), 50 U.S.C. §§1801–1811. Sometimes, tailor-made laws regulate by exemption; removing otherwise covered behavior from the law's reach. For example, 47 U.S.C. §230(c)(1) (exempting internet service providers from vicarious liability for libel otherwise available under common law).

[69] The Cybercrime Convention does not have universal aspirations, and its membership remains limited. As of June 2014, forty-two states had joined it, including the United States, Australia, Japan, and most European nations. "Status as of 17/6/2014" Convention on Cybercrime, at <http://conventions.coe.int/Treaty/Commun/ChercheSig.asp?NT=185&CL=ENG> accessed June 16, 2014. Several key states—including Russia—have refused to join, objecting to the treaty's cooperation requirements as invasive of state sovereignty.

[70] Article 2 requires states to adopt "legislative and other measures" to establish as criminal offences under their domestic law intentional access to the whole or any part of a computer system "without right." The accompanying Explanatory Memorandum clarifies that the "without right" caveat "leaves unaffected conduct undertaken pursuant to lawful governmental authority" including acts to "protect national security or investigate criminal offenses." Explanatory Report, Cybercrime Convention (November 8, 2001) para 38.

[71] Of course, international law's ambivalence towards cyber espionage does not (and has not) precluded states from treating foreign cyber espionage as a violation of their domestic criminal law. *Tallinn Manual* (n 4) 194.

[72] FH Easterbrook, "Cyberspace & the Law of the Horse" (1996) 1996 *University of Chicago Law Forum* 207 (arguing against a distinct field of cyberlaw by reference to Gerhard Casper's explanation for why the US legal system lacks a "law of the horse": "the best way to learn the law applicable to specialized endeavors is to study general rules. Lots of cases deal with sales of horses; others deal with people kicked by horses; still more deal with the licensing and racing of horses, or with the care veterinarians give to horses, or with prizes at horse shows. Any effort to collect these strands into a course on 'The Law of the Horse' is doomed to be shallow and to miss unifying principles").

the context from which the analogy originates shares sufficiently relevant similarities (and no relevant dissimilarities) with the targeted context.[73]

As with tailor-made law, analogical reasoning may draw lines to distinguish both the applicable field(s) of law as well as their contents. Take, for example, the recent *Tallinn Manual on the International Law Applicable to Cyber Warfare.* Written by a private—and distinguished—International Group of Experts (IGE) at the invitation of the NATO Cooperative Cyber Defence Centre of Excellence, the *Manual* purports to describe existing international law relating to cyberwar.[74] Unlike the specifically negotiated terms of the Cybercrime Convention, however, the *Tallinn Manual* relies on pre-existing fields of international law (e.g., the jus ad bellum, IHL) to populate this space. Various states, including the United States and (most recently) China, have adopted a similar approach, insisting that existing international law rules apply in cyberspace.[75]

Although the *Tallinn Manual* covers lots of topics—jus ad bellum, IHL, state responsibility, and so on—its scope has clear boundaries. It delimits rules that relate to regulating cyberwar while disavowing any intention to regulate other topics.[76] Thus, the *Tallinn Manual* does not elaborate rules for cybercrime.[77] It also declaims (like the Cybercrime Convention) any coverage of peacetime cyber espionage, whether pursued for industrial purposes à la China or US meta-data collection.[78] Instead, the *Manual*—and most states—appear to accept the analogy that since international law fails to prohibit espionage in real space it is absent in regulating espionage in cyberspace as well.[79]

The same logic that draws borders to separate the regulation of cybercrime from cyberwar (while leaving cyber espionage outside both camps) is at work *within* the cyberwar context to delimit different categories of applicable rules as well as the

[73] Sunstein (n 63) 745. Sunstein catalogs analogical reasoning by reference to four elements: (1) principled consistency; (2) a focus on particulars; (3) incompletely theorized judgments; and (4) principles operating at low or intermediate levels of abstraction. Sunstein (n 63) 746 et seq.

[74] Although IGE members participated in the *Tallinn Manual* in their individual capacity, a NATO affiliated organization—the NATO Cooperative Cyber Defense Center of Excellence—provided funding for the project. See *Tallinn Manual* (n 4) 1.

[75] See UN Group of Governmental Experts, "Report on Developments in the Field of Information and Telecommunications in the Context of International Security" UN Doc A/68/98 (June 24, 2013) 8.

[76] *Tallinn Manual* (n 4) 4 ("Cyber activities that occur below the level of a 'use of force'...like cyber criminality, have not been addressed in any detail. Nor have any prohibitions on specific cyber actions, except with regard to an 'armed conflict' to which the jus in bello applies").

[77] *Tallinn Manual* (n 4) 4 ("The Manual does not delve into the issue of individual criminal liability under either domestic or international law"). The *Manual* does acknowledge that international law allows states to exercise jurisdiction over cybercrimes in cases where they have jurisdiction to do so. *Tallinn Manual* (n 4) 18 (Commentary to Rule 1).

[78] *Tallinn Manual* (n 4) 18 ("Cyber espionage, theft of intellectual property, and a wide variety of criminal activities in cyberspace pose real and serious threats to all States, as well as to corporations and private individuals. An adequate response to them requires national and international measures. However, the Manual does not address such matters because application of the international law on uses of force and armed conflict plays little or no role in doing so."). Of course, the *Tallinn Manual* does include rules and commentary on cyber espionage where it occurs in the context of an armed conflict. See *Tallinn Manual* (n 4) 192–5 (Rule 66).

[79] See, for example, *Tallinn Manual* (n 4) 50 (noting "the absence of a direct prohibition in international law on espionage per se"); J Radsen, "The Unresolved Equation of Espionage and International Law" (2007) 28 *Michigan Journal of International Law* 595, 601–2.

requirements of the rules themselves. Most often, this is done by analogy to earlier statements and applications of law. Sometimes the analogic reasoning will be expressly acknowledged, while in other cases the analogy may be left implicit in the application of a general rule to a new context.[80] But whether acknowledged as law-by-analogy or not, the extent to which cyberspace and the context that generated the existing rule are similar (or dissimilar) serves to delimit the basic boundaries of the existing international law.

For example, as a category IHL only applies to a specific set of state (and to a more limited extent, non-state) behavior involving "armed conflicts."[81] For IHL to regulate a cyber operation, therefore, that operation must occur "in the context of and related to an armed conflict."[82] This poses no real barrier for cyber operations if they arise in the context of a conventional armed conflict such as the one between Russia and Georgia in 2008 (provided, of course, the operations are attributable to a state, an often difficult proposition).[83] Conventional conflicts encompass traditional means and methods of military confrontation—the deployment of troops and weapons—long regarded as "armed force" for purposes of triggering IHL's application.[84] Where states engage in *only* cyber operations, however, the analysis becomes murkier. Absent some tailor-made definition of "armed force" in cyberspace, analogies are necessary to explain which stand-alone cyber operations IHL will regulate.

Currently, where a cyber operation has "analogous effects" to kinetic force—significant death or injury, destruction or damage to infrastructure—states and scholars agree it

[80] Some, like Dinstein, object to the idea of an "analogy" in extending existing IHL rules to cyberspace, claiming that they are simply applying general principles to a particular context. Y Dinstein, "Cyber War and International Law" (2013) 89 *International Law Studies* 276, 283. But this critique may be one of form, not substance. Applying general principles in new contexts regularly requires analogical reasoning even if interpreters do not recognize it as such. Some interpreters may deduce meaning for general principles by reference to other principles or outside sources (e.g., morality). More often, however, the principle's meaning is rationalized in terms of previous iterations; that is, by analogy to what meaning(s) it received in the past. In articulating the idea of law as core and penumbra, HLA Hart noted the inevitability of such reasoning. HLA Hart, *The Concept of Law* (Oxford: Clarendon Press, 2nd edn 1994) 123, 124–8 (comparing familiar cases where "there is general agreement" on meaning to new variations that require considering whether the new case resembles familiar ones "'sufficiently' in 'relevant' respects"); Hart, *The Concept of Law*, 274–5 (describing the process by which judges, faced with indeterminacy, invoke a general principle and then choose among "competing analogies" to interpret it).

[81] IHL further sub-divides armed conflicts into two categories: international armed conflicts and conflicts not of an international character. See S Vite, "Typology of Armed Conflicts in International Humanitarian Law: Legal Concepts and Actual Situations" (2009) 91 *International Review of the Red Cross* 1, 82–6. Although controversial, the United States has advocated for a third category, known either as an "internationalized" or "transnational" armed conflict. See GS Corn, "Geography of Armed Conflict: Why it is a Mistake to Fish for the Red Herring" (2013) 89 *International Law Studies* 77, 80.

[82] C Droege, "Get Off My Cloud: Cyber Warfare, International Humanitarian Law, and the Protection of Civilians" (2012) 94 *International Review of the Red Cross* 533, 542; see also HH Dinniss, *Cyber Warfare and the Laws of War* (Cambridge: Cambridge University Press, 2012) 117.

[83] Indeed, despite claims that Russia bore responsibility for the 2008 attacks on Georgian websites, no formal attribution to Russia ever occurred. See Droege (n 82) 542; Dinniss (n 82) 127–9. For more on attribution see nn 100–2.

[84] According to the *Tadic* case, an international armed conflict arises "whenever there is a resort to armed force between States." International Criminal Tribunal for the Former Yugoslavia, *Prosecutor v Tadic*, Case No IT-94-1-A, Appeals Chamber Decision on the Defence Motion for Interlocutory Appeal on Jurisdiction, October 2, 1995 [70]. The "armed force" criterion has since become widely used to delineate IHL's application. See Droege (n 82) 546.

crosses into armed force territory.[85] The problem, as the *Tallinn Manual* acknowledges, lies in defining the quantum of harm required to constitute armed force. The IGE divided, for example, on whether the Stuxnet worm's damage—requiring the replacement of centrifuges at a nuclear fuel processing plant—sufficed.[86] In other words, IHL may not regulate stand-alone cyber operations that produce some, but not enough, damage analogous to the effects of non-cyber operations; cyber operations generating no damage presumably thus fall outside IHL's bounds entirely.[87] The problem is further complicated with respect to effects of cyber operations that may be severe but not analogous to effects now regarded as armed force (e.g., disruption without permanent damage to critical infrastructure like an electric grid).

Beyond delimiting which set of international laws govern a cyber operation, analogies also permeate efforts to delineate the borderlines drawn by specific rules. The *Tallinn Manual*, for example, adopts the law-by-analogy approach as its dominant method.[88] Sometimes the analogy is explicit. Rule 11 defines a cyber operation as constituting a use of force when "its scale and effects *are comparable* to non-cyber operations rising to the level of a use of force."[89] In other cases the analogy is implied by the rule's rationale. Rule 30 contains a critical definition—what constitutes an "attack" in cyberspace—which determines the reach of a host of related IHL rules on the conduct of hostilities. Rule 30 requires an attack to have violent consequences (injury or death to persons, damage or destruction to objects) and justifies that requirement by describing how it operates to define attacks with respect to non-cyber means (e.g., kinetic force, biological weapons). In other words, the line for a cyber attack is drawn according to the boundaries set by analogous prior effects.[90] Of course, this means non-analogous effects of cyber operations (e.g., those that do not produce physical damage or destruction to objects) do not cross the attack threshold. For example, a majority of the *Manual*'s IGE do not view a loss of functionality in a computer system as an attack absent a need to repair the system's physical hardware.[91]

(b) The limits of setting legal boundaries by analogy

The tension between the tailor-made and law-by-analogy approaches to regulating behavior in cyberspace mirrors competition over which boundaries delineate

[85] See Dinniss (n 82) 131; Droege (n 82) 546; see also MN Schmitt, "Classification of Cyber Conflict" (2012) 17 *Journal of Conflict & Security Law* 251.

[86] *Tallinn Manual* (n 4) 83–4.

[87] See, for example, Droege (n 82) 548–50 (proposing thresholds based on "severity" to define which cyber operations qualify as armed force or otherwise trigger IHL). But see *Tallinn Manual* (n 4) 136 (suggesting, without explanation, that IHL may apply to an operation "to degrade, deny, disrupt, or alter" global positioning satellite signals using "cyber means instead of conducting an operation that rises to the level of an attack (and that causes collateral damage)").

[88] There are, however, a few instances of proposals for new limits drawn from IHL's overarching purposes rather than analogy—for example, defining an "attack" to include severe illness and mental suffering. *Tallinn Manual* (n 4) 108.

[89] *Tallinn Manual* (n 4) 108 Rule 11.

[90] See *Tallinn Manual* (n 4) 108 Rule 30. In another example, the IGE found that introducing malware or production-level defects is analogous to laying mines, and thus qualify as attacks. *Tallinn Manual* (n 4) 108–9.

[91] *Tallinn Manual* (n 4) 108–9.

authority to do the regulating. The logic of boundaries dominates each discourse. In both contexts, moreover, there are claims that the novelty of cyberspace requires something "new" whether it's governance, a legal regime, or particular rules. Such claims are countered by claims cyberspace is not so new as to deserve *sui generis* treatment; existing authorities, legal regimes, and rules can regulate cyberspace without difficulty.

Given such similarities, it is not surprising that the same criticisms raised with respect to the boundaries for delimiting cyberspace governance extend to how international law draws boundaries between different legal regimes or within specific rules. The *lack of a unifying theory* appears at work in debates over using tailor-made law or law-by-analogy; those who seek a new regime or rules inevitably view cyberspace as something truly novel; a novelty denied by those who emphasize the capacity of existing rules or principles to operate by analogy. Functional problems with *accuracy*, *effectiveness*, and *completeness* are also visible. The one tailor-made set of rules—the Cybercrime Convention—fails to proffer an accurate set of lines around or within the field of cybercrime; it only binds the forty-two states parties, while many other states (e.g., China, Russia) draw very different lines, most notably by criminalizing "destabilizing" speech.[92] Nor has the tailor-made approach for cybercrime proved terribly effective; cybercrime remains a pervasive, and increasing, problem. And no matter how comprehensive in aspiration, tailor-made rules are never complete—over time, they inevitably encounter gaps or ambiguities requiring a new round of tailor-made law-making or law-by-analogy to clarify the situation. Of course, no one has made the mistake of suggesting that the Cybercrime Convention offers a complete response to cyber threats in any case.

The problems of legal boundaries are particularly exacerbated in the law-by-analogy context. On both theoretical and functional levels, the current emphasis on international law-by-analogy generates a distinct set of challenges. Thus, even if international law currently regulates *when* and *how* states may conduct a cyberwar, there are significant reasons to question *how well* it does so.

i. Theoretical problems: the fragmentation of international law in cyberspace

Aside from the lack of a unifying theory on the nature of cyberspace, efforts to analogize existing law to cyberwar present another theoretical challenge—fragmentation. Fragmentation refers to the phenomenon by which international law has splintered into a variety of "specialized and (relatively) autonomous rules or rule-complexes, legal institutions and spheres of legal practice."[93] International law's various regimes—for example, trade, human rights, international environmental law, the law of the sea, IHL—each now claim autonomy over a particular, defined area. The theoretical problem of fragmentation arises when more than one regime considers the same subject

[92] It is also inaccurate to treat law—whether domestic or international—as the sole purveyor of norms in cyberspace; other types of norms may also regulate cyber activities. See n 58 and accompanying text.

[93] See, for example, M Koskenniemi, "Report of the Study Group of the International Law Commission," *Fragmentation of International Law: Difficulties Arising from the Diversification and Expansion of International Law* (April 13, 2006) UN Doc. A/CN.4/L.682, 10–11.

matter to fall within its borders, giving rise to the prospect of competing proscriptions, interpretations, or remedies for the same conduct.[94] To the extent international law lacks general rules to resolve competition and conflict across these regimes, fragmentation suggests the international legal system lacks sufficient unity or coherence.[95]

Law-by-analogy implicates fragmentation in two respects. First, analogizing existing law into cyberspace replicates international law's existing fragmentation problems.[96] Analogical reasoning not only allows, but actively encourages, claims that specific cyber activities fall within the ambit of a pre-existing regime, such as IHL, international telecommunications law, or human rights law. In doing so, existing border disputes across regimes carry over into cyberspace. For example, IHL has long claimed to apply exclusively in geographic and temporal terms to armed conflicts vis-à-vis other fields of international law.[97] In recent years, human rights lawyers have challenged the impermeability of IHL's borders, arguing international human rights law applies alongside IHL.[98] By analogizing IHL *and* human rights to the cyber context, the competition between the two regimes may be perpetuated there as well. Similarly, attempts to devise proscriptions for "cyberterrorism" are likely to replicate the as-yet-unresolved debate over how law regards terrorism. Is it a crime? An act of war? Both? Or neither?[99]

Second, analogizing existing law into cyberspace may exacerbate international law's fragmentation problem by generating new instances of competition or conflict that did not exist previously. Aside from terrorism, criminal law has usually been separated

[94] See, for example, J Pauwelyn, *Conflicts of Norms in Public International Law: How the WTO Law Relates to Other Rules of International Law* (Cambridge: Cambridge University Press, 2003); C Borgen, "Treaty Conflicts and Normative Fragmentation," in DB Hollis (ed), *The Oxford Guide to Treaties* (Oxford: Oxford University Press, 2012) 448–50. This "normative" fragmentation is distinguishable from "institutional" fragmentation where different tribunals take jurisdiction over the same case or legal question. See, for example, Y Shany, *The Competing Jurisdictions of International Courts and Tribunals* (Cambridge: Cambridge University Press, 2003). Meanwhile, some view fragmentation as not a problem, but a pluralist outcome to be embraced. See, for example, M Koskenniemi and P Leino, "Fragmentation of International Law? Postmodern Anxieties" (2002) 15 *Leiden Journal of International Law* 553; JI Charney, "International Law and Multiple International Tribunals" (2002) 271 *Recueil des Cours* 101, 352–6.

[95] At one time, HLA Hart went so far as to suggest that international law consists of "rules which constitute not a system, but a simple set." Hart (n 80) 234. However, few today deny international law, with its unifying language and culture, exists as a legal system even with fragmentation. See, for example, M Proust, "All Shouting the Same Slogans: International Law's Unities and the Politics of Fragmentation" (2006) 17 *Finnish Yearbook of International Law* 131.

[96] Of course, it is worth noting that a tailor-made cyberspace legal regime would not necessarily escape a fragmentation critique. Developing an autonomous, tailor-made regime for cyberspace would add another regime to the rolls currently competing for normative priority. If, however, such tailor-made rules could achieve lex specialis status, they could alleviate fragmentation by providing a clear normative hierarchy in cases of conflict or tension with other legal regimes.

[97] IHL's borders are generally perceived as lex specialis, supplanting other laws when the conditions for IHL's application are met. See GS Corn, "Mixing Apples and Hand Grenades: The Logical Limit of Applying Human Rights Norms to Armed Conflict" (2010) 1 *International Humanitarian Law Studies* 52, 60.

[98] C Greenwood, "Human Rights and Humanitarian Law—Conflict or Convergence" (2010) 43 *Case Western Reserve Journal of International Law* 491, 495; A Orakhelashvili, "The Interaction between Human Rights and Humanitarian Law: Fragmentation, Conflict, Parallelism, or Convergence?" (2008) 19 *European Journal of International Law* 161.

[99] See, for example, DB Hollis, "Why States Need an International Law for Information Operations" (2007) 11 *Lewis & Clark Law Review* 1023, 1026–8.

quite easily from the boundaries claimed by laws on the use of force and IHL. Each regime regulates a different set of actors—criminal law regulates individuals; IHL and the prohibition on the use of force regulate states. In cyberspace, however, analogizing law based on the actor is complicated by the problem of attribution. Current information technology frequently makes it difficult (although not impossible) to identify the actual server from which an attack or exploit originates, let alone identify the actual perpetrators even if only by type.[100] Nor is this a transient problem—the very architecture of the internet enables sophisticated hackers to maintain anonymity if they so desire.[101] Attackers can even disguise their efforts to appear to originate with some other group or government.[102] Thus, cyberspace presents new opportunities for normative conflicts between criminal law and IHL that have not arisen in other contexts where the identities (or at least kinds) of perpetrators are often easier to discern.

ii. Inaccuracies in defining law's borders by analogy

Analogies may also generate inaccurate boundaries of behavior depending on the relationship between the analogy's origin and its target. Analogies work well when their origin and target have relevant similarities. But where there are relevant differences, an analogy may actually draw the wrong (or inaccurate) lines that require adjustment to serve the law's overarching values.

The *Tallinn Manual*'s approach to the jus ad bellum provides a paradigmatic example of the accuracy dilemma in using law-by-analogy. The UN Charter seeks to maintain international peace and security via jus ad bellum prohibitions on the use of force except when done in self-defense or pursuant to UN Security Council authorization under Chapter VII.[103] The *Manual*'s Rule 11 attempts to define a use of force in cyberspace by analogy; indicating that a cyber operation constitutes a use of force when its scale and effects are *comparable* to non-cyber operations rising to the level of a use of force.[104] But this line-drawing is open to challenge as either over- or under-inclusive. On the one hand, the UN Charter indicates that "complete or partial interruption of...

[100] Technical attribution is not necessarily impossible; over time, those with sufficient skill may be able to identify the origins of an attack. Or, attribution may come from other means (e.g., secondary intelligence, mistakes, luck). See DB Hollis, "An e-SOS for Cyberspace" (2011) 52 *Harvard International Law Journal* 373, 397–400.

[101] See L Greenemeier, "Seeking Address: Why Cyber Attacks Are So Difficult to Trace Back to Hackers" (2011) *Scientific American*; HF Lipson, *Tracking and Tracing Cyber-Attacks: Technical Challenges and Global Policy Issues* (CERT Coordination Center, 2002).

[102] MC Libicki, *Cyberdeterrence and Cyberwar* (Rand, 2009) 44. In some situations, a victim may not even realize an attack has occurred, attributing the threat to computer error or malfunction. JL Goldsmith, "The New Vulnerability," *The New Republic*, June 7, 2010; ET Jensen, "Computer Attacks on Critical National Infrastructure" (2002) 22 *Stanford Journal of International Law* 207, 212–13. Although some governments may assume state-sponsored origins for cyber operations based on their structure or method, such assumptions have proven incorrect in the past. See, for example, K Zetter, "Israeli Hacker 'The Analyzer' Indicted in New York—Update," *Wired*, October 29, 2008 (discussing US mistaken view that "Solar Sunrise" operation against US Department of Defense was state-organized when three teenagers perpetrated it).

[103] See UN Charter, Articles 2(4), 39–42, and 51.

[104] *Tallinn Manual* (n 4) Rule 11 ("A cyber-operation constitutes a use of force when its scale and effects are comparable to non-cyber-operations rising to the level of a use of force.").

telegraphic, radio, and other means of communication" are "measures not involving the use of armed force." This would seem to categorically deny cyber operations' "use of force" status if one focuses on the fact that cyber operations arise via interruptions and interference with electronic communications. But even if states regard that conclusion as untenable—and it seems most do—it does not follow that the effects of all cyber operations are so similar to non-cyber operations to justify drawing the use-of-force line entirely by analogy.[105]

Consider Stuxnet. The *Tallinn Manual* claims it constituted a use of force.[106] The US government would likely agree (at least if the target had been a US nuclear laboratory). The analogy at the root of Rule 11, however, does not appear to actually place Stuxnet beyond the use-of-force line. Rule 11 says cyber operations' effects must be "equivalent" to non-cyber uses of force to fall within the jus ad bellum. But Stuxnet was unique in its effects—it got so much attention precisely because it did something no one had previously thought possible. It caused damage to be sure, reportedly requiring repairs to 1,000 centrifuges.[107] But that damage was qualitatively different than the heat blast and fragmentation that would accompany a kinetic attack. Stuxnet caused its damage over a long period of time, and, remarkably, caused no apparent harmful effects on anything other than the centrifuges and related equipment. The objective of Stuxnet—delaying Iranian production of nuclear materials—may have been analogous to what military planners might have sought from a missile strike, but the direct effects by which it did so were not. Neither precision-guided weapons nor drones can achieve what Stuxnet did to Iranian operations over the course of months.

Stuxnet though at least had *some* physical effects, allowing a debate on just how comparable those effects and their scale are to the effects of kinetic or biological weaponry accepted as uses of force. But what happens if a cyber operation has *no* physical impacts? Does that mean it lies entirely outside the jus ad bellum's boundaries? Say—using hyperbole to make a point—a cyber operation randomly reassigns the decimal places in all Bank of America accounts and records. The data is irreversibly re-arranged, but no hardware needs to be replaced at all. Does this cross the line for a use of force? According to the Commentary accompanying Rule 11, it would not; absent analogous physical harm or damage no cyber operations may qualify as a use of force.

This result has some appeal if the goal is to ensure cyberspace does not become a legal loophole; that is, that international law should continue to treat effects analogous to those that crossed the use of force line in the past as still doing so regardless of their origins. And since cyber operations can do things analogous to the use of force by kinetic or biological weapons, it makes sense to reason that they also fall under the same prohibitions. Similarly, to the extent that cyber operations can do things analogous to those things that did not cross the use of force line previously (e.g., economic

[105] Hollis (n 99) 1041–2. For its part, the *Tallinn Manual* does not even address Article 41's implications for defining a cyber operation as a use of force.

[106] *Tallinn Manual* (n 4) 45 ("The clearest cases are those cyber-operations, such as the employment of the Stuxnet worm, that amount to a use of force"); see also n 5 and accompanying text.

[107] See n 5 and accompanying text.

sanctions), it makes sense to deny these operations use-of-force status. In both cases, law-by-analogy provides a way to accurately replicate earlier international law rules.

But cyber operations can do things that have no earlier analogue; things previously thought impossible—immediate, systemic consequences that generate no physical harm or bloodletting.[108] Deciding how to regulate such cases by analogy becomes problematic. Drawing the use-of-force line by reference only to what effects crossed that line in the past means excluding any new effects cyber operations may generate. But it is not clear that such a reflexive exclusion is rational any more than it would be to draw the use of force line by reference to what did not qualify as a use of force previously (i.e., any novel cyber capacities would be reflexively counted as uses of force). An accurate portrait of the jus ad bellum in cyberspace requires crafting a more tailor-made line for cyberspace, one drawn by assessing the various new capacities of cyber operations in light of the law's overarching object and purpose—maintaining international peace and security.

iii. The ineffectiveness of boundaries by analogy in regulating cyber conflicts

Even if the right analogies can be found for cyberspace, there is no guarantee the resulting law will work. The existing legal frameworks on which these analogies rely are ill-suited for the technology they target. As a result, much of the existing international law for cyberspace, including cyberwar, is unlikely to generate desired outcomes. Of course, law does not have to regulate every case or generate 100% compliance to be effective. But law should identify and regulate undesirable behavior with sufficient compliance to shape or deter future behavior. At present, however, the existing law-by-analogy approach suffers in terms of both coverage and compliance.

As a coverage matter, it is remarkable how many existing cyber threats, the law-by-analogy approach leaves unregulated. Cyber war may be covered by analogies, but, as discussed in section III(a), cybercrime is primarily a product of a tailor-made approach that is quite limited in its reach. Meanwhile, international law does little to nothing to regulate cyber espionage on the assumption that it is analogous to earlier forms of espionage, notwithstanding its vastly broader scale and scope.[109] Now, certainly some states desire this result—at least insofar as it concerns their own online surveillance and collection methods. Over the last several years, however, states and other actors have begun to question if international law should adjust to regulate cyber espionage in at least some instances. The United States for a time campaigned to differentiate industrial espionage (but not to prohibit it).[110] More recently, states have

[108] See, for example, L Kello, "The Meaning of the Cyber Revolution; Perils to Theory and Statecraft" (2013) 38:2 *International Security* 7.

[109] International law only directly addresses espionage during armed conflicts, differentiating prisoner of war treatment accorded scouts but not spies. Espionage may be prohibited by a victim state's domestic law, but international law is otherwise notoriously silent on regulating it. Radsen (n 79); John Murphy, "International Legal Process: Threat to U.S. Interests" (2013) 89 *International Law Studies* 309, 320.

[110] See DP Fidler, "Economic Cyber Espionage and International Law: Controversies Involving Government Acquisition of Trade Secrets Through Cyber Technologies," *ASIL Insights*, March 20, 2013.

sought no-spying agreements or even a global right to privacy in reaction to Edward Snowden's disclosures of US cyber espionage activities.[111] Without taking a stand on the merits of such proposals, it is enough to emphasize that law-by-analogy cannot regulate such behavior; any regulation, let alone effective regulation, requires a tailor-made approach.

However, even where law-by-analogy covers a subject matter, like cyberwar, the architecture of the internet makes it difficult for the existing law to ensure compliance. International law on the use of force and IHL (not to mention cybercrime law) operates by proscription—deterring or remedying violations by targeting the bad actors who perpetuate them. Proscription, however, requires attribution to work. And, as already detailed, attribution is notoriously difficult to obtain in cyberspace.[112] In most cases, the technology favors anonymity for those states and non-state actors with sufficient skill to hide or falsify their origins.[113]

Thus, no matter how clear the law is, compliance and enforcement may be problematic in a world where the law's subjects are so hard to identify. Cyber criminals rarely get caught.[114] To date, moreover, there is no consensus of any state conducting a cyber operation in violation of international law, let alone attribution of such a violation to a specific state(s).[115] As the Stuxnet case suggests, some states may still conform their cyber operations to IHL even if they think they will remain anonymous.[116] But it may be risky to generalize from one example.[117] States may still assume they can act with impunity; indeed, some cyber operations may not even be identified as such, but mislabeled as unfortunate accidents caused by computer error.

Where the authors of a cyber operation remain unknown, additional challenges to the law's effectiveness arise since it may be unclear what law applies. Today, we continue to be at risk of having one state assume a cyber operation lies within the boundaries of cybercrime, with any response required to respect state sovereignty, while

[111] See, for example, J O'Donnell & L Baker, "Germany, France demand 'no-spy' agreement with U.S.," *Reuters*, October 25, 2013; D Rushe, "UN Advances Surveillance Resolution Reaffirming 'Human Right to Privacy,'" *The Guardian*, November 26, 2013; K Roth, "NSA: Our Analogue Spying Laws Must Catch Up with the Digital Era," *The Guardian*, November 10, 2013.

[112] See nn 100–2 and accompanying text.

[113] To the extent that attribution may recently have become less of a problem, it is not clear that this will remain the status quo—the current push to establish cyber privacy protections suggests technology may again evolve to make it more—rather than less—difficult to identify bad actors in cyberspace.

[114] RA Grimes, "Everything is Hackable—and Cyber Criminals can't be Tracked," *InfoWorld*, May 10, 2011.

[115] The closest thing to a precedent came in 2013 when Mandiant, a private cyber security firm, published a report accusing the Chinese military of stealing data from over a hundred US companies, including *The New York Times*. See O Joy, "Mandiant: China is sponsoring cyber-espionage," *CNN*, February 20, 2013, at <http://edition.cnn.com/2013/02/19/business/china-cyber-attack-mandiant/>. Since then, the US government has charged five Chinese officers of the People's Liberation Army with violating US cybercrime laws, implicating the Chinese government in the process. J Finkle et al, "US Accuses China of Cyber-Spying on US Companies," *Reuters*, May 19, 2014.

[116] See, for example, J Richmond, "Note—Evolving Battlefields: Does STUXNET Demonstrate a Need for Modifications to the Law of Armed Conflict?" (2012) 35 *Fordham International Law Journal* 842, 883–893.

[117] This may be especially true where Stuxnet's existence came to light via an internal leak from the US Administration that reportedly sponsored it (along with Israel). See D Sanger, *Confront and Conceal: Obama's Secret Wars and Surprising Use of American Power* (New York: Broadway Books, 2012) Ch 10.

another views it as an "armed attack," for which self-defense is an available option. The fact that, technically speaking, it is difficult to distinguish the code used to perpetuate a cyber exploit or cyber espionage (which international law currently does not prohibit) from a cyber attack (which may trigger the jus ad bellum) only heightens this risk; states that discover an unexecuted cyber operation may have trouble discerning if its purpose is criminal, clandestine, or war-like.[118]

The near instantaneous speed at which cyber operations occur further complicates the effectiveness of law-by-analogy. In an effort to gauge relevant similarities (and dissimilarities) between a non-cyber use of force and its cyber equivalent, the *Tallinn Manual* introduces a multi-factored test for analogizing which cyber operations constitute uses of force.[119] The US State Department Legal Adviser Harold Koh offered a similar case-by-case standard.[120] Although the United States treats uses of force as equivalent to an "armed attack" (to which a state may lawfully respond in self-defense),[121] the *Tallinn Manual* views the latter criterion to have a higher threshold, requiring further analysis of a cyber operation's scale and effects.[122] Whether the two concepts are the same or different, however, may not matter much when victims of a cyber operation are asked to assess the situation. The relevance of so many factors inevitably muddies the jus ad bellum's boundaries; the more factors a legal question presents the more opportunities it offers for interpretative disagreements.[123] More

[118] Offensive cyber operations come in two basic forms: cyber attacks and cyber exploits. As Herb Lin details, "Cyber attack refers to the use of deliberate activities to alter, disrupt, deceive, degrade, or destroy computer systems or networks used by an adversary or the information and/or programs resident in or transiting through these systems or networks." In contrast, "cyber exploitation" refers to "deliberate activities designed to penetrate computer systems or networks used by an adversary, for the purposes of obtaining information resident on or transiting through these systems or networks." Herb Lin, "Cyber Conflict and International Humanitarian Law" (2012) 95 *International Review of the Red Cross* 515, 518–19. This chapter primarily focuses on cyber attacks. But given potential confusion between that term and "attack" as a term of art in IHL (not to mention armed "attacks" in the jus ad bellum), I opt to employ the more general "cyber operations" language throughout my discussion.

[119] *Tallinn Manual* (n 4) 48–51 (listing eight factors to "identify cyber-operations that are analogous to other non-kinetic or kinetic actions that the international community would describe as uses of force"—severity, immediacy, directness, invasiveness, measurability of effects, military character, state involvement, and presumptive legality).

[120] See C Borgen, "Harold Koh on International Law in Cyberspace," *Opinio Juris*, September 19, 2012 (quoting Koh: "Cyber activities that proximately result in death, injury, or significant destruction would likely be viewed as a use of force. In assessing whether an event constituted a use of force in or through cyberspace, we must evaluate factors: including the context of the event, the actor perpetrating the action (recognizing challenging issues of attribution in cyberspace), the target and location, effects and intent, among other possible issues").

[121] Borgen, "Harold Koh on International Law in Cyberspace" (n 120) (US position is that "the inherent right of self-defense potentially applies against any illegal use of force. In our view, there is no threshold for a use of deadly force to qualify as an 'armed attack'"). See also Y Dinstein, "Computer Network Attacks and Self-Defense" (2002) 76 *International Law Studies* 99, 103 (claiming armed attack must have "violent consequences"); M Waxman, "Cyber-Attacks and the Use of Force: Back to the Future of Article 2(4)" (2011) 36 *Yale Journal of International Law* 421, 431 (challenging Dinstein's view with respect to cyber attacks).

[122] *Tallinn Manual* (n 4) Rule 13 and Commentary. The originator of the attack may also be an additional factor given the international law controversy over whether "acts of non-State actors can constitute an armed attack absent direction by a State." See n 62.

[123] See DB Hollis, "The Existential Function of Interpretation in International Law," in A Bianchi (ed), *Interpretation in International Law* (Oxford: Oxford University Press, forthcoming 2015).

importantly perhaps, there may simply be no time to make the requisite analysis given the time pressures facing a state victim of a cyber operation.

If anything, the law-by-analogy approach to the jus ad bellum offers states a standard rather than an actual rule of behavior.[124] As such, the legality of a cyber operation (or a state's response to one) may only be determined ex post, instead of an assessment generated ex ante. Combined with problems in identifying the applicable law, the actors it targets, and even the behavior it covers, the law-by-analogy approach appears unlikely to effectively engender behavior, as opposed to providing grounds for arguing about it after the fact.

iv. The incompleteness of law-by-analogy to delineate boundaries of behavior

If the current law-by-analogy approach draws lines that are both inaccurate and ineffective, it necessarily suggests an incomplete regulatory response. Fortunately, there are alternatives that paint a more complete picture of law's regulatory capacities. For starters, borders could be tailor-made for cyber operations. How can international law do this? Broadly speaking, it may come via state practice (letting customary international law evolve as time passes) or agreement among states (and occasionally non-state actors). Both options are time tested, but in the cyber context a dialogue leading to agreement may be preferable. Between anonymity and the multitude of actors in cyberspace, there is relatively little incentive for states to disclose their behavior. As a result, state practice will be defined by leaks or technically inept actors, which does not suggest a solid framework on which to build rules for something as important as cyber warfare.[125]

In terms of their function, tailor-made laws may supplement or complement the boundaries generated by legal analogies. For example, to overcome the existing law's attribution problems, a number of proposals exist to assume attribution based on the type of cyber operation, its target or the nation-state from which it supposedly originates.[126] Despite extensive analysis—including much attention to when a state should be held responsible for non-state actor cyber operations of which it had (or should have had) knowledge, I remain skeptical of such approaches. The nature of the technology—both its widespread distribution and the potential for anonymity or false-flags—undercut the viability of attribution assumptions.[127] Nonetheless, future efforts to enhance the accuracy and efficacy of existing law may well bear fruit.

[124] Rules seek to bind parties to respond in specific, determinate ways when certain facts exist; once the facts are clear, so too is the expected behavior. Standards afford decision-makers discretion to decide (often ex post) what behavior satisfies an obligation by widening the range of relevant facts or authorizing application of some background policy or principle. Principles, in contrast, set forth broad considerations for evaluating future behavior without providing any particular norm for the behavior itself. See, for example, D Bodansky, "Rules vs. Standards in International Environmental Law" (2004) 98 *Proceedings of the American Society of International Law* 275; KM Sullivan, "The Justices of Rules and Standards" (1992) 106 *Harvard Law Review* 22, 57–9.

[125] See Hollis (n 99) 1054 [126] See Hollis (n 100) 400.

[127] Hollis (n 100) 400. Nor do I necessarily favor engineering a solution to the attribution problem for purposes of enhancing international law's effectiveness. The anonymity afforded by the current internet architecture reflects important values (e.g., privacy and freedom) that require as much—if not more—support than deterring or mitigating bad behavior.

Alternatively, tailor-made approaches could re-draw boundaries to supplant or substitute for those suggested by analogy. Examples of such efforts include the push to regulate cyber espionage or to redefine who constitutes a lawful combatant.[128] A third possibility involves drawing entirely new boundary lines, whether to distinguish some newly prohibited behavior or to identify new affirmative requirements for states. My earlier proposal for an e-SOS system falls within this latter category; the idea being states should agree on a duty to assist victims of the most severe cyber threats independent of any ability to identify those responsible.[129]

As in the governance context, a complete catalog of law's functions must extend beyond its capacity to produce boundaries.[130] Along with the sense of impermeability and definiteness, the nature of boundaries conveys a sense of limitation. And certainly limitations characterize many of the existing international legal rules for cyberwar, including the prohibition on states using force absent self-defense or UN Security Council authorization. Limitations may also address freedom of action; a duty to assist, for example, limits states (or other actors) by requiring them to take affirmative steps to help victims of cyber threats that they would otherwise be free not to take.

However, international law does not operate by limitation alone; it may perform other functions. As the governance context suggests, law may generate "secondary" rules (or "rules on rules") in addition to the "primary" rules that directly limit or constrain the behavior of states and others. International law may also empower or facilitate behavior without mandating it. The Geneva Conventions, for example, not only require or prohibit conduct, they also facilitate behavior—such as killing combatants—that would otherwise be legally impermissible.[131] In addition, international law may constitute new actors, such as IOs, who are assigned specific functions. They may, for example, provide a forum for tailor-made regulation (e.g., the United Nations Commission on International Trade Law (UNCITRAL)) or undertake managerial responsibilities in overseeing a regulatory regime adopted by states (e.g., the World Trade Organization (WTO), Montreal Protocol).[132] As a product of law,

[128] See nn 131–3 and accompanying text; S Watts, "Combatant Status and Computer Network Operations" (2010) 50 *Virginia Journal of International Law* 392; European Parliament Resolution of March 12, 2014 on the "US NSA surveillance programme, surveillance bodies in various Member States and their impact on EU citizens' fundamental rights and on transatlantic cooperation in Justice and Home Affairs 2013/2188(INI)," March 12, 2014.

[129] See Hollis (n 100) 400–1.

[130] One could, as in the governance context, extend this catalog to include avowedly non-legal sources that function in ways akin to legal norms. See nn 58–9 and accompanying text.

[131] See, for example, Geneva Convention for the Amelioration of the Condition of the Wounded and Sick in Armed Forces in the Field, August 12, 1949, 75 UNTS 31, Articles 3–18. Another example is the right of innocent passage states may—but are not required—to exercise under the law of the sea. UNCLOS (n 36) Article 17.

[132] See, for example, UN Commission for International Trade Law, at <http://www.uncitral.org/uncitral/en/index.html>; Montreal Protocol (n 38); Marrakesh Agreement Establishing the World Trade Organization, April 15, 1994, 1867 UNTS 154. In addition to providing the benefits of a permanent forum, expertise, and agency, IOs and other treaty bodies may generate specific costs, including forum-shopping, boot-strapping, and unforeseen impacts on state power. JE Alvarez, *International Organizations as Law-Makers* (Oxford: Oxford University Press, 2005) 627.

institutions can provide a mechanism for collecting and coordinating information and expertise on a subject such as cyberspace.[133]

In sum, legal boundaries offer a degree of certainty and constraint on state behavior. Analogical reasoning provides a useful—and, at present, necessary—basis for drawing such lines, both to distinguish between legal regimes like those for cybercrime and cyberwar as well as to delineate appropriate boundaries of behavior within each regime. But boundary projects and analogical methods have their limitations—whether in terms of *theory* or functional difficulties with *accuracy, effectiveness* and *completeness*. These limitations are very much on display with respect to international law's approach to cyberwar generally and cyber operations in particular. Given such problems, a re-thinking of the boundaries of law in cyberspace is warranted.

IV. The Duty to Hack in Military Operations? Rethinking IHL's Boundaries in Cyberspace

How should we rethink the boundaries of law in cyberspace? The task is a daunting one. As the discussion of cyberspace governance in section II suggests, theoretical and functional boundary problems counsel against further efforts to decide "who decides." The prospects appear equally dim for some comprehensive resolution of the boundaries of international law for cyberspace generally, or even cyberwar specifically. Certainly, a number of states and scholars have proposed drafting a new, global treaty that would establish tailor-made regulations for cyberwar.[134] But those proposals have gained little traction, whether due to a lack of political will or questions about the utility of any such treaty.[135] Moreover, the very complexity of cyberspace belies some "silver bullet" solution to the present difficulties, weighing against re-thinking the boundaries of law in cyberspace from the top down.

In contrast, there is more to recommend a bottom-up approach. For starters, the problems of fragmentation and the absence of a unifying theory are more distant when it comes to assessing lines drawn by specific rules *within* a particular international legal regime. Re-thinking IHL's rules, for example, does not automatically require resolving the relationship between IHL and international human rights law. As a functional matter, moreover, a specific regime's object and purpose can offer a basis for analyzing which rules provide accurate, effective, and relatively complete regulatory responses. Narrowing the scope of inquiry also lessens the complexity of analysis, allowing for a more focused examination of cyberspace's specific attributes with respect to the rules in question. Assuming some success can be achieved across

[133] See, for example, Alvarez (n 132); Geir Ulfstein, "Treaty Bodies and Regimes," in DB Hollis (ed), *The Oxford Guide to Treaties* (Oxford: Oxford University Press, 2012) 428, 432–42.

[134] OA Hathaway et al, "The Law of Cyber Attack" (2012) 100 *California Law Review* 817, 880; D Brown, "A Proposal for an International Convention to Regulate the Use of Information Systems in Armed Conflict" (2006) 47 *Harvard International Law Journal* 179; Hollis (n 100) 407 (discussing Russian proposals for a treaty banning cyberweapons).

[135] J Lewis, "A Cybersecurity Treaty is a Bad Idea," *US News*, June 8, 2012; CJ Dunlap Jr, "Perspectives for Cyber Strategists on Law for Cyberwar" (2011) 5 *Strategic Studies Quarterly* 82–4; PA Johnson, "Is it Time for a Treaty on Information Warfare" (2002) 76 *International Law Studies* 439.

a number of bottom-up projects, there is the further hope that the resulting experience, expertise, and good faith may lay a foundation for addressing the larger boundary disputes that remain.

Consistent with this volume's focus on cyberwar, I propose launching the bottom-up approach with an example from that area, specifically a re-assessment of IHL rules on weaponry and the conduct of hostilities. The state of the existing rules is (relatively) clear thanks, in no small part, to contributions like the *Tallinn Manual* and the work of ICRC officials.[136] IHL, moreover, benefits from relatively wide-spread acceptance of its core tenets—military necessity and humanity—along with the principles of distinction and proportionality.[137] As a result, there is a base-line from which to compare the functionality of existing rules with my own proposal. Specifically, I believe there is a normative case for having IHL re-cast its existing boundaries over the means and methods of warfare to include a Duty to Hack.

Simply put, the Duty to Hack would require that *states use cyber operations in their military operations when they are the least harmful means available for achieving military objectives.* This duty would thus mandate that—all other things being equal—militaries must employ cyber operations when doing so would generate *no* harm versus alternative means and methods that cause *some* harm. Similarly, cyber operations that cause *some* harm must take priority over alternatives that cause *more* harm.

IHL already imposes a version of this duty—the provisions on precautions—for cyber operations having violent consequences for civilians analogous to pre-existing means and methods of warfare.[138] But this law quickly runs into the same problems facing the use of analogies in drawing other international law lines in cyberspace. IHL-by-analogy exacerbates fragmentation and draws lines that ignore the novel capacities of cyberspace, leading to incoherent and ineffective results given IHL's overarching principles.[139]

In contrast, the Duty to Hack could alleviate (but not eliminate) the pressures of fragmentation and provide a clear rule for calibrating military necessity and humanity. Doing so would not, of course, be without costs or limits (in particular any Duty to Hack *must be* linked to the achievement of military objectives). Nonetheless, I submit

[136] Some, however, may contest that the content of these rules is sufficiently settled to allow for re-thinking; for such critics, a re-thinking project may be premature.

[137] See, for example, *Tallinn Manual* (n 4) Rule 20 ("Cyber-operations executed in the context of an armed conflict are subject to the law of armed conflict"); Rule 31 ("The principle of distinction applies to cyber attacks").

[138] See, for example, Protocol Additional to the Geneva Conventions of August 12, 1949, and relating to the Protection of Victims of Armed Conflict (Protocol I), June 8, 1977, 1125 UNTS 3, Article 57(a)(ii) (hereinafter, "AP I"). Much, but not all, of AP I is considered customary international law. See JM Henckaerts and L Doswald-Beck, 1 *Customary International Humanitarian Law* (ICRC, 2005) (hereinafter "*ICRC Study*"); MJ Matheson, "The United States Position on the Relation of Customary International Law to the 1977 Protocols Additional to the 1949 Geneva Conventions" (1987) 2 *American University International Journal of Law and Policy* 419.

[139] For discussion of IHL's overarching principles, see Schmitt (n 11) 802–3; International Committee of the Red Cross, *International Humanitarian Law and the Challenges of Contemporary Armed Conflicts* (ICRC, 2011) 36–8.

that adopting the Duty to Hack may better adapt IHL to cyberspace; it may even allow cyberspace to impact IHL in return.

(a) IHL-by-analogy: the law on weaponry and the conduct of hostilities

Assuming some form of armed conflict to which IHL applies, the fact that IHL lacks tailor-made rules for cyber operations is no bar to its application. IHL does not apply the *Lotus* principle (i.e., what international law does not prohibit, it permits).[140] Rather, the Martens Clause ensures that the absence of a specific prohibition does not mean new means and methods of warfare are automatically permitted.[141] Indeed, IHL's primary boundary lines involve distinguishing prohibited conduct from conduct that is authorized under certain conditions (but not required). For purposes of the Duty to Hack, two sets of requirements are relevant: (1) the law of weaponry; and (2) the conduct of hostilities.

The Law of Weaponry.

IHL attempts to balance the principle of military necessity (i.e., to secure the enemy's complete submission as soon as possible) and humanitarian considerations.[142] It does so through two prohibitions on certain means and methods of warfare.[143] First, IHL prohibits weapons "of a nature to cause superfluous injury or unnecessary suffering."[144] This rule prohibits weapons whose nature will necessarily cause more suffering or injury than another that offers the same or similar military advantage.[145] Second, IHL prohibits "indiscriminate" weapons.[146] Indiscriminate weapons include those incapable of distinguishing military and civilian targets as well as those whose harmful effects cannot be controlled and thus may spread among civilians and civilian objects.[147] The

[140] *The SS Lotus* (*France v Turkey*) [1927] PCIJ Ser A, no 10, 18–19.

[141] The clause, named after famed Russian international lawyer, Friedrich Martens, first appeared in the preamble to Hague Convention II with Respect to the Laws and Customs of War on Land of 1899. It has continued to appear in subsequent IHL treaties, including the Geneva Conventions and their Additional Protocols. See, for example, AP I (n 138) Article 1(2); Geneva Convention (IV) Relative to the Protection of Civilian Persons in Time of War, August 12, 1949, 75 UNTS 287, Article 158.

[142] See Schmitt (n 11); GS Corn et al, "Belligerent Targeting and the Invalidity of a Least Harmful Means Rule" (2013) 89 *International Law Studies* 536, 543 (noting military necessity cannot itself overcome IHL restrictions).

[143] "Means" refers to the weapon or instrument used to conduct an operation while "method" refers to how the operation is conducted. See *Tallinn Manual* (n 4) Rule 41.

[144] AP I (n 138) Article 35(1); *ICRC Study* (n 138) 237 (rule 70); see also W Boothby, *Weapons and the Law of Armed Conflict* (Oxford: Oxford University Press, 2009) Ch 5. This rule only reaches unnecessary suffering by combatants, members of organized armed groups, and civilians directly participating in hostilities; other individuals are immune from any attack. *Tallinn Manual* (n 4) 143.

[145] Dinniss (n 82) 254–5; *Tallinn Manual* (n 4) 143–4 (noting means or methods violate the prohibition if they necessarily will cause unnecessary suffering "regardless of whether it was intended to do so"). Where a weapon targets personnel, the ICJ has emphasized it must not "cause a harm greater than that unavoidable to achieve legitimate military objectives." *Legality of the Threat or Use of Nuclear Weapons* (Advisory Opinion) [1996] ICJ Rep [78].

[146] AP I (n 138) Article 51(4)(b)–(c); *ICRC Study* (n 138) Rules 12, 71; Boothby (n 144) Ch 6.

[147] *Tallinn Manual* (n 4) 144–5 (Rule 43).

harm, however, must involve some injury, loss of life, or damage requiring repair or replacement of physical objects.[148] States must review the weapons they acquire or use to ensure they comply with these and other IHL rules.[149] IHL further prohibits indiscriminate uses of otherwise discriminate means and methods of warfare.[150]

Which cyber operations constitute means and methods of warfare? Here again, the category is drawn by analogy. The *Tallinn Manual* defines cyber means as those weapons and weapon systems used, designed or intended to be used in an "attack." An "attack" involves "acts of violence," which are defined in the cyber context by analogizing to those acts previously regarded as violent, that is, not just violent acts releasing kinetic force but acts producing violent consequences such as those producing injury, death, destruction, or damage.[151] Damage, in particular is cast in terms of physical damage by analogy to its operation outside cyberspace. Thus, a majority of the *Tallinn Manual*'s authors accepted the view that a loss of functionality constitutes damage if it "requires replacement of physical components" but not if functionality could be restored by reinstalling the operating system.[152] As a result, a cyber operation not expected by design or use to generate damage akin to those previously falling within the "attack" concept (even if it produces "large-scale adverse consequences") likely lies outside the existing regulatory boundaries for means and methods of warfare.[153]

If a cyber operation is not an attack it cannot, by definition, constitute a means or method of warfare. It may thus have indiscriminate effects (e.g., the Stuxnet worm's world-wide infection of SCADA systems, albeit without triggering the payload targeting Iran's facilities). Moreover, such cyber operations need not undergo a legal review. Indeed, the US Army reportedly limits its legal reviews of new cyber capabilities to those "developed with the expectation or intent that its use will result in death, injuries, or damage, or destruction of property."[154]

The Conduct of Hostilities.

When it comes to the actual conduct of cyber operations, IHL provides three core sets of rules; those on (1) distinction; (2) proportionality; and (3) precautions.[155]

[148] *Tallinn Manual* (n 4) 145; see also n 87 and accompanying text (regarding the threshold for "damage").

[149] *Tallinn Manual* (n 4) 145 Rule 48; Geneva Convention (IV) (n 141) common Article 1. In addition, AP I requires states to study new weapons *and* new methods of warfare to determine that they are not prohibited by the Protocol or other international obligations, although the customary status of that rule is unsettled. *Tallinn Manual* (n 4) 154 (commentary to Rule 20).

[150] AP I (n 138) Article 51(4)(a); *ICRC Study* (n 138) Rules 11–12. The law of weaponry also prohibits specific means and methods of warfare such as biological, chemical and blinding laser weapons. C Greenwood, "The Law of Weaponry at the Start of the New Millennium," in M Schmitt and L Green (eds), *The Law of Armed Conflict: Into the Next Millennium* (Naval War College, 2012).

[151] *Tallinn Manual* (n 4) 141–2; see also nn 90–1 and accompanying text (discussing *Tallinn Manual* definition of "cyber attack").

[152] *Tallinn Manual* (n 4) 108–9. [153] *Tallinn Manual* (n 4) 108–9.

[154] P Walker, "Organizing for Cyberspace Operations: Selected Issues" (2013) 89 *International Law Studies* 341, 348. In contrast, the US Air Force reviews "all new, and newly-modified, cyber capabilities." Walker, 348–9.

[155] According to AP 1, proportionality is considered part of precautions. See AP I (n 138) Article 57. I choose to separate them here, however, just as the *Tallinn Manual* does. See *Tallinn Manual* (n 4) Rule 51 (Proportionality) and Rules 52–9 (Precautions). It is also worth emphasizing that, although these

Distinction is one of IHL's "cardinal principles," obligating states to distinguish civilians from combatants and civilian objects from military ones.[156] Such distinctions, in turn, enable states to implement the core of the principle—never deliberately attack civilians and civilian objects.[157]

Proportionality involves IHL's limits on how attacks impact civilians; it prohibits "an attack which may be expected to cause incidental loss of civilian life, injury to civilians, damage to civilian objects, or a combination thereof, which would be excessive in relation to the concrete and direct military advantage anticipated."[158] The principle is essentially a holistic equation, comparing the quantum of effects on civilians and their objects in relation to the expected military advantage and asking if the former is "excessive" given the latter.[159] Thus, "military advantage" includes not just suppression of a specific target, but other "concrete and direct" advantages as well (e.g., the benefits of the operation as a whole, not simply the attack in question) in the given circumstances.[160] At the same time, states must consider all qualifying incidental harm or "collateral damage" expected from an attack, whether as a direct result of the attack itself or via foreseeable indirect "knock-on" effects.[161] Not all harms qualify as collateral damage, however; only those implicated by the "attack" definition require attention—"inconvenience, irritation, stress, or fear" do not qualify, nor does a decline in civilian morale.[162]

Precaution refers to measures states must take in carrying out military operations and attacks.[163] For all military operations (i.e., not just attacks) IHL requires "constant care...to spare the civilian population, civilians and civilian objects."[164] For plans or decisions to attack, IHL requires compliance not only with the distinction and proportionality principles[165] but also (1) taking all feasible precautions in

are three of the most important IHL rules, IHL has additional requirements for attacks, not to mention treatment of combatants.

[156] *Nuclear Weapons Case* (n 145) para 78 (Higgins, J, dissenting); AP I (n 138) Article 48; *ICRC Study* (n 138) 3–8.

[157] AP I (n 138) Articles 48, 51(2), 52(2); *ICRC Study* (n 138) Rule 14.

[158] AP I (n 138) Article 51(5)(b); see also AP I Article 57(2)(ii); Dinniss (n 82) 205–8; *Tallinn Manual* (n 4) Rule 52 and Commentary; ET Jensen, "Cyber Attack: Proportionality and Precautions in Attack" (2013) 89 *International Law Studies* 198, 204–5.

[159] Corn et al (n 142) 547. Although the AP I Commentary suggests a ceiling on civilian harm that would automatically violate proportionality, most commentators view the principle as essentially relational. ICRC, *Commentary on the Additional Protocols of 8 June 1977 to the Geneva Conventions of 12 August 1949* (Martinus Nijhoff, 1987) para 1980 (hereinafter "*ICRC Additional Protocols Commentary*"); *Tallinn Manual* (n 4) 161. Thus, extensive civilian harm may not violate proportionality if the expected military advantage is great; the test is contextual and examines the reasonableness of the expected advantage and expected harm. See *Prosecutor v Stanislav Galić*, Case No IT-96-21-T, Trial Chamber Judgment (ICTY) (November 16, 1998) para 58.

[160] *Tallinn Manual* (n 4) 161–2. The advantage, however, must be "military" in nature; "exclusively economic, political, or psychological" advantages are not included in the calculus. *Tallinn Manual* (n 4) 131. In addition, whether military advantage accounts for sparing or maintaining one's own forces and capabilities is disputed. *Tallinn Manual* (n 4) 164.

[161] Dinniss (n 82) 207; *Tallinn Manual* (n 4) 160–1.

[162] See *Tallinn Manual* (n 4) 133, 160.

[163] IHL requires additional precautionary defensive measures against attacks. AP I (n 138) Article 58.

[164] AP I (n 138) Article 57(1).

[165] AP I (n 138) Article 57(2)(a)(i) and (iii). Attacks must be canceled or suspended if it becomes apparent that either one of these principles will be violated. AP I (n 138) Article 57(2)(b).

choosing a means and method of warfare to minimize "incidental loss of civilian life, injury to civilians, and damage to civilian objects"[166] and (2) selecting military objectives "expected to cause the least danger to civilian lives and to civilian objects" in cases where "a choice is possible between several military objectives for obtaining a similar military advantage."[167] Thus, even if an attack is proportional, IHL still requires minimizing collateral damage in planning what to attack and by what means and methods to do so.

The key criterion for precautionary measures is "feasibility," which is widely understood to require measures "that are practicable or practically possible, taking into account all the circumstances ruling at the time, including humanitarian and military considerations."[168] Thus, although there is no general obligation to employ precision capabilities in airstrikes, IHL may require using such capabilities over less precise means or methods when such capabilities are available and could minimize civilian harm.[169]

Two common themes stand out in IHL's existing requirements on the conduct of hostilities when extended to cyber operations. First, like the law of weaponry, IHL is generally regarded to only require distinction, proportionality, and precautions with respect to cyber operations that qualify as *attacks*.[170] IHL does not prohibit targeting or even harming civilians or civilian objects in a cyber operation so long as its effects are not analogous to those previously crossing the attack threshold (i.e., injury, death, destruction, or damage). The scope of IHL's precautions are similarly qualified in terms of attacks, meaning that military calculations need not consider non-attack cyber capabilities in choosing weapons, methods, or objectives for attacks.[171]

[166] AP I (n 138) Article 57(2)(a)(ii).

[167] AP I (n 138) Article 57(3). The customary international law status of this rule is disputed by some. See, for example, *Tallinn Manual* (n 4) 171. For additional precautionary requirements, see, generally, AP I (n 138) Article 57.

[168] See *Tallinn Manual* (n 4) 168; Dinniss (n 82) 211; Jensen (n 158) 209. In lieu of feasibility, a lower standard—"all reasonable precautions"—applies to attacks with effects only at sea or in the air. AP I (n 138) Article 57(4); *ICRC Additional Protocols Commentary* (n 159) 687–8.

[169] See CJ Markham and MN Schmitt, "Precision Air Warfare and the Law of Armed Conflict" (2013) 89 *International Law Studies* 669, 694. The requirement is not reciprocal; the fact that an enemy does not have the same capability does not alleviate the requirement to undertake all feasible precautions or choose a different objective. Markham and Schmitt, "Precision Air Warfare and the Law of Armed Conflict," 694.

[170] The one exception—the constant care requirement—remains underdeveloped; it applies only during hostilities and requires continuous respect or sensitivity to how military operations may affect civilians and their objects "to avoid unnecessary effects on them." *Tallinn Manual* (n 4) 166. Such aspirations comport with my Duty to Hack, but the current formulation is certainly looser and any practical examples of its operation or effectiveness are difficult to locate.

[171] *Tallinn Manual* Rule 54, for example, requires that "Those who plan or decide upon a cyber attack shall take all feasible precautions in the choice of means or methods of warfare employed in such an attack." *Tallinn Manual* (n 4) 168. By its terms, therefore, the rule is limited to cyber attacks, which Rule 30 defines in terms of those having violent consequences. *Tallinn Manual* (n 4) 106–7; nn 151–3 and accompanying text. Elsewhere, in discussing dual-use objects, the *Manual* suggests Rule 54 might cover cyber operations that do not qualify as attacks. *Tallinn Manual* (n 4) 136 (suggesting that, if feasible, IHL would require using an operation "to degrade, deny, disrupt, or alter" global positioning satellite signals via "cyber means instead of conducting an operation that rises to the level of an attack (and that causes collateral damage)"). However, it does so without explanation. As such, it is difficult to reconcile with the text of Rule 54 itself. This outcome would, in contrast, be required under the Duty to Hack I propose in sub-section (c).

Second, IHL's limitations on attack have an entirely *civilian orientation*. The restrictions on attacks in distinction, proportionality, and precautions all focus on limiting injury and death to civilians and destruction of, or damage to, civilian objects. When it comes to combatants and military objectives, however, the goal of military necessity prevails—seeking the enemy's complete submission as soon as possible. Recent proposals that IHL requires, when feasible, the capture rather than killing of enemy combatants generated an academic uproar, with the prevailing view denying such an obligation extends beyond those combatants rendered *hors de combat*.[172]

Similarly, IHL only protects civilian objects from damage and destruction. Civilian objects are defined as all objects that are not military objectives.[173] What constitute military objectives? IHL defines them in terms of "objects" which by their (1) nature, (2) geographic location, (3) future purpose, *or* (4) use "make an effective contribution to military action and whose total or partial destruction, capture or neutralization, in the circumstances ruling at the time, offers a definite military advantage."[174] Only "visible and tangible" things qualify as objects, leading a majority of the *Tallinn Manual*'s authors to deny that "data" resident on computers and computer networks are objects (unlike the computers and other tangible cyber infrastructure that contain such data).[175] Thus, neither IHL's protection of civilian objects nor its authority to attack military objectives apply to operations altering, disrupting, or deleting data unless they have indirect effects on some other objects in the process.[176]

With respect to cyber operations, moreover, the definition of military objects has one additional, important qualifier. IHL treats all "dual use" objects—objects used for both civilian and military purposes—as "military objects." For example, IHL treats an electrical grid supplying power to both military bases and civilian infrastructure as a military object, and thus subject to attack. This reasoning suggests IHL legitimizes attacks against much of cyberspace. Militaries regularly use the internet and information technology (IT) (e.g., Global Positioning Systems (GPS)). As such, these technologies qualify as military objects even if their civilian use is disproportionately greater. Treating cyberspace as

[172] For proponents of the "capture-but-do-not-kill" position, see, for example, ICRC, *Interpretive Guidance on the Notion of Direct Participation in Hostilities under International Humanitarian Law* (ICRC, 2008) 81–2; R Goodman, "The Power to Kill or Capture Enemy Combatants" (2013) 24 *European Journal of International Law* 819. For opposition, see, for example, Corn et al (n 142); MN Schmitt, "Wound, Capture, or Kill: A Reply to Ryan Goodman's 'The Power to Kill or Capture Enemy Combatants'" (2013) 24 *European Journal of International Law* 855; JD Ohlin, "The Duty to Capture" (2013) 97 *Minnesota Law Review* 1268, 1272–3; B van Schack, "The Killing of Osama Bin Laden and Anwar Al-Aulaqi: Uncharted Legal Territory" (2012) 14 *Yearbook of International Humanitarian Law* 255, 292.

[173] See AP I (n 138) Article 52(1) ("civilian objects are all objects which are not military objectives").

[174] AP I (n 138) Article 52(2); *ICRC Additional Protocols Commentary* (n 159) para 1979; *ICRC Study* (n 138) Rule 8; *Tallinn Manual* (n 4) 126 (citing military manuals). Objects that "make an effective contribution to military action" generally include "war-fighting" objects (e.g., materials used in combat) and "war-supporting" objects (e.g., factories producing war-fighting objects or infrastructure supporting military operations). Some states (e.g., the United States) also treat "war-sustaining" objects (e.g., an oil export industry that generates revenue to sustain the military) as military objects, but that is a minority view. *Tallinn Manual* (n 4) 130–1.

[175] *Tallinn Manual* (n 4) 127. A minority of the IGE would apply IHL to operations targeting as military objectives valuable and important civilian datasets. *Tallinn Manual* (n 4) 127. Dinniss would include "lines of code" in the definition of an object. Dinniss (n 82) 185.

[176] IHL does, however, protect certain types of data (e.g., personal medical data). *Tallinn Manual* (n 4) 206; *ICRC Study* (n 138) Rule 28.

a military objective means opening it up to attack; although proportionality and precautions must be taken into account in doing so.[177]

(b) Fragmentation and incongruities in IHL for cyber operations

Today, IHL clearly regulates cyber operations and their objectives, albeit indirectly and by analogy. It would be a mistake, however, to accept this lex lata as equivalent to the lex ferenda. Indeed, there are both theoretical and functional problems with the current approach that require further attention. For starters, by treating cyberspace as just another environment for IHL to regulate (and cyber weapons as just the latest in a long line of innovations in weaponry), IHL inadvertently takes a position on the contested question of what cyberspace "is" with all the baggage that question now generates.[178]

The problem of fragmentation also looms large in the IHL project. Explicating when and how IHL governs cyber operations is, if nothing else, a boundary project—mapping cyber operations by whether they do (or do not) fall within the corpus of IHL. And it is understandable that lawyers trained in IHL—whether they work for states, the ICRC, or in the academy—would approach that project via analogies to past applications of IHL. It is essentially a legal version of the "if all you have is a hammer everything looks like a nail" phenomenon.[179] Nonetheless, the IHL-by-analogy approach exemplifies the fragmentation problem identified already. It presupposes the propriety of using analogies to map out IHL's boundaries for cyber operations, without assessing whether stronger analogies emerge from other international legal regimes (e.g., human rights law). This sets the stage for transplanting many of the existing contests over IHL's boundaries with respect to conventional conflicts into cyberspace.[180] Problems with attribution and attempts to extend the idea of "armed conflicts" to stand-alone cyber operations may also generate new tensions with those who would assume other regimes (e.g., criminal law) apply instead. The continuing disagreement over whether Stuxnet falls within IHL's boundaries demonstrates, moreover, that this is not simply a theoretical dilemma.

Problems with IHL-by-analogy, however, do not end with that regime's relationship to the rest of international law; they include the lines drawn by analogy *within* IHL as well. Using analogies to pre-existing operations to delimit the lawfulness of cyber operations cannot be justified to the extent cyber operations are not analogous to anything in the past. As with the *jus ad bellum*, IHL-by-analogy can justify the lines it draws to ensure that previously unlawful behavior (e.g., indiscriminate attacks, excessive civilian losses, choosing weapons without regard to minimizing civilian damage) remains unlawful in the cyber context.[181] And those lines make similar sense when cyber

[177] *Tallinn Manual* (n 4) 134–135; Dinniss (n 82) 194. Questions persist, however, on whether doubt as to the military/civilian status of an object defaults to require its treatment as a civilian object. *Tallinn Manual* (n 4) 137–8.

[178] See section II (c).

[179] Nor is this limited to IHL; other disciplines (e.g., intellectual property, human rights) each approach and conceptualize the borders of law and cyberspace via analogies to their respective, pre-existing disciplines.

[180] See nn 96–9 and accompanying text.

[181] See nn 103–8 and accompanying text.

operations are analogous to behavior that IHL either considered lawful previously (e.g., employing feasible methods to minimize civilian damage) or excluded from its reach (e.g., dissemination of propaganda, psychological operations).

However, what is to be done with cyber operations that have no earlier analogue? Cyber operations may now cause damage, but as Stuxnet reveals, the quality is of an entirely different sort from earlier weaponry, whether kinetic or non-kinetic (e.g., biological or chemical) in nature. Cyber operations can also do things that neither military *nor* political or economic measures could do before; the paradigmatic example is a cyber operation that remotely turns off an electric grid for a period of time without permanent harm to the grid or the personnel who run it.[182] IHL-by-analogy, however, largely excludes such novel cyber capacities from its reach. Military objectives still require the tangible and visible quality akin to previous objectives notwithstanding the transformative and revolutionary utility of data, big, meta-, or otherwise. Cyber operations require no review unless they will cause (directly or indirectly) injury, death, destruction, or damage akin to earlier weaponry. And attacks are defined to only include cyber operations with the same violent consequences.

Incongruous results flow, however, from such analogical reasoning precisely because some of cyber's novel capacities belong within IHL's reach (just as some do not). For example, the bombing of a single house clearly involves the physical damage necessary to label it as an attack, but a temporary disruption of an electrical grid affecting tens of thousands of people via malware, which does not require replacing any physical parts (or even the operating system), would not.[183] For such malware to qualify as an attack, its indirect effects must generate the requisite damage.[184] And while such damage seems reasonably foreseeable from a power outage, this will not always be the case. Disrupting the ability of a stock market, factory, or warehouse to operate may generate no physical damage or injury. And if such hacks are not attacks, then IHL does not require any assessment of the malware's lawfulness as a weapon in terms of unnecessary suffering or discrimination.[185]

[182] See, for example, Dinniss (n 82) 197; Droege (n 82) 539.

[183] See Droege (n 82) 558–9 for the original hypothetical. The *Tallinn Manual* commentary suggests that a majority of the IGE regarded physical replacement after a loss of functionality to qualify as damage for purposes of labeling a cyber operation as an attack, while some of that majority would have regarded the need to restore an operating system as sufficient to do so. But it is possible to conceive of malware that might not even require restoring the operating system; if, for example, the malware would bring the shutdown to an end and wipe itself clean from the system at that time. Accord N Lubell, "Lawful Targets in Cyber-operations: Does the Principle of Distinction Apply?" (2013) 89 *International Law Studies* 252, 266 (noting how a cyber operation damaging a computer system that can be repaired in under an hour by replacing a part would constitute an attack but not "a cyber-operation that incapacitates a whole system for two days" if there is no physical damage or repair "other than waiting for the operation to be over").

[184] See *Tallinn Manual* (n 4) 109 ("a cyber-operation might not result in the requisite harm to the object of the operation, but cause foreseeable collateral damage... Such an operation amounts to an attack").

[185] There is some question whether the principle of distinction covers more than attacks, to include either military operations or hostilities. See, for example, Droege (n 82) 553–5 (reviewing literature); see also MN Schmitt, "Cyber-Operations and the jus in bello: Key Issues" (2011) 87 *International Law Studies* 91 (distinction, proportionality and precaution rules extend only to attacks); Dinniss (n 82) 196–202 (extending IHL rules on conduct of hostilities to "military operations"); N Melzer, *Cyberwarfare and International Law* (UNIDIR Paper 2011) 28 (extending IHL rules to hostilities). For purposes of the present analysis, I assume IHL's rules extend only to attacks, although the Duty to Hack would broaden the reach of IHL's core principles in similar (although not identical) ways.

Nor would IHL's principle of distinction apply; the military or civilian status of the stock market, factory, or warehouse would not play into any decision to target it.[186]

Perhaps most ironically, if shutting down factories or warehouses via cyber means do not comprise attacks and a military *did* want to actually attack them (say, as war-supporting objectives), IHL does not require that military to deploy the malware in lieu of the attack no matter how feasible it might be to do so.[187] Instead, IHL appears to authorize attacks—kinetic or otherwise—that cause physical damage and loss or injury of human life so long as they compare favorably to losses from other potential "attacks."[188] Assuming malware and a kinetic attack could achieve the same or similar military advantage in such cases, this result flies in the face of the humanity value at the heart of IHL.

Other scholars have noted the incongruities I describe and have sought to try and wrestle the analogies into a more coherent form. Thus, Knut Dormann has recommended regarding cyber operations that produce the "neutralization" of a military objective as attacks even if no damage results.[189] Cordula Droege, meanwhile, flags a criterion of inconvenience to differentiate cyber operations that are not attacks from those that are, while acknowledging a more tailor-made approach may need to emerge from state practice.[190] But such efforts remain nascent at present and the prevailing view continues to bound IHL's definitions of damages and attacks by analogy.

All of this of course presupposes that having IHL draw lines will constrain the conduct of conflicts in or otherwise involving cyberspace. But even if states followed all the directives of IHL-by-analogy, it would do little to stem the vast, vast majority of existing cyber threats, namely cybercrime and cyber espionage. Of course, IHL's ambitions are narrower, namely to regulate "armed conflicts." As already noted, however, it will not always be clear what cyber operations constitute "armed conflicts" where they cannot be attributed to a state or actors under a state's control or where the quantum of harm is not analogous to prior armed conflicts (e.g., Stuxnet).[191] Such uncertainty undermines IHL's effectiveness. It is not clear that states will follow rules where any compliance (or non-compliance) can be hidden. And even if states were

[186] No proportionality analysis would be necessary either, because such a hack would not involve any incidental injury, death, destruction, or damages. The only limitation involves the requirement that constant care be taken in such operations for their civilian effects. See n 170.

[187] This assumes that precautionary measures are only required for attacks. For alternative views see n 185. For purposes of this hypothetical, assume the second state has met its obligations to take precautions against the effects of attacks. AP I (n 138) Article 58.

[188] The same scenario might befall a stock market that could be cataloged as a military objective if war-sustaining objectives constitute military objectives. See n 174.

[189] K Dormann, *Applicability of the Additional Protocols to Computer Network Attacks* (ICRC, 2004) 4 (relying on the definition of military objective in AP I Article 52(2) as including one "whose total or partial destruction, capture or *neutralization*, in the circumstances ruling at the time, offers a definite military advantage" (emphasis added)).

[190] Droege (n 82) 556–0.

[191] See, for example, nn 100–2 and accompanying text (re attribution); nn 106–8 and accompanying text (re Stuxnet as an armed conflict). Whether a state is required to exercise "overall" or "effective" control is disputed. Compare *Case concerning Military and Paramilitary Activities in and against Nicaragua (Nicaragua v United States)* (Merits, Judgment) [1986] ICJ Rep 14, 64–5 [115] (test is "effective control") with *Prosecutor v Dusko Tadic aka "Dule"* (Judgment) ICTY, No IT-94-1-A (July 15, 1999) [131], [145] (test is "overall control").

inclined to comply with IHL, the ambiguity in its application may lead to inconsistent or even conflicting behavior in doing so.

Taken together, the current problems with the IHL-by-analogy approach suggest it is, at best, an incomplete regulatory response, a situation only exacerbated by the rate at which the militarization of cyber capabilities are expanding. But what alternatives can the law offer? Obviously, one solution would be a tailor-made effort (by practice or agreement) to clarify the boundaries IHL currently draws; that is, to settle which cyber means constitute weapons, what operations constitute attacks, and which cyber resources constitute military objectives. But, there is nothing inherent in IHL that requires blind adherence to its existing boundaries. To the extent cyberspace presents novel capacities, it can—and perhaps must—adopt equally novel regulatory responses to accommodate IHL's core principles of military necessity and humanity. The existing rhetoric of IHL draws boundaries around prohibitions, leaving outside behavior that is authorized only by implication.[192] It is worth asking, however, whether IHL can—and should—accommodate a different rhetoric where the boundary marks a duty, thus drawing a line between what is required and what is merely permitted. The Duty to Hack idea adopts this latter formula.

(c) A duty to hack in military operations?

My Duty to Hack proposal is straightforward: IHL should require states to use cyber operations in their military operations when they are expected to be the least harmful means available for achieving military objectives. This duty departs from the current law in two key respects.

First, it removes the "attack" threshold on precautionary measures. Current law only requires choosing among means and methods and objectives of an "attack" on a military objective to minimize civilian injury, death, destruction, or damage. In doing so, IHL presumes that the only way to achieve a military objective (and the definite military advantage it offers) is via an attack that directly or indirectly causes physical harm. But the novel and wide-ranging capacities of cyber operations unsettle such a presumption. A cyber operation may be able to achieve a military objective (e.g., shutting down a factory for some desired period of time) without causing *any* physical harm. Rather than leave such cyber operations outside the requirements of precaution because they do not meet the definition of an "attack," a Duty to Hack would require that they be part of any choice in means, methods, and objectives. Thus, a cyber operation that can achieve a particular military objective without an attack should be required in lieu of any "attack" on that same objective, by other means or methods, whether cyber, kinetic, or non-kinetic in nature.[193] Similarly, in choosing between

[192] *ICRC Additional Protocols Commentary* (n 159) 689 ("The law relating to the conduct of hostilities is primarily a law of prohibitions: it does not authorize, but prohibits certain things").

[193] The fact that cyber operations may cause no harm or harm qualitatively different than pre-existing violent consequences differentiates my argument from earlier claims that IHL should require the use of precision-guided munitions. Precision-guided weapons may cause less violent consequences than earlier weapons but, in contrast to cyber operations, even the most precise weapon must still cause *some* violent consequence.

two military objectives offering a similar military advantage, the cyber operation that achieves its objective without an attack must be selected over means and methods for achieving another military objective via an attack. In other words, so long as the military objective is achievable, the Duty to Hack requires employing cyber operations generating *no* physical harm over those means and methods of warfare that, by definition, must generate *some* physical harm.

Second, the Duty to Hack addresses all forms of physical harm from cyber operations, not just those of a *civilian* character. Existing IHL—distinction, proportionality, and precautions—only require efforts to avoid, limit, or minimize *civilian* harm. Absent the harmful civilian impacts protected by these and other IHL rules, militaries are free to employ destructive and lethal force against military objects and belligerents. This approach furthers military necessity—complete submission of the enemy as soon as possible—and makes sense where military objectives were usually military in character and dual-use objects qualified as military objects only on occasion.

However, the nearly universal dual-use conditions in cyberspace now means that most cyber operations will employ or target as military objects, objects that are also—often primarily—civilian in nature, use or purpose. As military objects, dual-use objects may be attacked (and damaged or destroyed) without regard to any questions of distinction, proportionality, or precautions vis-à-vis those objects themselves; IHL's cardinal principles only apply with respect to any collateral damage to civilians or civilian (i.e., non-dual use) objects that result from such attacks. Thus, in cyberspace, the fundamental goal of distinction—to protect the civilian population and their objects—is likely undercut by the specific iteration of IHL-by-analogy. One way to mitigate this problem is to require more careful segregation of what exactly constitutes the military object in cases where it is situated within or among civilian objects.[194]

A Duty to Hack, however, takes a different, and simpler, approach. In military operations, it requires using cyber operations that cause the least harm to achieve a military objective. For example, assuming disruption of Iran's nuclear processing plant was a lawful military objective, the prospect of deploying Stuxnet to achieve that objective would take priority over doing so by an airstrike if that airstrike—even a precise one—would foreseeably involve greater risks of injury, death, damage, or destruction than spinning centrifuges out of control periodically. Similarly, where shutting down an electric grid (or a factory, or an airport, or some portion of the internet itself) is a lawful military objective and the indirect effects of doing so are not excessive, the Duty to Hack requires using a cyber operation to achieve that objective in lieu of traditional means like bombs or missiles where the latter may be expected to cause greater harm overall via direct effects even if the indirect effects of both operations are identical.

[194] See *Tallinn Manual* (n 4) 135 (treating military servers in server farm as distinct from civilian servers such that an attack on the facility's cooling system must employ proportionality and precautions in its effect on those civilian servers).

To be clear, the Duty to Hack is not designed to impact military necessity. The duty only applies to those *available* cyber operations expected to *achieve military objectives*. States are not required to acquire new cyber capacities if they do not already have them nor create ones that do not yet exist.[195] Moreover, the Duty does not mandate deploying any cyber operation—irrespective of whether and how much harm it may generate—if that operation cannot achieve the set objectives. The plural—objective*s*—is significant; it requires conceptualizing a cyber operation's capacities in light of military operations as a whole, rather than any individual operation or attack.

Thus, the Duty to Hack is impacted by desired objectives; what constitutes the least harmful means can—and should—vary depending on the specific objective (i.e., to disrupt, to damage, to destroy, and so on) and its relationship with other military objectives. Even where the objective is total destruction of a military object, however, the Duty to Hack may still mandate using a cyber operation if doing so will generate fewer indirect harms to civilians or even other military objects (many of which should be dual use) than available alternatives. In doing so, therefore, the Duty to Hack is designed to preserve the principles of distinction and proportionality as they exist; IHL would continue to prohibit direct attacks on civilians and their objects by cyber operations or otherwise, just as any military operation that does constitute an attack must not generate excessive civilian harm. Nor would the Duty to Hack override the requirement to comply with the principles of discrimination and avoidance of unnecessary suffering when it comes to developing or deploying cyber operations.

How would a duty to hack compare to the prevailing difficulties with IHL-by-analogy? Quite well it turns out. A Duty to Hack adopts a middle-ground approach to what cyberspace "is" and thus what type of law it requires. The Duty privileges *some* cyber operations where the capacities are truly novel while preserving IHL-by-analogy when they are not. Similarly, it tries to accommodate the difficulties cyberspace poses for the civilian/military distinction given the extensive range of dual-use objects, without necessarily dispensing with it altogether.

When it comes to fragmentation, the Duty to Hack adopts a more constructive approach as well. IHL-by-analogy has tended to treat cyberspace as a one-way street, where the only question is how does IHL map onto cyberspace and what rules does it apply there.[196] In contrast, the Duty to Hack is constructed not just in light of what IHL has to say about cyberspace but what cyberspace may have to say about IHL. As a result, this project may not just re-cast IHL's boundaries in cyberspace but IHL's own boundaries within international law. Doing so could limit carrying over the continuing tensions between IHL's boundaries geographically (e.g., the controversy over

[195] In this respect, the Duty to Hack mirrors existing IHL with respect to certain precision capacities in the right context. See Markham and Schmitt (n 169) 687.

[196] R Geiß and H Lahmann, "Cyber Warfare: Applying the Principle of Distinction in an Interconnected Space" (2012) 45 *Israel Law Review* 381, 382 ("legal discussions have focused primarily on the question of when the laws of war are applicable in relation to military cyber-operations... this has been a line-drawing exercise").

transnational armed conflicts) or across regimes (e.g. its relationship to human rights) into cyberspace.[197]

More specifically, the Duty to Hack attempts to engage more accurately with the similarities *and* differences cyberspace poses for armed conflicts. Instead of drawing artificial lines according to some pre-existing concepts of weapons, attacks and military objects, it tries to draw boundaries in light of the capacities available in cyberspace consistent with IHL's overarching goal of balancing military necessity and humanity. It recognizes that this balance is not fixed or static, but can change with the times, and that new capacities—particularly those as novel as cyberspace offers—may require recalibrating the balance. A Duty to Hack does this by requiring the most humanitarian means available—those least harmful to civilians and their objects (even if they may also bear the "military" label)—to achieve objectives the military identifies as necessary to achieve its goal(s) whether it is the enemy's complete submission or some other change in its behavior.

But could a Duty to Hack actually constrain state (let alone non-state actor) behavior? To be sure, nothing in such a duty will fix the technical problems with attribution nor will it alleviate ongoing uncertainty about IHL's scope or meaning. Indeed, by limiting it to military operations, it continues to ignore the large swath of cyberthreats actually occurring today.[198] Yet, the Duty to Hack has the potential to be more effective than the status quo. It offers a more accurate line for implementing IHL's balance of military necessity and humanity and a more complete response than IHL-by-analogy to cyberspace's unique characteristics.

In any case, the Duty to Hack's effectiveness need not turn solely on its rationality, let alone reciprocity.[199] Its very formulation as a "duty" is designed to convey the potential that exists in IHL rules for self-enforcement.[200] The Duty to Hack aspires to take advantage of IHL's roots in professionalism; in obtaining compliance with rules by having militaries internalize those rules as part of their very identity.[201] This is particularly important in the cyber context. The difficulties of attributing responsibility for cyber operations make it difficult for victims or third parties to observe—let alone enforce—compliance with IHL. If internalized over time by states and militaries, however, the Duty to Hack could generate such compliance where states implement it regardless of whether their behavior is attributed to them by other states or actors.

[197] An alternative approach would be to sketch out the legal regimes applicable to cyber operations beyond IHL and assess how they inter-relate. For example, MN Schmitt, "'Below the Threshold' Cyber-operations: The Countermeasures Response Option and International Law" (2014) 54 *Virginia Journal of International Law* (forthcoming).

[198] A more radical Duty to Hack proposal might extend its application beyond military operations (which require the existence of an armed conflict and hostilities) to cover all government-sponsored cyber operations under a least-harmful means rule. The relative costs and benefits of such a proposal require additional study, and, thus, I do not advocate for it here.

[199] See Hollis (n 100) 418.

[200] Indeed, IHL has its origins, among other things, in the Lieber Code—a set of regulations for Union forces in the Civil War designed to be self-regulating. See R Hartigan (ed), "General Orders 100," in *Lieber's Code and the Law of War* (Chicago: Precedent, 1983).

[201] Or, states may comply with international rules because they have become acculturated to them, instead of being persuaded or coerced into doing so. See R Goodman and D Jinks, *Socializing States—Promoting Human Rights through International Law* (Oxford: Oxford University Press, 2013).

(d) The benefits of a duty to hack would outweigh its limitations and costs

In appreciating the potential of the Duty to Hack, it is equally important to accept its limitations. Just as the technology may make new, less harmful, military operations possible, so too will it constrain whether, when, and how often cyber operations can do so. For all the hypotheticals about what cyber operations *can* do, there is not yet proof of concept behind each and every variation. Unlike the science-fiction charges first levied at the cyberwar discourse in the 1990s, there are certainly now concrete examples demonstrating what cyber operations can do in principle (causing a generator to self-destruct) and in the wild (Stuxnet). But, one may need to approach with some skepticism the idea that if one can conceive of it, it can be done in cyberspace.[202]

i. *Logistical challenges and opportunities in the Duty to Hack*

Assuming the capacity exists, actually crafting and deploying a cyber operation is a very complicated process. Time and money are required (which may be relatively quick and cheap or lengthy and expensive depending on the point of comparison).[203] Three logistical hurdles loom large: (1) finding a path to *access* the target; (2) identifying a *vulnerability* in the target system; and (3) exploiting that vulnerability to deliver a *payload* that performs the desired operation.[204] Although Stuxnet demonstrated that internet access is not required to access a target, computers and computer networks that are air-gapped from the internet do pose access issues (requiring an insider or some proximity to the target to proceed).[205] Vulnerabilities, meanwhile, whether by accident or design, are the entry point to get into the targeted system or network itself. And while some vulnerabilities are inevitable given the millions of lines of code employed in information technology, the most effective ones are hard to come-by. "Zero-day exploits"—previously unknown exploits—can be costly to acquire and use precisely because once a vulnerability becomes known patches and defenses may be developed.

Finally, there is the payload itself, which may require tailoring to the specific target (as Stuxnet did to ensure it only disrupted certain Iranian SCADA systems and not similar SCADA systems world-wide). As a result, cyber operations and the Duty to Hack they involve will only implicate military objectives where the operation can be *planned in advance* or where some *pre-positioning* has occurred (e.g., a rootkit) that allows the cyber operation to be deployed on-call. Unplanned or unanticipated targets

[202] See, for example, T Rid, "Think Again: cyberwar" (2012) *Foreign Policy* 5 et seq; Droege (n 82) 539–40.

[203] A cyber operation will likely cost less than a ballistic missile but more than the $22,000 Joint Direct Attack Munition (JDAM) used to give unguided bombs precision capabilities. See, for example, Markham and Schmitt (n 169) 673 (describing JDAMs). Under IHL, however, the expenditure or savings associated with using a particular means or method of warfare should be irrelevant to principles of precaution and proportionality. Markham and Schmitt (n 169) 682–8.

[204] Lin (n 118) 517–18.

[205] But see DE Sanger and T Shanker, "N.S.A. Devises Radio Pathway Into Computers," *The New York Times*, Jan 14, 2014.

will rarely be amenable to cyber operations and thus the Duty to Hack will not present itself.

Although challenging, the logistics of developing cyber operations does not undercut the Duty to Hack idea. One of the features of cyberspace is the cat-and-mouse game between offensive and defensive capacities. Thus, as defenses emerge to block or remedy old cyber operations, new types of cyber operations will inevitably emerge. Given the requirement that a Duty to Hack only applies where it is expected to be capable of achieving a military objective, this feature of cyberspace should not deter its implementation. On the contrary, one thing the Duty to Hack might do is incentivize particular types of cyber operations, namely those that are less harmful than earlier cyber (or kinetic) alternatives.

Deploying cyber operations introduces a separate set of risks. The inter-connected nature of the internet facilitates dissemination in unheralded ways. As a result, malware of various types can quickly be distributed world-wide. Without careful planning, cyber operations may be indiscriminate or cause harms greater than expected or necessary to achieve the military objective. And even with planning, mistakes can still happen. Many assume, for example, that Stuxnet's indiscriminate distribution to civilian nuclear SCADA systems world-wide was not a design feature of the operation, but an unexpected consequence. The Moscow Theatre incident also serves as a poignant reminder that even non-violent means can be deployed with unexpected lethality.[206]

The fact that cyber operations may spread indiscriminately and may cause unexpected harms should not, however, forestall the adoption of the Duty to Hack. Such risks must be counter-balanced by the potential of cyber operations to cause less harm than existing alternatives. Moreover, although only cyber operations expected to cause physical harm are now subject to legal review, the Duty to Hack would require review of *all* cyber operations capable of achieving a military objective.[207] Such reviews (and the experience that comes with a duty to use more cyber operations in the future) would hopefully mitigate unexpected distributions or harms from the cyber operations developed and deployed.

ii. The operational costs and benefits of the Duty to Hack for IHL and cyberspace

Adopting the Duty to Hack would undoubtedly represent a departure from IHL's traditional method of regulation. When it comes to weapons, IHL either prohibits or permits; it has never before mandated the use of a particular weapon. That said, the requirements of precaution have in effect dictated a requirement that as specific capacities become available—such as precision capabilities—IHL may require their use.[208] In essence, the Duty to Hack would establish a similar, albeit explicit, requirement for

[206] In 2002, Russian forces sought to overtake Chechen rebels holding 850 hostages in a Moscow Theater with a non-lethal (chemical) weapon. Dozens of rebels and more than one hundred hostages died as result of adverse reactions to the gas. DP Fidler, "The Meaning of Moscow: 'Non-lethal' weapons and international law in the early 21st century" (2005) 87 *International Review of the Red Cross* 859.

[207] See Walker (n 154) 348.

[208] See, for example, Markham and Schmitt (n 169) 694.

cyber operations. The Duty to Hack is limited to *available* capacities that can accomplish military objectives.

This last caveat is an important one. IHL has recently witnessed an extensive debate over whether a least-harmful-means rule applies to targeting of belligerents or, more succinctly, a "capture but do not kill" requirement when feasible.[209] Although similarly titled, my proposal is distinct from that one in four key respects. First, that debate is over what IHL currently requires, while the Duty to Hack proposed here is clearly *de lege ferenda*. Second, the capture–kill debate would purportedly muddy and complicate an IHL that has clarity and simplicity in its directives to combatants. In contrast, as this chapter has shown repeatedly, IHL-by-analogy is presently complicated and confusing. A Duty to Hack would actually provide the sort of clear rule that opponents say already exists among combatants.

Third, the capture–kill debate focuses on combatants, whereas the duty to hack addresses cyber operations planning and execution. These are two very different candidates for a least-harmful means rule. Combatants are trained to fight in the intensive and violent environment of the battlefield, where they must put their lives at risk. Those who plan cyber operations are, in contrast, not facing combat or (usually) risks to their own lives. The options open to individual soldiers in combat are limited; as Fritz Kalshoven famously observed, the individual soldier has no military equivalent to a bag of golf clubs from which to select an appropriate weapon for the task at hand.[210] Cyber operations, in contrast, do precisely this—their effectiveness depends on being matched with careful calibration to their targets. And while the "corporate aspect of the enemy" may reason against obligations that dilute treating enemy combatants as individuals rather than a collective so long as they are not hors de combat, those reasons are less persuasive in cyberspace, where de jure military objects are often de facto civilian objects as well.[211] In any case, the Duty to Hack is conditioned on achieving military objectives which undercut objections that it somehow would force militaries to sacrifice their initiatives.

Fourth, and finally, the soldiers at the heart of the capture–kill debate are not equivalent to the decision-makers who will plan and deploy cyber operations. The Duty to Hack idea requires its least-harmful-means calculus at a strategic and operational level. Absent dramatic shifts in technology, it is unlikely that a Duty to Hack will implicate tactical questions or Rules of Engagement.

It is the Duty to Hack's impact on cyberspace that may sound the one cautionary note for further thought. To implement the duty, states will need an array of data about various means and methods of deploying cyber operations as well as the various targets against which they might operate. Thus, states will have an incentive to seek out and collect that information, including presumably (and most often) through cyber means. In other words, the Duty to Hack might encourage a continuation (or even an increase) in cyber espionage. Of course, given recent revelations about the

[209] See n 172 and accompanying text.

[210] F Kalshoven, "The Soldier and His Golf Clubs," in C Swinarski (ed), *Studies and Essays on International Humanitarian Law and Red Cross Principles, in Honour of Jean Pictet* (Leiden: Martinus Nijhoff, 1984) 369.

[211] See, for example, Corn et al (n 142) 566.

extent of existing government surveillance and data collection in cyberspace, it may be hard to imagine an increase. And even if we could, that result may not be troubling to those who accept espionage as either unregulated or unregulatable. But for those pressing for new international law rules on such behavior, further clarification may be necessary to balance the perpetuation of the least harmful forms of cyber operations in military operations with the need to limit over-reaching in government surveillance and intelligence collection in peace-time.

V. Conclusion

This chapter calls for a re-thinking of the boundaries of law in cyberspace. So much of the cyberspace literature, particularly on cyberwarfare, has sought to carve out a particular piece of cyberspace or a particular area of international law and analyze it as if in a silo. More troublesome still has been the tendency to construct these silos almost exclusively by analogies to how they were constructed in other times and contexts.

I have illustrated some of the problematic logic in using boundaries and analogies from other contexts without adjusting for the distinctive aspects of cyberspace. Such thinking lies at the root of many of cyberspace's ongoing debates about who should set its laws, which sets of laws to apply, and what the particular rules themselves actually mean. Whether the context is governance, the jus ad bellum, or IHL, the resulting boundaries generate theoretical and functional problems without offering effective or complete regulatory solutions.

Why should a book on cyberwar spend so much time unpacking the admittedly conceptual framework of boundaries and the law? Isn't governance a distinct issue from what the rules are? To ask that question, however, is only to reinforce the intellectual power of bounded thinking. By implying that a border separates these topics, the question itself purports to justify treating them separately and in isolation.

However, "deciding who decides" what the law is for cyberspace *is* tremendously important to issues of when or how the jus ad bellum and IHL apply there. At the most extreme, boundaries could justify cyberspace's autonomy from nation-states, which would undermine the force there of the international law on which those states agreed. The increasing prevalence of national boundaries and state authority in cyberspace provides some reassurance that this will not happen; that international law does—and will continue to—regulate cyber operations. It would be a mistake, however, to accept the status quo as a complete assurance. States and their borders may have authority to govern with respect to some rules and some behavior, but other actors have important voices, just as technical standards and other norms may mediate what international law actually says in this new environment.

Less obviously, the boundaries of the jus ad bellum and IHL may be equally important to the cyberspace governance project. Deciding what behavior is prohibited, permitted, or required in cyberspace provides an opportunity to circumvent the current governance boundary stalemate. Projects like the *Tallinn Manual* and widespread agreement that IHL regulates certain cyber operations begin the work of constructing uniform expectations of future behavior that may, in turn, serve as the basis for developing a more robust governance model for cyberspace, even absent any agreement on what cyberspace "is."[212]

Whatever systemic benefits the laws for cyberwarfare may have for cyberspace governance or vice versa, it would be a mistake to assume those laws are themselves sufficient or effective regulations. The same problems of *theory, accuracy, effectiveness*, and *completeness* on display in using boundaries to produce law are apparent in using law to produce boundaries whether the context is the jus ad bellum or IHL. Fragmentation problems are inherent in efforts to analogize existing fields of international law into cyberspace without attempting to redress pre-existing tensions among them or to avoid new ones. The rationale for line-drawing by analogy is further flawed by how it treats non-analogous, novel cyber capacities. Thus, there are serious reasons to question the accuracy of current claims as to the boundaries for a use of force under the jus ad bellum or an attack under IHL. Add to that the problems of effectiveness created by the attribution problem and the speed of cyber operations, and the current rules become a patently incomplete response to their objectives, whether that is maintaining international peace and security or balancing military necessity with humanity.

To correct such errors, tailor-made solutions are required. This chapter offers one solution in particular—the Duty to Hack—as a response to the particular problems of IHL-by-analogy in regulating the conduct of cyber operations during hostilities. This duty would unsettle boundaries within IHL to better accommodate the law's overarching goals of military necessity and humanity given cyberspace's unique aspects. It would avoid copying IHL's old lines where their rationales do not mesh with current conditions or capabilities, most notably by removing the attack and civilian qualifiers that otherwise bind the choices among means and methods for cyberwarfare. The result is a duty for states to use cyber operations in their military operations when they are expected to be the least harmful means available for achieving military objectives.

The Duty to Hack is not a total panacea for the law's current insufficiency and ineffectiveness. But it offers the promise of increasing and even incentivizing humanitarian protections without sacrificing the need for militaries to do their job—defeating the enemy. Cyber operations will not always be easy or without risks. And although a Duty to Hack may have operational implications for IHL and cyberspace itself, its benefits favor adoption. The capacity of cyber operations to cause significant harm has occupied the public (and military) consciousness for some time. But the capacity of cyber operations to avoid or lessen the harm military operations would otherwise produce deserves attention as well. The Duty to Hack accommodates both possibilities. It presents a unique opportunity to re-calibrate how IHL bounds behavior in cyberspace while also using cyberspace to re-cast the boundaries of IHL itself.

In the end, laws and boundaries need each other. By its existence, a boundary can delineate who is in charge and what law applies to a particular setting. And law exists, if only in part, to draw boundaries. For all its unique attributes, cyberspace is

[212] See n 60 (discussing cyberspace governance via incompletely theorized agreements).

no exception to this phenomenon. And while the law-by-analogy approach has performed an important—and necessary—function to date, it clearly requires adjustments. Cyberspace requires new thinking, not just about tailor-made proposals like the Duty to Hack, but more comprehensive and critical reflections on how information technology relates to legal regimes, governance, and law. It is a significant project, but given how cyberspace now touches nearly every human endeavor, including warfare, it is a necessary one.

8

Cyber Espionage or Cyberwar?

International Law, Domestic Law, and Self-Protective Measures

*Christopher S Yoo**

I. Introduction

The academic discourse on cyberspace followed a pattern that is now well recognized. Early scholarship embraced cyber exceptionalism, pronouncing that the internet's inherently transnational character transcended traditional notions of sovereignty and made it inherently unregulable by nation states.[1] Others disagreed, arguing that the internet fell comfortably within established legal principles.[2] Although President Clinton once confidently stated that China's attempt to crack down on the internet was "like trying to nail Jell-O to the wall,"[3] history has largely proven him wrong. As of today, belief in the internet's unregulability is now essentially defunct, at least with respect to law.[4]

The same debate over internet exceptionalism is currently playing out in the scholarship on international law, albeit with a much more skeptical tone. Early scholarship generally took the position that established international law principles governing when it is appropriate to go to war (jus ad bellum) and the appropriate ways that war may be conducted (jus in bello) applied to cyber conflicts.[5]

* The author would like to thank Jonathan Smith, who has been instrumental in shaping my thinking about cyberwar, and Jean Galbraith and Bill-Burke White for comments on earlier drafts. The chapter also benefitted from presentations at the 8th Annual Symposium of the Global Internet Governance Academic Network held in conjunction with the Internet Governance Forum, as well as the Roundtable on "Cyberwar and the Rule of Law," the Philadelphia Area Cyberlaw Colloquium, and the Conference on "Invisible Harms: Intellectual Property, Privacy, and Security in a Global Network," all held at the University of Pennsylvania.

[1] See, for example, John Perry Barlow, "A Declaration of the Independence of Cyberspace," Electronic Frontier Foundation, February 8, 1996, at <https://projects.eff.org/~barlow/Declaration-Final.html>; David Johnson and David Post, "Law and Borders—The Rise of Law in Cyberspace," (1996) 48 *Stanford Law Review* 1367, 1375.

[2] See, for example, Frank H. Easterbrook, "Cyberspace and the Law of the Horse," (1996) *University of Chicago Legal Forum* 207; Jack L. Goldsmith, "Against Cyberanarchy," (1998) 65 *University of Chicago Law Review* 1199.

[3] William J. Clinton, Remarks at the Paul H. Nitze School of Advanced International Studies, March 8, 2000, in *Public Papers of the President of the United States: January 1 to June 26, 2000* (2000) 404, 407.

[4] See, for example, Alex Kozinski and Josh Goldfoot, "A Declaration of the Dependence of Cyberspace," (2009) 32 *Columbia Journal of Law & the Arts* 365; Tim Wu, "Is Internet Exceptionalism Dead?" in Berin Szoka and Adam Marcus (eds), *The Next Digital Decade: Essays on the Future of the Internet* (Washington, DC: TechFreedom, 2010) 179, 179–82, at <http://nextdigitaldecade.com/ndd_book.pdf>.

[5] See, for example, Michael N Schmitt, "Computer Network Attack and the Use of Force in International Law: Thoughts on a Normative Framework," (1999) 37 *Columbia Journal of Transnational Law* 885; Walter Gary Sharp, Sr, *CyberSpace and the Use of Force* (Falls Church, VA: Aegis Research Corp, 1999).

By September 2012, US State Department Legal Advisor Harold Koh could confidently declare that established principles of international law apply to cyberspace.[6] More recently, scholarship has begun to raise doubts about this conclusion.[7] Even those confident about the application of international law to cyber conflicts acknowledge that some types of cyber operations do not fit easily into the traditional categories.[8]

This chapter will explore how these two bodies of international law governing conflicts between states applies to cyber operations, using as its lens prominent examples that have been in the news of late. These include the surveillance programs alleged by Edward Snowden to have been conducted by the National Security Agency (NSA), the 2007 and 2008 distributed denial of service attacks launched by Russian hackers against Estonia and Georgia that disrupted their communications networks without damaging any property, and the 2008 Stuxnet virus introduced into a key Iranian nuclear facility that caused the centrifuges used to enrich uranium to accelerate and decelerate unexpectedly and eventually to destroy themselves. In addition, this chapter will examine the type of practices described in a 1999 book entitled *Unrestricted Warfare*, in which two Colonels of the Chinese People's Liberation Army (PLA) describe ways that a country in the position of China in the late 1990s could defeat an opponent that was in a stronger economic and technological position. The tactics described in the book include lawfare (ie, the use of international and multilateral organizations to subvert an opponent's policies), manipulation of trade, manipulation of financial transactions, as well as network warfare (cyberwarfare) targeted at a nation's financial and communications systems, just to name a few.[9]

Section II applies current principles as understood through the *Tallinn Manual on the International Law Applicable to Cyber Warfare* produced by a special Independent Group of Experts (IGE) convened by NATO, concluding that jus ad bellum and jus in bello as well as the customary international law of non-intervention will not reach much, if any, of this behavior. Instead, such conduct is relegated to the law of espionage, described in section III, which is governed almost entirely by domestic law. The absence of an overarching legal solution to this problem heightens the importance of technological self-protective measures, described in section IV.

[6] Harold Hongju Koh, Legal Advisor, US Department of State, "International Law in Cyberspace," Remarks, before the USCYBERCOM Inter-Agency Legal Conference, Ft. Meade, Maryland, September 18, 2012, at <http://www.state.gov/s/l/releases/remarks/197924.htm>.

[7] See Susan W. Brenner and Leo L. Clarke, "Civilians in Cyber war: Conscripts," (2010) 43 *Vanderbilt Journal of Transnational Law* 1011; Jack M. Beard, "Legal Phantoms in Cyberspace: The Problematic Status of Information as a Weapon and a Target Under International Humanitarian Law," (2014) 47 *Vanderbilt Journal of Transnational Law* 67.

[8] See, for example, Sharp (n 5); Koh (n 6).

[9] Qiao Liang and Wang Xiangsui, *Unrestricted Warfare: Assumptions on War and Tactics in the Age of Globalization* (Beijing: PLA Literature and Arts Publishing House, 1999).

II. The Applicability of Jus ad Bellum, Jus in Bello, and Non-Intervention

The law of war is dominated by two bodies of law that govern hostile activities among states.[10] Jus ad bellum determines when a use of force is justified. Jus in bello governs how states should conduct themselves during periods of armed conflict. The literature exploring these concepts is too voluminous to cite comprehensively and will only be sketched here, primarily by referring to the leading primary sources of international law as well as the interpretation offered by the *Tallinn Manual* as to how these principles will apply to cyber operations. Admittedly, the exposition presented in the *Manual* glosses over many important ongoing debates. As the leading source on how the law of war applies to cyber conflicts, it does provide an important cornerstone for analysis. As we shall see, the threshold determination for the applicability of jus ad bellum and jus in bello has been whether the damage to persons or property is analogous to those inflicted by traditional kinetic war. This standard leaves many of the types of cyber operations that have raised the greatest concern outside the scope of the law of war.

(a) Jus ad Bellum

Although the principles on when international law permits nations to go to war have a long history, the starting point for modern analysis is the Charter of the United Nations. Although the Charter is by no means the only relevant source of law, the International Court of Justice (ICJ) has recognized that the Charter's restrictions on the use of force have become part of the customary international law of jus ad bellum.[11]

Two provisions of the UN Charter have particular importance. Article 2(4) of the Charter specifically bans the use of force against other states when it provides, "All Members shall refrain in their international relations from the threat or use of force against the territorial integrity or political independence of any state, or in any other manner inconsistent with the Purposes of the United Nations." The Charter recognizes certain exceptions to the prohibition of the use of force. In particular, Article 51 recognizes that nothing in the Charter abrogates states' "inherent right of individual or collective self-defense if an armed attack occurs against a Member of the United Nations, until the Security Council has taken the measures necessary to maintain international peace and security."

In addition to jus ad bellum, nations are also subject to the duty of nonintervention. Article 2(1) recognizes the "principle of sovereign equality" among UN members,

[10] To date the scholarly debate has focused primarily on the application of these two areas of law to cyber war. That said, they do not represent the only relevant bodies of international law. For example, some scholars have begun to explore the extent to which human rights treaties govern electronic surveillance. See, for example, Marko Milanovic, "Human Rights Treaties and Foreign Surveillance: Privacy in the Digital Age," (forthcoming) 56 *Harvard International Law Journal*, draft available at http://ssrn.com/abstract=2418485.

[11] Military and Paramilitary Activities in and Against Nicaragua (*Nicaragua v US*), 1986 ICJ 14, paras 188–90 (June 27).

which implies a principle of nonintervention preventing other states from taking actions that deprive another state of the ability to control governmental matters implicit in being a state. Although the precise contours of what constitutes intervention have long been the subject of extensive debate, the ICJ has recognized that intervention is wrongful when it constitutes coercion.[12]

The key concepts are thus the conduct that is prohibited by Article 2(4) (threat or use of force) and the occurrence that triggers the Article 51 exception to that prohibition (armed attack) as well as the threshold for the principle of nonintervention implicit in Article 2(1) (coercion). Although the most aggressive forms of cyber operations would fall within the scope of these terms, many forms of cyber surveillance and interference would not.[13]

i. Use of force

We begin our analysis with the use of force, which is the primary conduct prohibited by Article 2(4). No treaty provides a definition of the use of force, although the UN Charter does provide some guidance. For example, other Charter provisions offer guidance as to the types of conduct that fall within Article 2(4). The Charter's preamble clearly stated the signatory nations' determination "to save succeeding generations from the scourge of war" by "ensur[ing] . . . that armed force shall not be used, save in the common interest." This language is generally regarded as establishing that armed military operations within another country would constitute a use of force prohibited by Article 2(4). Furthermore, in describing measures that the Security Council may take that do "not involve[e] the use of armed force," Article 41 specifically includes the "complete or partial interruption of economic relations and of rail, sea, air, postal, telegraphic, radio, and other means of communication." This suggests that not all disruptive operations constitute the use of force.

The *travaux préparatoire* of Article 2(4) reveals that the San Francisco Conference specifically rejected an amendment that would have included economic coercion within the scope of the use of force prohibited by Article 2(4).[14] Moreover, when considering a draft containing language that paralleled Article 2(4), the UN Commission on Friendly Relations rejected arguments that all forms of political and economic pressure fell within its scope.[15]

[12] ICJ *Nicaragua* judgment, para 205.

[13] For other enlightening discussions, see also Watts (Chapter 4, this volume), Blank (Chapter 5, this volume), and Hollis (Chapter 6, this volume).

[14] 6 UN CIO Docs. 334, 609 (1945); Doc. 2, 617 (e) (4), 3 UN CIO Docs. 251, 253–4 (1945). This drafting history also rebuts arguments that the use of the term "force" without the word "armed" in Article 2(4) suggested that Article 2(4) prohibits economic coercion that does not rise to the level of military action. This underscores that the primary impetus for the creation of the UN was to outlaw armed conflict as a legitimate instrument for effectuating national policy rather than to insulate countries from all imbalances in economic bargaining position. See Tom J Farer, "Political and Economic Coercion in Contemporary International Law," (1985), 79 *American Journal of International Law* 405, 410; Schmitt (n 5) 905; Marco Roscini, "World Wide Warfare—Jus ad Bellum and the Use of Cyber Force," (2010) 14 *Max Planck Yearbook of United Nations Law* 85, 105.

[15] UN GAOR Special Comm. on Friendly Relations, UN Doc. A/AC.125/SR.110 to 114 (1970); Rep. of the Special Comm. On Friendly Relations and Cooperation among States, 1969, UN GAOR, 24th Session, Supp. No 19, at 12, UN Doc. A/7619 (1969).

This interpretation is reinforced by principles of customary international law, particularly the aspect of ICJ's *Nicaragua* judgment rejecting Nicaragua's claim that US funding of the contras constituted an impermissible use of force. Although the ICJ lacked the jurisdiction to determine whether those actions violated Article 2(4) of the UN Charter, it did have the authority to determine whether the actions taken by the US violated customary international law. The ICJ concluded that "organizing or encouraging the organization of irregular forces or armed bands... for incursion into the territory of another State" and "participating in acts of civil strife... in another State" did constitute an impermissible use of force. At the same time, the ICJ ruled that "the mere supply of funds to the *contras*... does not in itself amount to a use of force."[16]

The *Tallinn Manual* drew guidance from the ICJ's focus on scale and effects when determining whether conduct constituted an armed attack. Moreover, the IGE that authored the *Tallinn Manual* took note of the historical materials indicating that the use of force excluded mere economic and political pressure and concluded that cyber operations analogous to such activities, such as psychological operations or funding a hacktivist group, did not constitute uses of force, nor would providing sanctuary or safe haven for those mounting cyber operations unless coupled with other acts.[17] Actions taken by a state's intelligence agency or a contractor whose conduct is attributable to a state can constitute a use of force despite the fact that it was not undertaken by the state's armed forces.[18]

The heart of the *Tallinn Manual*'s proposal was to identify cyber operations with kinetic operations that the international community would clearly recognize as uses of force, including all conduct that rose to the level of armed attack and acts that injure or kill persons or damage or destroy objects.[19] For other cases, the IGE put forward eight nonexclusive factors to guide the inquiry: severity, immediacy, directness, invasiveness, measurability of effects, military character, state involvement, and presumptive legality.[20] The *Manual* observed that "actions such as disabling cyber security mechanisms in order to monitor keystrokes would, despite their invasiveness, be unlikely to be seen as a use of force."[21]

Measured against this standard, the type of surveillance described in the classified documents disclosed by Edward Snowden, lacking as it did any injury to people or property, would not likely rise to the level to constitute a use of force. Similarly, the use of lawfare or economic warfare through the erection of trade barriers or financial transactions would almost certainly not constitute uses of force. Presumably neither would the partial disruption of communications recognized by Article 41 of the UN Charter. And, as we shall see in subsection ii, intrusion into another state's systems by breaching firewalls and cracking passwords fails to violate nonintervention; because that standard is lower than the standard governing the use of force, *a fortiori* this type of conduct does not constitute use of force.

[16] ICJ *Nicaragua* judgment, para 228.

[17] Michael N Schmitt (ed), *The Tallinn Manual on the International Law Applicable to Cyber Warfare* (New York: Cambridge University Press, 2013) 47–9.

[18] Schmitt (n 17) 46.

[19] Schmitt (n 17) 49.

[20] Schmitt (n 17) 49–52.

[21] Schmitt (n 17) 50–1.

Attacks on a nation's financial or cyber infrastructure of the type envisioned by *Unrestricted Warfare* are a matter of degree. Temporary denial of service would be insufficient to constitute a use of force even if highly invasive. At the same time, "some may categorize massive cyber operations that cripple an economy as a use of force even though economic coercion is presumptively lawful."[22] The *Tallinn Manual* does, moreover, presume that Stuxnet constituted a use of force.[23]

ii. *Self-defense*

Another key concept under traditional international law are the conditions under which a state may act in self-defense. As noted earlier, a nation may exercise its inherent right of self-defense under Article 51 of the UN Charter when confronted with an armed attack.

Because the UN Charter does not provide a definition of armed attack, customary international law and treaty law coexist side by side.[24] The ICJ's *Nicaragua* judgment distinguished between "the most grave forms of the use of force (those constituting an armed attack) from other less grave forms" and noted that "measures which do not constitute an armed attack... may nevertheless involve a use of force."[25] In so doing, the ICJ recognized that armed attack is a subset of the use of force. Thus, while all forms of armed attack constitute uses of force, not all uses of force constitute armed attacks. This in turn implies that states may be the targets of operations that are sufficiently severe to constitute uses of force that are illegal under Article 2(4), but not sufficiently severe to justify responding in kind under Article 51.[26]

The ICJ further noted that in addition to sending regular forces across an international border, armed attack also includes "'sending... armed bands, groups, irregulars or mercenaries, which carry out acts of armed force against another State of such gravity as to amount to' *(inter alia)* an actual armed attack conducted by regular forces, 'or its substantial involvement therein.'"[27] The ICJ further observed that some instances of sending armed bands into the territory might represent "a mere frontier incident" rather than an armed attack.[28] In short, whether particular conduct constitutes an armed attack depends on its "scale and effects."[29] The ICJ seemed to entertain the possibility that a series of smaller actions might constitute an armed attack if considered together.[30] Any actions taken in self-defense in response to an armed attack are subject to the customary international law requirements of necessity and proportionality.[31]

[22] Schmitt (n 17) 55–6. [23] Schmitt (n 17) 47. [24] ICJ *Nicaragua* judgment, para 176.

[25] ICJ *Nicaragua* judgment, paras 191, 210; see also Oil Platforms (*Iran v US*), judgment, 2003 ICJ 161, paras 51, 64 (November 6).

[26] Michael N Schmitt, "'Attack' as a Term of Art in International Law: The Cyber Operations Context," in Christian Czosseck, Rain Ottis, and Katharina Ziolkowski (eds), *Proceedings of the 4th International Conference on Cyber Conflict (CYCON 2012)* (Tallinn, Estonia: NATO CCD COE Publications, 2012), 283, 286–7.

[27] ICJ *Nicaragua* judgment, para 195 (quoting Declaration on the Definition of Aggression, GA Res. 3314 (XXIX), UN Doc A/RES/29/3314 (December 14, 1974)).

[28] ICJ *Nicaragua* judgment, para 195. [29] ICJ *Nicaragua* judgment, para 195.

[30] ICJ *Oil Platforms* judgment, para 64.

[31] ICJ *Nicaragua* judgment, paras 176, 194; Legality of the Threat or Use of Nuclear Weapons, Advisory Opinion, 1996 ICJ 226, para 41 (July 8); ICJ *Oil Platforms* judgment, paras 43, 74, 76.

In applying these principles to cyberspace, the commentary on Rule 13 governing self-defense against armed attack noted that the IGE "unanimously concluded that some cyber operations may be sufficiently grave to warrant classifying them as an 'armed attack' within the meaning of the Charter."[32] A majority of the IGE agreed that the term "armed" did not require the employment of weapons.[33] The IGE concurred that to constitute an armed attack, conduct must exceed the scale and effects needed to qualify as a use of force.[34] On the one hand, the IGE agreed that any force that injures or kills persons or damages or destroys property would have sufficient scale and effects to constitute an armed attack. On the other hand, the IGE also agreed that acts of cyber intelligence, cyber theft, and brief or periodic interruptions of nonessential cyber services do not constitute armed attacks.[35]

How to characterize intermediate cases divided the IGE. Some members held that some harm to persons or property is necessary for an incident to be considered an armed attack, while others focused on the severity of the broader effects.[36] For example, some would describe a cyber assault on the New York Stock Exchange as mere financial loss that did not rise to the level of armed attack, while others focused on the catastrophic effects of such an attack.[37] The IGE also disagreed as to the role of intent, with a majority focusing exclusively on scale and effects and a minority refusing to characterize cyber espionage that unexpectedly inflicted significant damage to another state's cyber infrastructure as an armed attack.[38]

The *Tallinn Manual* included rules imposing principles of necessity, proportionality, imminence, and immediacy.[39] But application of these rules to cyber conflicts proved controversial. The debate over imminence provides an apt illustration. Although Article 51 on its face applies when "an armed attack occurs," a majority of the IGE held that a state may defend itself against armed attacks that are "imminent," while a minority rejected the concept of anticipatory self-defense.[40] While the IGE agreed that the speed of cyber operations dictated that a state need not wait until an attack had already been launched before responding, a minority would have required that an attack be about to be launched, while a majority "rejected this strict temporal analysis" in favor a "last feasible window of opportunity" standard that permits self-defense when failure to act would reasonably leave a state unable to defend itself effectively once the attack commences.[41]

Although the *Tallinn Manual* found it "indisputable" that actions of non-state actors may constitute an armed attack if undertaken under the direction of a state, the IGE divided over the legal implications of such actions in the absence of any direction by a state. A majority concluded that state practice supported the exercise of the right of self-defense against non-state actors such as terrorists and rebel groups, which in the cyber context would extend to actions taken by information technology corporations, while a minority disagreed.[42]

[32] Schmitt (n 17) 54. [33] Schmitt (n 17) 54. [34] Schmitt (n 17) 54.
[35] Schmitt (n 17) 55. [36] Schmitt (n 17) 55. [37] Schmitt (n 17) 55.
[38] Schmitt (n 17) 56. [39] Schmitt (n 17) 59–63. [40] Schmitt (n 17) 60–61.
[41] Schmitt (n 17) 61. [42] Schmitt (n 17) 56–57.

If the state from which the non-state actors are launching their cyber armed attack (which the *Tallinn Manual* calls the territorial state) consents, the victim state may take self-defensive actions against non-state actors within the territorial state.[43] The IGE divided on how to address situations in which the territorial state does not consent to self-defensive actions by the victim state, but remains unable or unwilling to stop the cyber armed attack. A majority of the IGE concluded that self-defensive actions by the victim state within the territorial state are permissible as a result of the duty of each state to ensure that its territory is not used to violate international law.[44] When confronted with such an attack, the victim state may ask the state from which the non-state cyber operations constituting a cyber armed attack are emanating (called the territorial state) to address the situation.[45] A minority disagreed, arguing that such actions were impermissible absent an action based on a plea of necessity.[46]

How would these principles apply to the real world examples under consideration? The *Tallinn Manual* concluded that as of 2012, no cyber incidents had occurred that had generally been recognized as constituting an armed attack, including the 2007 cyber operations against Estonia.[47] Under these standards, because the surveillance programs described in the documents disclosed by Edward Snowden did not represent a use of force, *a fortiori* it also did not constitute an armed attack. If, as noted already, monitoring keystrokes did not rise to the lower standard of the use of force, it necessarily did not constitute an armed attack. Moreover, courts would likely not categorize the types of operations suggested by *Unrestricted Warfare*, lawfare and economic warfare through manipulation of trade or the financial transactions, as armed attacks. Whether direct actions against the financial infrastructure itself that caused it to crash would meet the standard of armed attack divided the IGE.

Most interestingly, the IGE did not regard Stuxnet, often regarded as the strongest real-world candidate for being classified as an armed attack, as completely clear. Although the *Tallinn Manual* indicates that some IGE members believed that the damage to the Iranian centrifuges was sufficient to reach the level of armed attack,[48] the clear implication is that other IGE members thought otherwise.

(b) Jus in Bello

During periods of armed conflict, state conduct is governed by jus in bello. The primary sources of law are the Geneva Conventions of 1949 and the Additional Protocols I and II of June 8, 1977, as well as customary international law. Specifically, these include the principles of distinction, which requires parties to target only those engaged in fighting, and proportionality, which requires parties to forbear from acting when the likely civilian casualties would exceed the anticipated military advantage.

Jus in bello applies only if certain threshold considerations are met; it applies only under circumstances constituting "armed conflict;" it places restrictions on states when taking actions deemed to constitute "attacks;" and it only applies to actions by

[43] Schmitt (n 17) 26, 58.
[44] Schmitt (n 17) 58.
[45] Schmitt (n 17) 59.
[46] Schmitt (n 17) 59.
[47] Schmitt (n 17) 56.
[48] Schmitt (n 17) 56.

state actors and non-state actors under their control. Many common types of cyber operations that are of public concern fall outside the scope of jus in bello.

i. Armed conflict

As noted earlier, jus in bello applies only in the presence of an armed conflict. The Geneva Conventions distinguish between two types of armed conflicts: international armed conflict, defined as armed conflict among two or more states, and non-international armed conflict, defined as armed conflict within a state between the armed forces of a state and one or more armed groups or between organized armed groups.[49]

Although the treaties do not define armed conflict, Article 1(2) of Additional Protocol II to the Geneva Conventions established that "situations of internal disturbances and tensions, such as riots, isolated and sporadic acts of violence and other acts of a similar nature" do not constitute armed conflict.[50] The International Criminal Tribunal for the Former Yugoslavia (ICTY) has also ruled that armed conflict arises "whenever there is a resort to armed force between States."[51] In making this determination, the Tribunal focused on two key criteria—the intensity of the hostilities and the organization of the parties[52]—and has developed factors by which to evaluate them.[53] The actions of a non-state organized group may be attributed to a state if that state exercises "overall control" over that group.[54] Support, such as through financing, training, equipping, and providing operational assistance, is not sufficient to establish overall control.[55] The ICJ has mentioned the overall-control test favorably without adopting it, instead applying a test that focuses on effective control.[56] Although some IGE members argued that an international armed conflict can exist between a state and a non-state armed group whose actions cannot be attributed a state, a majority of the IGE rejected that position. Instead, the *Tallinn Manual* regards such a situation as a non-international armed conflict. The IGE incorporated the intensity and organization standard established by ICTY for determining when a non-international

[49] Geneva Convention for the Amelioration of the Condition of the Wounded and Sick in Armed Forces in the Field, August 12, 1949, 75 UNTS 31, Articles 2 and 3.

[50] Protocol Additional to the Geneva Conventions of August 12, 1949, and Relating to the Protection of Victims of Non-international Armed Conflicts, June 8, 1977, 1125 UNTS 609, Article 1(2).

[51] Prosecutor v. Tadić, Case No. IT-94-1-A, Appeals Chamber Decision on the Defence Motion for Interlocutory Appeal on Jurisdiction, October 2, 1995, para 70.

[52] ICTY *Tadić* decision, para 70. Although *Tadić* initially also included factors such as geographic scope and temporal duration as separate criteria, later decisions incorporated these concepts into intensity. *Prosecutor v Haradinaj*, Case No IT-04-84-T, Trial Chamber Judgment, April 3, 2008, para 49.

[53] On intensity, see *Prosecutor v Delalić/Mucić*, Case No IT-96-21-T, Trial Chamber Judgment, November 16, 1998, para 187; *Prosecutor v Milošević*, Case No IT-02-54-T, Decision on Motion for Judgment of Acquittal, June 16, 2004, paras 28–31 *Prosecutor v Limaj*, Case No IT-03-66-T, Trial Chamber Judgment, November 30, 2005, paras 135–67; *Prosecutor v Hadžihasanović*, Case No IT-01-47-T, Trial Chamber Judgment, March 15,2006, para 22; *Prosecutor v Mrkšić*, Case No IT-95-13/1-T, Trial Chamber Judgment, September 27, 2007, paras 39–40, 407–8, 419; ICTY *Haradinaj* judgment, para 49. On organization, see ICTY *Limaj* judgment, paras 94–129; see also *Prosecutor v Akayesu*, Case No ICTR-96-4-T, Trial Chamber Judgment, September 2, 1998, paras 619–21.

[54] ICTY *Tadić* decision, paras 131, 145, 162.

[55] ICTY *Tadić* decision, para 137.

[56] Application of the Convention on the Prevention and Punishment of the Crime of Genocide (*Bosnia and Herzegovina v Serbia and Montenegro*), Judgment 2007 ICJ 108, para 404 (February 26).

armed conflict exists.[57] When these criteria are met, these situations are subject to many aspects of the law of armed conflict, including criminal responsibility of commanders, the principles of distinction and proportionality, the obligation to respect medical and U.N. personnel, journalists, cultural property, and diplomatic archives, and to protect detained persons, although principles such as combatant status and belligerent immunity do not apply.[58]

The *Tallinn Manual* recognizes that jus in bello in the context of cyber operations requires the presence of an armed conflict.[59] The IGE concluded that armed conflict requires the existence of hostilities, which in turn "presuppose the collective application of means and methods of warfare, consisting of kinetic and/or cyber operations."[60] IGE members disagreed as to whether a single cyber incident was sufficient to satisfy the requisite threshold of violence to constitute international armed conflict.[61] Consistent with Article 1(2) of Additional Protocol II, they did agree that "[s]poradic cyber incidents, including those that directly cause physical damage or injury," as well as that "cyber operations that incite incidents such as civil unrest or domestic terrorism" do not constitute non-international armed conflict.[62] Similarly insufficient are "network intrusions, the deletion or destruction of data (even on a large scale), computer network exploitation, and data theft" and "[t]he blocking of certain internet functions and services."[63] The IGE divided over whether such cyber operations conducted during civil disturbances or combined with other acts of violence or nondestructive but severe cyber operations might satisfy the criterion.[64]

Turning now to our motivating cases, the IGE noted that no cyber operation has been publicly characterized as an international armed conflict.[65] The *Tallinn Manual* makes clear that none of the surveillance practices described in the classified documents disclosed by Edward Snowden even arguably constituted armed conflict. The damage-free denial of service attack against Estonia and calls by the Russian minority for civil unrest in 2007 also did not rise to the level of armed conflict either in terms of harm or organization, but the similar cyber operations against Georgia in 2008, which took place in conjunction with military operations, did.[66] Neither lawfare nor the economic warfare through trade barriers or financial transactions described in *Unrestricted Warfare* would constitute armed conflict, nor would presumably more active attempts to overwhelm the financial or cyber infrastructure.[67] The IGE could not even come to agreement as to whether Stuxnet inflicted sufficient damage to constitute an armed conflict.[68]

[57] Schmitt (n 17) 72–3, 76, 77.

[58] Schmitt (n 17) 81, 84, 88, 90, 95, 97–9, 106–7, 115, 120–2, 124–5, 127, 130–2, 137, 139–44, 149, 155, 158, 168–71, 173–4, 176–7, 180, 187, 192.

[59] Schmitt (n 17) 68.

[60] Schmitt (n 17) 74.

[61] Schmitt (n 17) 74–5.

[62] Schmitt (n 17) 77.

[63] Schmitt (n 17) 78.

[64] Schmitt (n 17) 78.

[65] Schmitt (n 17) 75.

[66] Schmitt (n 17) 68, 74, 77.

[67] Cordula Droege, "Get Off My Cloud: Cyber Warfare, International Humanitarian Law, and the Protection of Civilians," (2012) 94 *International Review of the Red Cross* 533, 548.

[68] Schmitt (n 17) 75.

ii. Attacks

In addition, Additional Protocol I requires distinction and proportionality when states engage in certain activities. For example, Article 48 implements distinction by requiring that "the Parties to the conflict shall at all times distinguish between the civilian population and combatants and between civilian objects and military objectives and accordingly shall direct their operations only against military objectives."[69] More specifically, Articles 51(2) and (4) and 52(1) prohibit signatory parties from launching an "attack" on a civilian population or on civilian objects or from indiscriminate attacks.[70] Article 51(5)(b) similarly implements proportionality by prohibiting any "attack which may be expected to cause incidental loss of civilian life, injury to civilians, damage to civilian objects, or a combination thereof, which would be excessive in relation to the concrete and directly military advantage anticipated."[71]

Article 49(1) defines attacks as "acts of violence against the adversary, whether in offence or defence."[72] Accompanying commentaries emphasize that attack refers to "combat action"[73] and "physical force" and does not apply to the "dissemination of propaganda, embargoes, or other non-physical means of psychological or economic warfare" or to "military movement or maneuver as such."[74] It bears mentioning that the concept of attack in jus in bello plays a role that is far different from the concept of armed attack in jus ad bellum. The former is the trigger for a range of legal protections for civilian populations, whereas the latter serves to authorize the use of force in self-defense against another state.

Rule 31 of the *Tallinn Manual* provides that "[t]he principle of distinction applies to cyber attacks," while Rule 32 incorporates the principle that civilians and civilian objects may not be attacked.[75] Rule 30 defines a cyber attack as "a cyber operation, whether offensive or defensive, that is reasonably expected to cause injury or death to persons or damage or destruction to objects."[76]

The *Tallinn Manual* commentary on attacks emphasizes the concept of violence used in Article 49(1) of Additional Protocol I as the key concept that distinguishes attacks from other military operations and lists psychological cyber operations and cyber espionage as nonviolent operations that do not qualify as attacks.[77] As the precedents on chemical, biological, and radiological weapons indicate, attacks do not have to have a kinetic effect in order to be violent so long as they foreseeably have the consequences set for in the rule.[78] For example, a cyber operation targeted at an electrical grid that starts a fire would qualify; *de minimis* damage or destruction would not.[79]

[69] Protocol Additional to the Geneva Conventions of August 12, 1949, and Relating to the Protection of Victims of International Armed Conflicts (Protocol I), June 8, 1977, 1125 UNTS 3, Article 48.
[70] Additional Protocol I, Articles 51(2), (4), and 52(1).
[71] Additional Protocol I, Article 51(5)(b). [72] Additional Protocol I, Article 49(1) and (2).
[73] Yves Sandoz, Christophe Swinarski, and Bruno Zimmerman (eds), *Commentary on the Additional Protocols of 8 June 1977 to the Geneva Conventions of 12 August 1949* (Geneva: ICRC, 1977), para 1880.
[74] Michael Bothe, Karl Josef Partsch, and Waldemar A. Solf, *New Rules for Victims of Armed Conflicts* (Boston: Martinus Nijhoff Publishers 1982), 289.
[75] Schmitt (n 17) 95, 97. [76] Schmitt (n 17) 92. [77] Schmitt (n 17) 92.
[78] Schmitt (n 17) 92–3. [79] Schmitt (n 17) 92.

The IGE agreed that cyber operations targeted at data could constitute an attack. The IGE divided over whether cyber interference with the functionality of an object could constitute a cyber attack, with a majority opining in the affirmative and a minority in the negative.[80] While an attack on a computer-based control system for an electrical grid that requires the replacement of components would constitute an attack, IGE members disagreed over whether an attack that required reinstallation of the operating system or the restoration of data might qualify as an attack.[81] The IGE agreed that cyber operations that do not cause damage, but result in nationwide blocking of email or other similar large-scale adverse consequences would not constitute an attack.[82] Introduction of malware or latent defects that are capable of launching an attack would, however, as would attacks that are successfully intercepted by firewalls, antivirus software, or other protective systems.[83]

The surveillance programs described in the classified documents disclosed by Edward Snowden would not rise to the level of injury to persons, damage to objects, or violence sufficient to constitute an attack under the criteria laid out in the *Tallinn Manual*, nor would apparently the type of denial of service attack directed against Estonia in 2007, although a similar attack that is part of a wider operation, as was the case in Georgia in 2008, would.[84] The same would apply to the lawfare and economic warfare through trade barriers and financial transactions described in *Unrestricted Warfare*, nor would presumably more active attempts to disrupt the financial or cyber infrastructure so long as they did not cause physical damage. The IGE did not offer an assessment as to the application of these principles to Stuxnet, although one could surmise that if introducing a bug into the control system for an electric grid that caused a fire would constitute an attack, so would introducing a bug into a control system that led to the destruction of centrifuges.

iii. Attribution to the state

In addition to constituting an armed conflict, the acts must be attributable to a state for jus in bello to apply. According to the International Law Commission's Articles on Responsibility of States for Intentionally Wrongful Acts, states are responsible for private actors who are operating under their direction and control.[85] The ICJ and ICTY have ruled that the conduct of non-state actors may be attributed to the state if the state exercises "effective control" or "overall control."[86]

As an initial matter, the inherent anonymity of internet-based communications makes the true source of a packet hard, if not impossible, to verify. Furthermore, many attacks rely on botnets where a bot controller can use thousands or millions of infected computers to launch an attack without their owners' permission or awareness. The fact that many cyberwar capabilities are developed and executed by private

[80] Schmitt (n 17) 93. [81] Schmitt (n 17) 93–4. [82] Schmitt (n 17) 94.
[83] Schmitt (n 17) 94. [84] Schmitt (n 17) 94.
[85] Int'l L Comm'n, Responsibility of States for Internationally Wrongful Acts, GA Res 56/83 annex, UN Doc. A/RES/56/83, December 12, 2001.
[86] ICJ *Nicaragua* judgment, para 115; ICJ *Genocide* judgment paras 399–401; ICTY *Tadić* decision, paras 131, 145.

companies raises questions when this conduct may be fairly attributed to the state. Such groups must be organized; collective activity of "hacktivists" acting independently is not sufficient.[87]

(c) Non-intervention/coercion

Customary international law has long recognized the principle of non-intervention to be implicit in the principles of sovereignty and the equality among nations. In its *Nicaragua* judgment, the ICJ recognized that non-intervention "forbids all States... to intervene directly or indirectly in internal or external affairs of other States." It also ruled, "Intervention is wrongful when it uses methods of coercion in regard to" a nation's "choice of a political, economic, social and cultural system, and the formulation of foreign policy." The relevant level of coercion "is particularly obvious in the case of an intervention which uses force, either in the direct form of military action, or in the indirect form of support for subversive or terrorist armed activities within another State."[88]

Under this standard, the ICJ found that the US's "financial support, training, supply of weapons, intelligence and logistic support" for the contras "constitute[d] a clear breach of the principle of non-intervention."[89] Such conduct clearly met the relevant standard of coercion.[90] Moreover, although simply funding the contras did not constitute a use of force, it was "undoubtedly an act of intervention in the internal affairs of Nicaragua."[91] On the other hand, humanitarian assistance would not constitute intervention,[92] nor would cessation of economic aid or the imposition of a trade quota or an embargo.[93]

The *Tallinn Manual* recognized that cyber conduct that does not rise to the level of a use of force may nonetheless constitute a violation of non-intervention.[94] At the same time, "not all cyber interference automatically violates the international law prohibition on intervention," including cyber espionage and cyber exploitation lacking a coercive element as well as mere intrusion into another state's systems even when such intrusion requires abrogating virtual barriers such as breaching firewalls or cracking passwords.[95] Whether other cases violate non-intervention depends on the circumstances. For example, attempts to achieve regime change by manipulating elections or public opinion in advance of elections constitute improper intervention, while lesser forms of political and economic interference do not.[96]

Measured against these standards, the surveillance practices described in the classified documents disclosed by Edward Snowden represented nothing more than the type of surveillance that the *Tallinn Manual* did not regard as constituting intervention. Lawfare similarly lacks a coercive element, and economic warfare through the erection of trade barriers or financial transactions would represent little more than the selection of trade partners that the ICJ deemed permissible. Whether a more

[87] Schmitt (n 17) 38, 73–4.
[88] ICJ *Nicaragua* judgment, para 205.
[89] ICJ *Nicaragua* judgment, para 242.
[90] ICJ *Nicaragua* judgment, para 241.
[91] ICJ *Nicaragua* judgment, para 228.
[92] ICJ *Nicaragua* judgment, para 242.
[93] ICJ *Nicaragua* judgment, para 245.
[94] Schmitt (n 17) 46.
[95] Schmitt (n 17) 47.
[96] Schmitt (n 17) 47.

direct assault on a nation's financial or cyber infrastructure of the type envisioned by *Unrestricted Warfare* would constitute intervention is less clear. Because non-intervention represents a lower threshold than the use of force, the fact that the *Tallinn Manual* presumes that Stuxnet constituted a use of force means that Stuxnet must necessarily constitute a violation of non-intervention as well.[97]

III. The Applicability of the Law of Espionage

With a few narrow exceptions, jus ad bellum and jus in bello do not govern the type of information gathering or interference that characterize cyber operations such as the type of surveillance described in the confidential documents disclosed by Edward Snowden, the denial of service attacks directed at Estonia, or the type of measures described in *Unrestricted Warfare*. Indeed, the initial section regarding the scope of the *Tallinn Manual* explicitly notes that it does not address such matters as "[c]yber espionage, theft of intellectual property, and a wide variety of criminal activities in cyberspace" because "the international law on uses of force and armed conflict plays little or no role" with respect to those subjects.[98] It explains why the IGE members did not regard practices such as monitoring keystrokes, breaching firewalls, cracking passwords, or intruding into another state's systems as a use of force for purposes of jus ad bellum.[99] It also explains why they did not regard "network intrusions, the deletion or destruction of data (even on a large scale), computer network exploitation,... data theft," "[t]he blocking of certain internet functions and services, and the nationwide disruption of email as constituting an armed conflict for purposes of *jus in bello*."[100]

Instead, this type of conduct falls within the province of the law of espionage. Sometimes called the second oldest profession,[101] espionage traces its roots back to ancient Egypt, Greece, Rome, and China.[102] Indeed, the great seventeenth-century legal scholar Hugo Grotius noted that "there is no doubt, but the law of nations allows anyone to send spies, as Moses did to the land of promise, of whom Joshua was one."[103] The law of espionage has remained relatively undeveloped, with what little work that exists focusing on espionage during times of war, with the most salient example being the Hague Regulations provision providing that "a spy who, after re-joining the army to which he belongs, is subsequently captured by the enemy, is treated as a prisoner of war, and incurs no responsibility for his previous acts of spying."[104] Thus, Falk's

[97] Schmitt (n 17) 47. [98] Schmitt (n 17) 18. [99] Schmitt (n 17) 47, 50.

[100] Schmitt (n 17) 78, 94.

[101] Allison Ind, *A Short History of Espionage* (New York: David West Co., 1963); Phillip Knightley, *The Second Oldest Profession: Spies and Spying in the Twentieth Century* (New York: W.W. Norton & Co., 1986).

[102] Allen Dulles, *The Craft of Intelligence* (Guilford, CT: Globe Pequot, 1961); Richard A Falk, "Forward," in Roland J. Stranger (ed), *Essays on Espionage and International Law* (Columbus, OH: Ohio State University Press, 1962); Adrienne Wilmoth Lerner, "Espionage and Intelligence, Early Historical Foundations," in Lee Lerner and Brenda Wilmoth Lerner (eds), *Encyclopedia of Espionage, Intelligence, and Security*, (New York: Thomson Gale, 2004) at <http://www.encyclopedia.com/doc/fullarticle/1G2-3403300282.html>.

[103] Hugo Grotius, *The Rights of War and Peace, Including the Law of Nature and of Nations* (New York: Cosimo, Inc., 2007) 331.

[104] Convention (IV) Respecting the Laws and Customs of War on Land and Its Annex: Regulations Concerning the Laws and Customs of War on Land, October 18, 1907, Article 31, 36 Statute 2277, 2304.

admonition that "[t]raditional international law is remarkably oblivious to the peacetime practice of espionage" continues to hold true today.[105] Indeed, the ICJ appears to be going out of its way to avoid creating opinio juris with respect to peacetime espionage.[106]

Scholars have divided sharply on the legality of peacetime espionage. The majority of scholars assert that it *is* legal, pointing to the predominance of the practice and asserting that better information about what other countries are doing promotes stability and is implicit in the right of preemptive self-defense.[107] Others disagree, condemning it as an impressible violation of the spied-upon country's territorial integrity.[108] Still others assert that the legal status of espionage remains ambiguous, with international law neither condemning nor condoning it.[109] In the absence of any clear principles, with the exception of a handful of exceptions such as interference with diplomatic communiques, espionage remains the province of domestic law and falls outside the province of jus ad bellum and jus in bello.

A literature has only begun to emerge regarding cyber espionage.[110] There are aspects of cyber espionage that may change the optimal outcome. On the one hand, as was the case with satellite surveillance,[111] the lack of territorial invasion makes cyber surveillance less problematic.[112] On the other hand, the dramatic drop in the cost of surveillance and the lower likelihood of apprehension may lead firms to engage in it when before the benefits did not justify the costs. Some regard this as a benefit, as the nonlethal aspects of cyber capabilities make them preferable to other means.[113] Others

[105] Falk (n 107) v.

[106] Dieter Fleck, "Individual and State Responsibility for Intelligence Gathering," (2007) 28 *Michigan Journal of International Law* 687.

[107] Lassa Oppenheim, *International Law* (New York: Longmans, Green, and Co., 2nd edn, 1912); Julius Stone, "Legal Problems of Espionage in Conditions of Modern Conflict," in Stranger (n 107); Beth M Polebaum, "National Self-defense in International Law: An Emerging Standard For a Nuclear Age," (1984) 59 *New York University Law Review* 187; John Kish and David Turns, *International Law and Espionage* (Boston: Martinus Nijhoff, 1995); Geoffrey Demarest, "Espionage in International Law," (1996) 24 *Denver Journal of International Law & Policy* 321; Roger D. Scott, "Territorial Intrusive Intelligence Collection and International Law," (1999) 46 *Air Force Law Review* 217; Simon Chesterman, "The Spy Who Came in From the Cold War: Intelligence and International Law," (2006) 27 *Michigan Journal of International Law* 1071; Fleck (n 111); Luke Pelican, "Peacetime Cyberespionage: A Dangerous, but Necessary Game," (2012) 20 *CommLaw Conspectus* 363.

[108] Falk (n 107); Quincy Wright, "Espionage and the Doctrine of Noninterference in Internal Affairs," in Stranger (n 107); Manuel R Garcia-Mora, "Treason, Sedition and Espionage as Political Offenses Under the Law of Extradition," (1964) 26 *University of Pittsburgh Law Review* 65; Ingrid Delupis, "Foreign Ships and Immunity for Espionage," (1984) 78 *American Journal of International Law* 53; A John Radsan, "The Unresolved Equation of Espionage and International Law," (2007) 28 *Michigan Journal of International Law* 595; John F. Murphy, "Cyber War and International Law: Does the International Legal Process Constitute a Threat to U.S. Vital Interests?," (2013) 89 *International Law Studies* 309.

[109] Christopher D Baker, "Tolerance of International Espionage: A Functional Approach," (2004) 19 *American University International Law Review* 1091; Daniel B Silver, "Intelligence and Counterintelligence," in John Norton Moore and Robert F Turner (eds), *National Security Law* (Durham, NC: Carolina Academic Press, 2nd edn, 2005) (updated and revised by Frederick P Hitz and J E Shreve Ariail), 965.

[110] Sean P Kanuck, "Information Warfare: New Challenges for Public International Law," (1996) 37 *Harvard International Law Journal* 272; Pelican (n 112); Murphy (n 113); David Weissbrodt, "Cyber-conflict, Cyber-crime, and Cyber-espionage," (2013) 22 *Minnesota Journal of International Law* 347.

[111] Stone (n 112). [112] Pelican (n 112).

[113] Jeffrey TG Kelsey, "Hacking into International Humanitarian Law: The Principles of Distinction and Neutrality in the Age of Cyber Warfare," (2008) 106 *Michigan Law Review* 1427.

take the contrary view, arguing that because cyber espionage increases the scale of intelligence-gathering capability, it should be curbed by treating it more severely than traditional espionage.[114] Still others support doing nothing and allowing state practices to evolve.[115] In any event, general agreement never emerged with respect to traditional espionage, and there seems little reason to expect that consensus is more likely to appear in the cyber context.

The law of war, thus, has little to say about the types of surveillance that are generating the most concern. That is why Rule 66(a) of the *Tallinn Manual* explicitly states, "Cyber espionage and other forms of information gathering directed at an adversary during an armed conflict do not violate the law of armed conflict."[116] The commentary similarly notes that cyber espionage does not constitute the use of force or armed attack for purposes of jus ad bellum, armed conflict or attack for purposes *of jus in bello*, or a violation of the principle of non-intervention.[117] It also explains why the commentary notes at several points that international law does not address espionage per se and that, as such, conduct related to espionage is presumptively legal as a matter of international law.[118]

Even more tellingly, the *Tallinn Manual* commentary distinguishes between cyber espionage, which necessarily takes place in territory controlled by one of the parties to the armed conflict, from computer network exploitation and cyber reconnaissance, which are conducted from outside enemy controlled territory. Such conduct is not cyber espionage at all.[119]

The type of activities described in the classified documents disclosed by Edward Snowden, the denial of service attack on Estonia, and the tactics described in *Unrestricted Warfare* are, thus, not governed by international law at all. The legality of such conduct is consigned instead to domestic law.

IV. Potential Cyber Warfare Defense Strategies

Given the unlikely prospect of legal solutions to the problems posed by cyber war, those confronting the risk of cyber attacks must necessarily undertake self-protective measures to protect themselves. In this section of the article we examine defenses and defensive strategies intended to preserve the functioning of a society's information infrastructures and the systems controlled by them.

(a) "Air gaps"

As many of the problems of information infrastructures exploitable by adversaries are enabled by internet connectivity, one defensive strategy that suggests itself is to

[114] Anna Wartham, "Should Cyber Exploitation Ever Constitute a Demonstration of Hostile Intent That May Violate UN Charter Provisions Prohibiting the Threat or Use of Force?" (2012) 64 *Federal Communications Law Journal* 643; Alexander Melnitzky, "Defending America Against Chinese Cyber Espionage Through the Use of Active Defenses," (2012) 20 *Cardozo Journal of International and Comparative Law* 537.

[115] Pelican (n 112). [116] Schmitt (n 17) 158. [117] Schmitt (n 17) 47, 50–1, 52, 56, 75, 92.

[118] Schmitt (n 17) 36, 50, 52, 159. [119] Schmitt (n 17) 159.

keep machines off the global internet. For example, a network intended to carry sensitive military traffic might be created, with its own internet protocol (IP) address and domain name system (DNS) infrastructures and web servers; as long as there is no interchange with the conventional global internet, this can be very effective.[120] There are at least two difficulties, however. First, replicating an internet is logistically challenging and, therefore, expensive, and to truly "air gap" one must prevent all accidental interconnections (e.g., with a laptop connected to the military network with a cable and connected to the global internet with WiFi).

A second issue, also logistical in essence, is that data transfers and updates of software must be done in a way that preserves the "air gap," for example, via storage media such as USB memory sticks and CDs or DVDs. As Stuxnet demonstrates, the data transfers must be carried out carefully to ensure that there is no hidden malware on the medium, and the individuals involved must be trusted to ensure such checking is performed 100% of the time.[121]

(b) "Kill switches"

The "kill switch" idea is that a national authority would have the capability to disconnect their nation from the global internet, in order to disable network-based attacks of various types. As an example, consider the directed denial of service (DDoS) attacks against Estonia. If Estonia had a kill switch capability, then the floods of external traffic would be cut off. Some engineering would be required to accomplish this, notably ensuring that the DNS would continue to work via caches and redirections.

To engineer such a defensive solution requires that all, or a vast majority, of the network resources incorporate a control mechanism by which the national authority can exercise the cut-off decision. Economics and reliability suggest that minimizing the number of such control points makes sense, but this has the unfortunate consequence of creating fewer points that can fail before a catastrophe. These control points also create an attractive target for an adversary, and by the nature of their role are difficult to test previous to their engagement.

(c) Special treatment of information systems for critical infrastructure

Consistent with the observations in subsections (a) and (b), connectivity to the global internet enables attacks on critical infrastructures. Imagine, for example, that a bored employee at a control point in the electrical power grid connects a personal laptop, tablet, or WiFi-enabled smartphone into the building network to access entertainment such as online gaming. There are various scenarios, for example, a cellular network/

[120] Defense Advanced Research Projects Agency, "Military Networking Protocol (MNP)," FedBizzOps.gov, October 28, 2008, at <http://www.fbo.gov/index?s=opportunity&mode=form&id=01886bf13926063b1cc0e996b223440f&tab=core&_cview=1>.

[121] Nicholas Falliere, Liam O Murchu, and Eric Chien, "Symantec W32.Stuxnet Dossier, Version 1.4," Symantec, February 2011, at <http://www.symantec.com/content/en/us/enterprise/media/security_response/whitepapers/w32_stuxnet_dossier.pdf>.

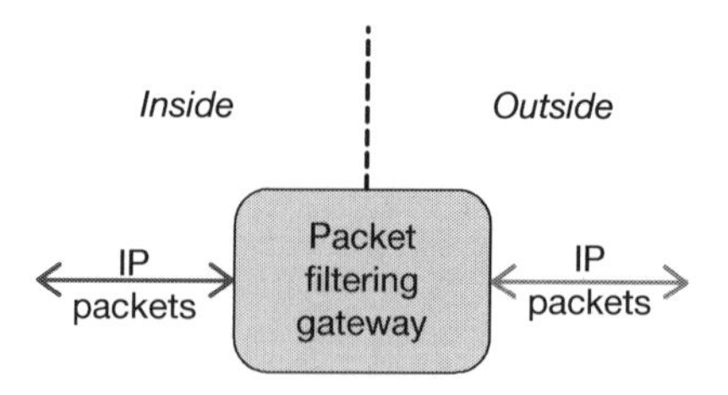

Fig. 8.1 Packet-filtering gateways used for network defense

building network bridge, that conceivably could provide malware or malicious actors with access to the power grid.

Here, the right strategy is isolation of the grid control interfaces from the internet and machines connected to it. While it is conceptually possible (see subsection (d)) to have application-specific gateways, complexity seems to inevitably creep in and create opportunities for malware to overcome the gateway's role (isolation) in the system design.

(d) Network-embedded perimeter defenses

An alternative to physical air gaps is the idea of a packet-filtering gateway or firewall, as illustrated in Figure 8.1. A "firewall" is a packet-filtering gateway.[122] The typical role of such a device is to segregate packet traffic into an "inside" and "outside," where certain activities are allowable if originating from the inside but not if originating from the outside.

The decisions about what activities are allowable and not allowable is encoded in a packet-filtering policy implemented by the firewall. The problem with firewalling as a strategy is that many security threats have moved into applications, for example, into interpreters embedded into applications such as document formatters and renderers, as well as browsers. "Port 80" (used by the Web's HTTP) cannot be closed, yet much of today's dangerous material flows in from the web. This suggests application gateways that provide extremely constrained interfaces to applications, perhaps complemented by one or more packet-filtering gateways.

(e) Improved software engineering

A major issue in cyberwarfare is software that can be exploited to perform unwanted actions. The very fact that such unwanted actions are feasible is a sign that there are mistakes in the software's design. The discipline of software engineering is intended to produce correct software.[123] Software correctness in the context of security means that the right action is performed for the right person in the right place at the right time, with no missing functions, errors or "extras." Modern design techniques, such

[122] WR Cheswick and SM Bellovin, *Firewalls and Internet Security: Repelling the Wily Hacker* (Boston: Addison-Wesley, 1994).

[123] Daniel M Hoffman and David M Weiss (eds) *Software Fundamentals: Collected Papers by David L. Parnas* (Boston: Addison-Wesley, 2001).

as software development with the aid of theorem provers,[124] and use of modern programming languages, such as Haskell and OCaml, preclude many of the most common errors (e.g., buffer overflows) that pervade software implementations.[125] While such tools do not guarantee a correct or appropriate design, they remove a great deal of low-hanging fruit exploitable by the attacker and allow the defending programmer to focus more attention on the software logic, interfaces, and overall design.

Software engineering has a great deal to offer as a discipline,[126] yet it is underutilized as a defensive mechanism, not least because market forces pressure developers towards a strategy of release early and often that may or may not be wise. Cyber security considerations have very limited traction in the marketplace, as there seems to be little in the way of financial gain or penalty for writing software that is more secure or less secure. One policy, for better or worse, that would create financial incentives for correct functioning would be to create a legal doctrine of software liability, involving torts for failures of the software, as well as the inevitable documentation of "best-practices" defenses against such torts.[127]

V. Conclusion

At several points in the discussion, the disturbing impression arises that the technology has run ahead of the policy thinking in the domain of cyberwar. We make here a few suggestions on questions and directions that might lead to informed policy-making in this area.

International policies and agreements should clarify how cyber actions fit into existing international law. Although the *Tallinn Manual* is a step in the right direction within its scope, the more problematic area of espionage remains largely unaddressed. Cyber war's greater ability to deploy latent disruptive capabilities and a greater ability to conduct surveillance will make guidelines to determine the propriety of this type of conduct increasingly important in years to come.[128]

On the technological side, engineers should develop means to ensure reliable attribution of actions in cyberspace. In addition, they should develop the technical means to distinguish combatants from non-combatants in order to honor distinction and neutrality. If developed, these capabilities would make it easier to reconcile cyber war with existing principles of international law.

On a broader level, any proposals should account for the huge social and economic value inherent in cooperation, as exemplified by the success of the internet's federated

[124] Benjamin C Pierce, Chris Casinghino, Marco Gaboardi, Michael Greenberg, Cătălin Hriţcu, Vilhelm Sjöberg, and Brent Yorgey, "Software Foundations," Penn Engineering: Computer and Information Science, July 2014, at <http://www.cis.upenn.edu/~bcpierce/sf/>.

[125] Graham Hutton, *Programming in Haskell* (Cambridge: Cambridge University Press, 2007); Xavier Leroy, Damien Doligez, Alain Frisch, Jacques Garrigue, Didier Rémy, and Jérôme Vouillon, "The OCaml System (Release 4.01): Documentation and User's Manual," The Caml Language, September 12, 2013, at <http://caml.inria.fr/pub/docs/manual-ocaml/>.

[126] Hoffman and Weiss (n 128); John Viega and Gary McGraw, *Building Secure Software: How to Avoid Security Problems the Right Way* (Boston: Addison-Wesley, 2002).

[127] Jonathan M Smith, "At Issue: Should Software Manufacturers be Liable for Vulnerabilities in Their Software?" (2003) 13 *CQ Researcher* 811.

[128] Baker (n 114); Demarest (n 112).

but cooperative architecture. At the same time, they should recognize how a lack of alignment in underlying interests, the increase in the number of participants, and the reduced ability to verify others' compliance have the tendency to cause cooperation to break down. In the absence of such cooperation, nation-states may use lawfare as a tool of obstruction in the course of waging total war.[129]

These are not concrete proposals as such, but rather directions that might result in productive discussions by technologists interested in policy issues and policy-makers interested in the risks and management challenges associated with a defensive posture in cyber space.

[129] Qiao and Wang (n 9).

9

Deception in the Modern, Cyber Battlespace

William H Boothby

I. Introduction

If movement, and thus maneuver, characterized many of the twentieth century's developments in the conduct of warfare, information and its manipulation seem destined to be the features critical to success in the contests of the twenty-first century. Conventional methods of warfare will continue to feature. The rifle, the bayonet, mortars, bombs, missiles, and mines will remain critically important tools in the conduct of hostilities in many future, conventional armed conflicts. The oldest fighting domains of land and sea, together with the additions of the twentieth century, airspace and outer space, will remain the centerpiece of the conduct of most hostilities. But cyberspace will, as the author of this chapter sees it, become the environment in which adversaries employing some degree of operational sophistication will seek to gain and to maintain military advantage by leveraging their own hostile activities while impeding the enemy's capacity to organize and operate. Some see cyberspace as a domain in its own right—a fifth domain for military operations. Others see it as the way in which advantage may be obtained in the four established domains. That is a controversy that is largely irrelevant to the matters discussed in the present chapter and which we can, therefore, now safely ignore.

If the manipulation of information is likely to be the cornerstone of success in many future operations, that suggests that cyber deception operations will be a recurring and potentially decisive feature of future warfare. The purpose of the present chapter is, therefore, to consider what the law has to say about deception operations in warfare and to work out what the implications are for the lawful exploitation of cyber deception methods. The analysis suggests that certain foreseeable types of cyber deception operation will cause the cyber hacker to become the "attacker" and thus to have the attendant responsibilities, for example, under Article 57 of Additional Protocol I. Other kinds of cyber interference may not cause the hacker to displace the attacker in this way, but may obstruct adverse party compliance with the principle of distinction. If, as the International Court of Justice (ICJ) has observed, that principle is intransgressible, the Article questions the acceptability of such cyber operations and the adequacy of contemporary legal rules that would currently permit them.

In the second section, we consider whether deception operations in the cyber sphere seem likely and, if so, what form they are liable to take. In the third section, we take a brief look at the history of the use of deception in warfare. In section IV we trace the development of the relevant law, explaining what the modern law permits and, respectively, prohibits. In section V we seek to apply those legal rules to different kinds of cyber deception and ask ourselves whether the law challenges, or is challenged by, the foreseeable use of such technologies. In the sixth and final section we shall try to draw some conclusions.

The discussion in this chapter will, sensibly, be conducted by reference to the approach taken in the *Tallinn Manual*.[1] It will, therefore, be useful to state at the outset the meanings that the *Manual* gives to certain pivotal terms. Its Glossary describes "cyber" as connoting a relationship with information technology so that "cyber operations" are the "employment of cyber capabilities with the primary purpose of achieving objectives in or by the use of cyberspace."[2]

II. Likelihood and Form of Cyber Deception Operations

In this section, we consider whether cyber deception operations are a likely development in the way in which warfare is conducted, and if so, the kinds of cyber operation that seem likely to be employed for such purposes.

The UK Cyber Security Strategy[3] notes that "a growing number of adversaries are looking to use cyberspace to steal, compromise or destroy critical data"[4]; that "other states....seek to conduct espionage with the aim of spying on or compromising our government, military...assets"; that in "times of conflict, vulnerabilities in cyberspace could be exploited by an enemy to reduce our military's technological advantage"[5]; that "with the borderless and anonymous nature of the Internet, precise attribution is often difficult and the distinction between adversaries is increasingly blurred"; and that "[s]ome states regard cyberspace as providing a way to commit hostile acts 'deniably'."[6]

The US Department of Defense Strategy document[7] notes that "[t]oday, many foreign nations are working to exploit DoD unclassified and classified networks, and some foreign intelligence organisations have already acquired the capacity to disrupt elements of DoD's information infrastructure.... We recognize that there may be malicious activities on DoD networks and systems that we have not yet detected."[8] More specifically:

> [p]otential US adversaries may seek to exploit, disrupt, deny, and degrade the networks and systems that DoD depends on for its operations. DoD is particularly concerned with three areas of potential adversarial activity: theft or exploitation of

[1] This is a reference to Michael N Schmitt (ed), *Tallinn Manual on the International Law Applicable to Cyber Warfare* (New York: Cambridge University Press, 2013). The author was a member of the International Group of Experts who prepared the *Manual*.

[2] Cyberspace is the "environment formed by physical and non-physical components, characterized by the use of computers and the electro-magnetic spectrum, to store, modify, and exchange data using computer networks."

[3] The UK Cyber Security Strategy, Protecting and Promoting the UK in a digital world, November 2011 at <https://www.gov.uk/government/uploads/system/uploads/attachment_data/file/60961/uk-cyber-security-strategy-final.pdf > accessed June 24, 2014 (UK Cyber Security Strategy).

[4] UK Cyber Security Strategy (n 3) para 2.3.

[5] UK Cyber Security Strategy (n 3) para 2.5.

[6] UK Cyber Security Strategy (n 3) paras 2.8 and 2.14.

[7] US Department of Defense Strategy for Operating in Cyberspace, July 2011 at <http://www.defense.gov/home/features/2011/0411_cyberstrategy/docs/DoD_Strategy_for_Operating_in_Cyberspace_July_2011.pdf> accessed June 24, 2014 (US DoD Strategy).

[8] US DoD Strategy (n 7) 1.

data; disruption or denial of access or service that affects the availability of networks, information or network-enabled resources; and destructive action including corruption, manipulation, or direct activity that threatens to destroy or degrade networks or connected systems.[9]

The US Army Cyberspace Concept and Plan goes on to emphasize the vulnerability to attack, degradation, or destruction of network-enabled systems and services on which the US Army relies to provide a communications infrastructure linking soldiers and platforms with global information sources.[10] It is, therefore, reasonable to conclude from all of this that the use of cyber capabilities to distort or disrupt the enemy's critical military information systems is not only foreseeable, even likely, but has indeed been foreseen and drives the development of strategic concepts as to future operations.[11] Distortion and disruption are, of course, rather different activities and the legal significance of that difference will be considered later.

If disruption and distortion operations in cyberspace seem likely in the future, we now look to the recent past to consider some of the forms that such activities may be expected to take.[12]

On April 27 and 28, 2007 a series of cyber denial of service operations, involving an apparent degree of co-ordination, started to affect websites in Estonia during its dispute with Russia. Simple ping requests were followed by malformed web queries against governmental and media websites. In 2007, from April 30 until May 18, more coordinated cyber operations occurred, involving distributed operations aimed at producing a directed denial of service (DDos) from targeted websites. Cyber operations were scheduled to take place at particular times to maximize their effect and targeted sites were rendered inaccessible for periods of time. The indication was that botnets, discussed further, were being used to achieve a precise, concentrated impact.[13] Some Estonian websites were defaced by so-called "patriotic hackers." Of vital significance

[9] US DoD Strategy (n 7) 3. The US Army's Cyberspace Operations Concept Capability Plan 2016–2028, dated February 22, 2010, TRADOC PAM -525-7-8 (US Army Cyberspace Concept and Plan) explains the importance of this contest. It points out that a significant advantage goes to the party to the conflict that "gains, protects and exploits advantage in the contested and congested cyberspace" while "the side that fails in this contest, or that cannot operate effectively when their systems are degraded or disrupted, cedes a significant advantage to the adversary"; para 1–1.c.

[10] US Army Cyberspace Concept and Plan, para 3.2.

[11] The scale of the threat is laid bare in para 3.4: "The Army cannot adequately identify, attack, exploit, and defeat the expanding cyber-electromagnetic threats or mitigate the increasing vulnerability of its own networks."

[12] There are also reports that the US considered the possible use of cyber capabilities in the preparatory phase before air operations commenced over Libya in 2011, clearly demonstrating that cyber warfare is considered to have moved from a theoretical construct to a realistic option. See E Nakashima, "US Cyberweapons Had Been Considered to Disrupt Gaddafi's Air Defences," *Washington Post*, October 18, 2011, at <http://www.washingtonpost.com/world/national-security/us-cyber-weapons-had-been-considered-to-disrupt-gaddafis-air-defenses/2011/10/17/gIQAETpssL_story.html> accessed June 24, 2014. It appears, however, that time constraints did not allow the use of cyber capabilities.

[13] E Tikk, K Kaska, and L Vihul, *International Cyber Incidents: Legal Considerations* (2010) published by Cooperative Cyber Defence Centre Of Excellence, Tallinn and available at http://www.ccdcoe.org/publications/books/legalconsiderations.pdf, 18–25.Note also the DDoS operation on April 26–8, 2008, which targeted the website of Radio Free Europe/Radio Liberty's Belarus service reported and discussed at Tikk et al *International Cyber Incidents*, 39–48 and the cyber operation that targeted Lithuania on June 17, 2008. See Tikk et al, 51–64.

for the purposes of the current chapter, however, it was never formally determined whether any particular state was responsible for sponsoring these cyber operations against Estonia.[14]

On 19 July 2008, a computer located at a US ".com" Internet Protocol (IP) address was used to achieve remote command and control of malware hosted on that computer, which initiated a DDoS operation against the website of Mr Mikheil Saakashvili, President of Georgia, putting the website out of operation for twenty-four hours.[15] Then, on August 8, 2008, there was a second wave of DDoS operations against Georgian websites that spread to computers throughout the Georgian government. It is clear that these cyber operations coincided with the movement by Russia of its military forces into Georgia,[16] but an important feature of these events was the ease with which private citizens, motivated as loyal citizens, could become involved as cyber warriors.[17] While the individuals so involved are liable to be characterized as direct participants in the hostilities and, thus, to lose their protection from attack while so involved,[18] again the important aspect of the matter for the purposes of this chapter is the difficulty that has been encountered in determining definitively and provably which state, if any, was responsible for the cyber operations.

However, a relatively recent cyber event that perhaps most clearly demonstrates the ability to mount cyber deception operations and the form that they can take involved the use of an integrated set of components, known as "Stuxnet," to undertake computer network attacks. During the Stuxnet attack against Iranian nuclear centrifuges discovered in 2010, malware affected the computerized control of centrifuges apparently associated with an Iranian nuclear program so as to cause the latter to revolve at excessive speed, resulting in the reported damage of the centrifuges. It appears that

[14] W A Owens, K W Dam, and H S Lin, *Technology, Policy, Law and Ethics Regarding US Acquisition and Use of Cyberattack Capabilities* (2009) Washington DC: National Academies Press, 173–6.

[15] J Markoff, "Georgia takes a Beating in the Cyberwar with Russia," *The New York Times Bits Blog*, August 11, 2008, at <http://bits.blogs.nytimes.com/2008/08/11/georgia-takes-a-beating-in-the-cyber-war-with-russia/> accessed June 24, 2014; European Union Independent International Fact Finding Mission on the Conflict in Georgia, Report (2009) and see also Tikk et al (n 13) 67–79. Consider also operations against Kyrgyzstan in 2009; J Hunker, *Cyber War and Cyber Power: Issues for NATO Doctrine* (NATO Research Paper No 62, November 2010) 3.

[16] J Markoff, "Cyber Attacks Disable Georgian Websites," *The New York Times Bits Blog*, August 11, 2008, at <http://bits.blogs.nytimes.com/2008/08/11/georgia-takes-a-beating-in-the-cyberwar-with-russia/> accessed June 24, 2014.

[17] Evgeny Morozov, "An Army of Ones and Zeroes: How I Became a Soldier in the Georgia-Russia Cyberwar," *Slate*, August 14, 2008, at <http://www.slate.com/articles/technology/technology/2008/08/an_army_of_ones_and_zeroes.html>.

[18] The events mentioned so far in this section are referred to as cyber operations. Rule 30 of the *Tallinn Manual* defines a cyber attack as "a cyber operation, whether offensive or defensive, that is reasonably expected to cause injury or death to persons or damage or destruction to objects." This definition is based on the definition of "attack" in Article 49(1) of the Protocol Additional to the Geneva Conventions of August 12, 1949, and Relating to the Protection of Victims of International Armed Conflicts, Geneva, June 8, 1977. The Commentary to Rule 30 makes the point that the notion of "acts of violence" in Article 49(1) "should not be understood as limited to activities that release kinetic force" and explains that it is the effects that are caused that lie at the crux of that notion; *Tallinn Manual*, commentary accompanying Rule 30, para 3 and note that Walker comes to the conclusion that the cyber operations against Georgia and Estonia did not reach this threshold and were not, therefore, cyber attacks. PA Walker, "Rethinking Computer Network 'Attack', Implications for Law and US Doctrine", 1 *National Security Law Brief* 33, 50–3.

the cyber attack was so conducted that the sensors and computers responsible for monitoring the performance of the centrifuges were provided with false status data and thus had, and presented, a false overview of the status of the centrifuges, namely that they were operating normally. This seems to suggest that cyber deception operations of the sort we are discussing may involve pretence, for example, that no attack has occurred, or that no military cyber operation has taken place, when in fact it has.[19] The operations against Estonia and Georgia, referred to earlier, may further suggest that cyber operations may be undertaken by persons whose actions are not in fact attributable to a state, or may be so arranged by a state as to cause the operations to appear to be non-attributable to it. It would also seem sensible to deduce that a cyber operation by one state may be so arranged as to cause it to appear to have been undertaken by or on behalf of another state.

Cyber deception operations may also mask or falsely describe the nature of the operation that is being undertaken. For example, military logistics activities might be falsely labeled as medical processes in order to protect them unlawfully from attack. Cyber means might be used during an armed conflict by one of the parties to give false information as to the disposition of its forces or a party may use cyber means to intrude into enemy computer networks and insert false information in substitution for correct data. Enemy military computer messages may be corrupted or forged, false orders may be issued, out of date or otherwise inaccurate sensor data may be inserted into computer networks that support targeting decision-making, pattern of life data may be corrupted, and the common operational picture on which the commander's decision-making critically depends may be distorted or destroyed. Naturally, there are other possible kinds of cyber deception operations that are likely to be employed in the course of future armed conflicts, but those mentioned in the present paragraph appear to be likely examples.

It follows from this that future armed conflict will almost certainly witness cyber intrusions into and interference with the enemy's networks, and this will likely include those concerned with targeting decision-making, using techniques deliberately so

[19] Stuxnet is delivered in part by a computer worm. It inserts itself onto air-gapped networks by means of such devices as a thumb drive, a DVD, or a CD-ROM. It checks for a distinct set of conditions, such as the presence and configuration of specific industrial control software, before activating. It then operates in a designed way. Sophisticated malware delivered in a similar way was used to effect the Stuxnet attack on Iran in July 2010; see J Fildes, "Stuxnet Worm Attacked High Value Iranian Assets," *BBC News*, September 23, 2010, at <http://www.bbc.co.uk/news/technology-11388018> accessed June 24, 2014; and W J Broad, J Markoff, and D E Sanger, "Israeli Test on Worm Called Crucial in Iran Nuclear Delay," *The New York Times*, January 15, 2011, at <http://www.nytimes.com/2011/01/16/world/middleeast/16stuxnet.html?pagewanted=all> accessed June 24, 2014. As to the likelihood of future clandestine operations in cyberspace, consider UK Ministry of Defence, DCDC, Future Land Operating Concept, JCN 2/12 dated May 2012 para 349, which emphasizes the need to resource and exercise the integrated exploitation of deception operations, and of defensive measures against deception, particularly through the use of emerging technologies and cyberspace. The need to use military and civilian intelligence agencies to coordinate such deception-related activities is noted at para 350. Current UK air power doctrine describes offensive counter space (OCS) operations in terms of preventing adversaries from exploiting space "by attacking their capabilities through deception, disruption, denial, degradation and destruction. As adversaries become more dependent on space, OCS Operations will become increasingly important in affecting their ability to organize and orchestrate military campaigns." British Air and Space Power Doctrine, Air Publication 3000, 4th edn, at <http://www.raf.mod.uk/rafcms/mediafiles/9E435312_5056_A318_A88F14CF6F4FC6CE.pdf> accessed June 24, 2014.

designed as to remain concealed from the enemy. Consequently, the danger is that data that are essential to accurate and lawful targeting decision-making will be rendered unreliable or factually wrong without the intended users becoming aware of that unreliability. A party to the conflict that is trying to comply with international law obligations in attack, including the principles of distinction and discrimination and the precautions rules, must of necessity base its decisions on the information that it has at its disposal. This will critically include the data on its own support systems. In the event of such cyber operations, that party will be basing its decisions on what turns out to be false information about what is going on in the battlespace. So attacks that that party believes are targeting combatants or military objectives may in fact be attacking protected persons and/or objects. This may be happening despite the fact that that party has taken all feasible precautions as required by Article 57 of API. Indeed, the deception may go further in that the attacker's battle damage assessments may also be falsified so as to indicate, incorrectly, that the attack did indeed engage the intended target in the planned way. It is the purpose of the remainder of this chapter to consider whether such cyber deception activities are coherent with the law as it applies to deception operations in general during armed conflict. In section III, therefore, we consider the evolution of the use of deception during warfare.

III. The Use of Deception in Warfare

The extensive history of the use of deception operations in armed conflict demonstrates conclusively that there is nothing unlawful, per se, in their employment. The mere fact of deception does not pre-suppose illegality. Virgil refers in the *Aeneid* to the use of a wooden model of a horse to infiltrate a Greek unit into the city of Troy after ten years of siege.[20] Very much more recently, Operation Mincemeat[21] during World War II involved the use of a dead body bearing false papers describing a fictitious attack plan with the purpose of deceiving the German High Command into believing that the focus of the allied attack in 1943 would be on Sardinia and Greece whereas Sicily was intended to be the focus of the planned attack.

Spaight describes the use of false nationality marks on aircraft during World War I. As he comments:

> The inadmissibility of the use of such marks was established, first, by the accusations which the belligerents made against one another of resorting to the practice, secondly by their indignant denials of any complaints that they had done so themselves.[22]

Simply to feign death, however, in order to avoid being attacked and with a view thereafter to escaping from a difficult tactical situation is an established and

[20] Virgil, *Aeneid*, Bk II, lines 13–16 describes the crafting of "a horse, thanks to Pallas's divine skill, high as a mountain, built on a framework of ribbing with interlocked sections of pitch pine." A captured Greek, Sinon, tells the Trojans falsely that the wooden horse is an offering to the Gods and if they take it into their city, the Greeks will be defeated. They do so and the soldiers concealed within the structure are released at night, upon a signal, by Sinon and kill the Trojan guards;Virgil, *Aeneid*, lines 63–267.

[21] B Mcintyre, *Operation Mincemeat: The True Spy Story that Changed the Course of World War II* (London: Bloomsbury, 2010).

[22] J M Spaight, *Air Power and War Rights* (3rd edn, Longmans, 1947) 85–6.

legitimate tactic.[23] Similarly, making use of dummy communications to pretend to the enemy that fighter aircraft were active when they were not[24] was also considered to be lawful.

This shows that deception operations the purpose of which was to mislead the enemy about the military strength, the identity, the military plans, the military equipment or the operational objectives of the party employing the deception were considered to be lawful. Conversely, the use of false or enemy nationality marks in air warfare, for example, was regarded as prohibited. In the section IV we will consider how the modern law on these matters has evolved.

IV. The Evolving Law as to Deception Operations in Warfare

When considering the evolution of the modern law on deception operations, we should take as our starting point the Code written by Dr Francis Lieber.[25] The Lieber Code asserts that military necessity "admits of deception, but disclaims acts of perfidy."[26] As we shall see this is a distinction that has stood the test of time and that is reflected in the modern law of armed conflict.

The broader context of these legal rules is reflected in the Hague Regulations 1907 provision that "the right of the belligerents to adopt means of injuring the enemy is not unlimited."[27] More specifically, the Regulations especially prohibit "kill[ing] or wound[ing] treacherously individuals belonging to the hostile nation or army."[28] Maintaining the distinction that Dr Lieber made, Article 24 stipulates: "Ruses of war and the employment of measures necessary for obtaining information about the

[23] Spaight (n 22) 173. Spaight also discusses events involving Lieutenant L G Hawker, who was attempting to attack a German airship shed at Gontrode in April 1915. It appears that Hawker used "an occupied German captive balloon to shield him from fire whilst manoeuvring to drop the bombs." Spaight, 174, citing *London Gazette*, May 8, 1915.

[24] Spaight (n 22) 176–8 cites numerous examples of such ruses in both World Wars.

[25] Instructions for the Government of Armies of the United States in the Field, April 24, 1863 (Lieber Code). This Code was written for President Lincoln in 1861 and was issued as General Orders 100 to the Union side in the American Civil War in 1863. It does not have the formal status of a source of international law, but is an authoritative statement by an informed observer of what the law was then considered to comprise.

[26] Lieber Code, Article 16. Article 101 spells out the position as follows: "While deception in war is admitted as a just and necessary means of hostility, and is consistent with honorable warfare, the common law of war allows even capital punishment for clandestine or treacherous attempts to injure an enemy, because they are so dangerous, and it is difficult to guard against them." For an appreciation of the likely intended meaning of these provisions, see S Watts, "The Structure of Law-of-War Perfidy," February 11, 2013, at <http://papers.ssrn.com/sol3/papers.cfm?abstract_id=2220380> accessed June 24, 2014, 18–21 where, among other aspects, the significance of Articles 15, 63, 65, and 117 of the Code and of other contemporary understanding and writings is discussed. Tracing the evolving arrangements in the Brussels Declaration and in the Oxford Manual, Sean Watts comes to the view that these instruments represent "an early effort to evolve perfidy from generally prohibited conduct to a specific and technically proscribed method of warfare." 25.

[27] Annex to Hague Convention IV Respecting the Laws and Customs of War on Land, Article 22, The Hague, October 18, 1907. This text superseded a similar but not identical set of Regulations annexed to Hague Convention II of 1899. The Regulations are widely regarded as having customary status and thus as binding all states, irrespective of their adoption of the Convention.

[28] For the purposes of the present chapter, the author will treat the words "treachery" and "perfidy" as, in practical terms, synonymous. *Tallinn Manual*, para 1 of the commentary accompanying Rule 60. For a discussion of historical examples of treacherous and perfidious conduct, see Watts (n 26) 1–9.

enemy and the country are considered permissible."[29] The important legal distinction that is emerging here, and that binds, of course, states that are not party to API, is between treacherous and, therefore, unlawful killing or wounding, and ruses and espionage both of which are explicitly described as lawful.

The modern law on deception operations is contained in Articles 37 and 38 of API. The first of these Articles deals with the perfidy/ruses distinction as follows:

> (1) It is prohibited to kill, injure or capture an adversary by resort to perfidy. Acts inviting the confidence of an adversary to lead him to believe that he is entitled to, or is obliged to accord, protection under the rules of international law applicable in armed conflict, with intent to betray that confidence, shall constitute perfidy. The following acts are examples of perfidy:
>
> (a) the feigning of an intent to negotiate under a flag of truce or of a surrender;
>
> (b) the feigning of an incapacitation by wounds or sickness;
>
> (c) the feigning of civilian, non-combatant status; and
>
> (d) the feigning of protected status by the use of signs, emblems or uniforms of the United Nations or of neutral or other States not Parties to the conflict.
>
> (2) Ruses of war are not prohibited. Such ruses are acts which are intended to mislead an adversary or to induce him to act recklessly but which infringe no rule of international law applicable in armed conflict and which are not perfidious because they do not invite the confidence of an adversary with respect to protection under the law. The following are examples of such ruses: the use of camouflage, decoys, mock operations and misinformation.

Of course perfidy[30] and lawful ruses[31] have much in common. Both have the purpose of misleading the enemy so as to cause him to act contrary to his interests. However, the vital distinction lies in the nature of the false belief that the operation seeks to induce. A ruse cannot relate to protected status under the law of armed conflict whereas perfidy has that protected status at the core of the deception, whether as to entitlement to receive protection or as to the obligation to accord it. The Program on Humanitarian Policy and Conflict Research at Harvard University (HPCR) *Air and Missile Warfare Manual*[32] comments that a "typical example of perfidy would be to

[29] Sean Watts considers the Hague Regulations formulation of the perfidy rule to be "a nearly rote reproduction of mid-to-late nineteenth century nascent positivism." Watts (n 26) 30.

[30] Sean Watts takes the view that, for states party to API, the formulation of the perfidy rule in Article 37 essentially replaces the corresponding rules in the Hague Regulations. Watts (n 26) 42–3. As to the meaning of perfidy, see S Oeter, "Methods and Means of Combat", in D Fleck (ed), *The Handbook of International Humanitarian Law* (Oxford: Oxford University Press, 2nd edn, 2008) 119, 227–9.

[31] Noting the API Commentary at para 1521, Stefan Oeter confirms that transmitting misleading messages for example by using the enemy's radio wavelengths, passwords or codes and infiltrating his command chain to transmit false orders are established elements of traditional tactics. Oeter (n 30) 227.

[32] Program on Humanitarian Policy and Conflict Research at Harvard University, *Manual on International Law Applicable to Air and Missile Warfare*, published with Commentary, March 2010, at <http://ihlresearch.org/amw/HPCR%20Manual.pdf> accessed June 24, 2014 (*AMW Manual*).

open fire upon an unsuspecting enemy after having displayed the flag of truce, thereby inducing the enemy to lower his guard."[33]

It is important to appreciate that the perfidy rule is only breached if the perfidious act has an adverse impact on the enemy. Furthermore, that adverse impact must come within the categories listed in Article 37(1), that is, death, injury, or capture. The Rule is not, therefore, broken if the adverse consequences of the perfidious act are limited to annoyance, inconvenience, or damage.[34] Moreover, the perfidious act must be the proximate cause of the death, injury, or capture.[35] The time interval between the perfidious act and the death, injury or capture may be short or long. That time interval is not the relevant issue. What matters is, quite simply, whether the former causes the latter.[36]

The International Group of Experts (IGE) who prepared the *Tallinn Manual* considered whether deception as to protected status that does not operate upon a human mind but that influences the operation of a computer thereby causing death, injury, or capture is capable of amounting to perfidy. They considered as an example a cyber operation that causes an enemy commander's pacemaker to malfunction causing his death. A majority of them felt that if such a cyber operation betrays the confidence of the computer controlling the pacemaker, it would breach the perfidy rule. The minority view was that the "notion of confidence presupposes human involvement, such that influencing a machine's processes without consequently affecting human perception falls outside the Rule."[37] In the author's view, the minority view represents the lex lata position, given that, in the stated example, the computer controlling the pacemaker is not an intelligent agent of the kind that must be deceived for a breach of the Rule to be established. As a matter of lex ferenda, it is possible that the Tallinn majority view will prevail among states, although the author suspects that deception of computer processes is likely to be and to become such a fundamental and prevalent aspect of cyber operations in general that states are unlikely to regard it as amounting to perfidious conduct.

In order to reinforce this point, we should consider a further example. A cyber operation is undertaken against, say, the computer system that controls the air traffic control facility at an enemy military airfield where employees of a civilian company provide air traffic services. The military cyber attacker uses the password of a civilian

[33] *AMW Manual* (n 32), commentary accompanying Rule 111(a), para 8. The *Tallinn Manual*'s perfidy rule is expressed in similar terms to Article 37. *Tallinn Manual*, Rule 60. It reflects the customary law Rule by limiting Rule 60 to perfidious killing or injuring and by omitting the API reference to capture. Hague Regulations Article 23(b) does not mention capture. The Rome Statute of the International Criminal Court, 1998, at Article 8(2)(b)(xi) also does not refer to capture. Article 8(2)(e)(ix) of the Rome Statute, which relates to non-international armed conflicts, refers to "[k]illing or wounding treacherously a combatant adversary", a clear reflection of the Hague Regulations formulation. Generally, on the applicability of the perfidy rule to non-international armed conflict, see R B Jackson, "Perfidy in Non-International Armed Conflict" (2012) 88 *International Law Studies* 237–54. In the author's view, in application during non-international armed conflict the Rule will require some adjustment. It would, for example, have to be made clear that a feigning that the individual is not involved in the non-international armed conflict as a fighter and is entitled to protection under the law of non-international armed conflict would be an example of potentially perfidious activity in the context of such a conflict.

[34] *AMW Manual* (n 32), commentary accompanying Rule 111(a), para 7.

[35] M Bothe, K J Partsch, and W A Solf (eds), *New Rules for Victims of Armed Conflicts* (Martinus Nijhoff Publishers,1982) 129, 204.

[36] *Tallinn Manual* (n 1), commentary accompanying Rule 60, para 6.

[37] *Tallinn Manual* (n 1), commentary accompanying Rule 60, para 9.

employee of the company to gain access to the computerized air traffic control system. He interferes with the normal operation of that system in such a way as to cause an aircraft incident in which deaths or injuries to enemy personnel are caused. The confidence of the computer has been induced by the misuse of the password. That misuse takes the form of feigning the civilian status of an employee of the air traffic control company. The majority of the *Tallinn Manual* experts would consider this sort of cyber operation perfidious. Indeed, in their view any cyber operation that feigns civilian status by falsely using a civilian's password to gain access to a closed system and which, having falsely obtained that access, causes death, injury, or capture of persons belonging to the adverse party is going to breach the perfidy rule, a position with which, as previously noted, the author does not agree.

The *AMW Manual* points out that the fact "that a person is fighting in civilian clothing does not constitute perfidy"[38] and lawful ruses that give rise to the death, injury or capture of enemy personnel do not as a result of so doing become unlawful.[39] The *AMW Manual* goes on to list the following examples of lawful ruses, namely: "(a) mock operations[40]; (b) disinformation[41]; (c) false military codes and false electronic, optical or acoustic means to deceive the enemy (provided that they do not consist of distress signals, do not include protected codes, and do not convey the wrong impression of surrender)[42]; (d) use of decoys and dummy-construction of aircraft and hangars; and (e) use of camouflage."[43]

Examples of perfidious conduct during air operations would include: "(a) the feigning of the status of a protected medical aircraft, in particular by the use of the distinctive emblem or other means of identification reserved for medical aircraft; (b) the feigning of the status of a civilian aircraft; (c) the feigning of the status of a neutral aircraft; (d) the feigning of another protected status; and (e) the feigning of surrender."[44]

Moreover, aircraft may not improperly use distress codes, signals, or frequencies, nor may aircraft other than military aircraft be used as a means of attack.[45] Distress

[38] *AMW Manual* (n 32), commentary accompanying Rule 111(b), para 4; the individual may, however, not be entitled to combatant immunity and may thus be liable to prosecution and punishment under applicable domestic law. Consider *Mohamed Ali et al v Public Prosecutor*, Privy Council, [1969] AC 430, 449; a regular soldier committing an act of sabotage when not in uniform loses entitlement to prisoner of war status.

[39] *AMW Manual* (n 32), commentary accompanying Rule 113, para 3.

[40] The *AMW Manual* refers to the following as examples of lawful ruses of war: air attacks on the Pas de Calais during the weeks leading up to D-Day in 1944, the movement of, for example, an aircraft carrier to create a false impression as to the likely nature of an attack and the use of simulated air attacks to persuade the enemy to activate its ground based air defences.

[41] Examples of disinformation might include operations to induce the enemy to surrender by creating the false impression that they are surrounded, or that an overwhelming attack is about to occur; *AMW Manual* (n 32), commentary accompanying Rule 116(b), para 2. Note the distinction between such permissible disinformation and the unlawful use of false information as to civilian, neutral, or other protected status. *AMW Manual*, para 3.

[42] Lawful ruses include the use of enemy IFF codes, or the use of the enemy's password to avoid being attacked when summoned by an enemy sentry or inducing a false return on the enemy radar screen indicating the approach of a larger force than is the case. *AMW Manual* (n 32), commentary accompanying Rule 116(c), paras 2–3.

[43] *AMW Manual* (n 32), Rule 116. [44] *AMW Manual* (n 32), Rule 114.

[45] *AMW Manual* (n 32), Rule 115. IFF codes are not included for these purposes within distress codes signals and frequencies; *AMW Manual*, commentary accompanying Rule 115(a), para 5.

signals may only be used for humanitarian purposes[46] and use of a distress signal to facilitate an attack is prohibited. So while sending a false distress signal is prohibited, flying an aircraft in a manner that induces the enemy falsely to assume that the aircraft has been damaged does not as such breach the Rule.[47]

A helpful list of lawful ruses is set forth in the *UK Manual* as follows:

> Transmitting bogus signal messages and sending bogus despatches and newspapers with a view to their being intercepted by the enemy; making use of the enemy's signals, passwords, radio code signs, and words of command; conducting a false military exercise on the radio while substantial troop movements are taking place on the ground; pretending to communicate with troops or reinforcements which do not exist. . . . ; and giving false ground signals to enable airborne personnel or supplies to be dropped in a hostile area, or to induce aircraft to land in a hostile area[48]

The other kind of deception operation that is of most likely relevance to the current discussion involves the misuse of particular flags, emblems, and indicators. The current law on these matters is to be found in Articles 38 and 39 of API that prohibit, respectively, the misuse of recognized emblems and of emblems of nationality. Making "improper use" of the distinctive emblem of the Red Cross or Red Crescent[49] or of other emblems, signs, or signals provided for in the Conventions or in the Protocol is prohibited. Similarly prohibited is the deliberate misuse of other internationally recognized protective emblems, signs, or signals.[50] There is no requirement that death, injury, or capture result from the misuse. Improper use, by which is meant any use other than that for which the emblem, sign or signal was intended, is sufficient to establish the breach of the Rule.[51] There was, however, no agreement among the Tallinn Experts as to whether the misuse of a domain name alone, for example, "icrc.org," without the misuse of any associated emblem, sign, or signal would breach the rule.[52]

[46] *AMW Manual* (n 32), commentary accompanying Rule 115(a), para 1.

[47] It should be recalled that a damaged aircraft is not necessarily disabled and is not necessarily surrendering; *AMW Manual* (n 32), commentary accompanying Rule 115(a), para 3. The simulation by a pilot of an aircraft deploying paratroopers of a situation of distress with the intention of giving a false impression that the deploying personnel are entitled to protection under Article 42 of API, "this could amount to prohibited perfidy if it leads to the killing, injuring (or capturing) of an adversary"; *AMW Manual*, commentary accompanying Rule 115(a), para 4 and note API, Article 42(3): "Airborne troops are not protected by this Article."

[48] *UK Manual on the Law of Armed Conflict, UK Ministry of Defence*, Oxford: Oxford University Press, 2004, para 5.17.2. Sean Watts discusses lawful ruses and analyses where legitimate camouflage ends and perfidious conduct starts. Watts (n 26) 51–6. Note the reference here to misuse of enemy passwords as being a lawful ruse, an interpretation that rather supports the author's view that deception directed at a computer's decision-making processes is not, as a matter of lex lata, capable alone of constituting a breach of the perfidy rule.

[49] Article 2(1) of Additional Protocol III to the Geneva Conventions extends this prohibition to the red crystal adopted by that instrument as a distinctive emblem.

[50] This would extend to the distinctive signs for cultural property and for civil defence, to the flag of truce and to the electronic protective markings set out in Annex I to API; Cultural Property Convention, Articles 16 and 17, API, Article 66, Hague Regulations, Article 23(f) and API, Annex I, para 9; *Tallinn Manual*, Commentary accompanying Rule 62, para 2 and see AMW, Rule 112(a) and (b).

[51] *Tallinn Manual* (n 1), commentary accompanying Rule 62, paras 3 and 4, in the latter case citing the *ICRC Study*, commentary accompanying Rule 61.

[52] *Tallinn Manual* (n 1), Commentary accompanying Rule 62, paras 6 and 7.

Article 38(2) of API provides that it is "prohibited to make use of the distinctive emblem of the United Nations, except as authorised by that Organization."[53] Unauthorized use of the emblem by electronic means is, therefore, also prohibited, but, again, the Tallinn Experts could not agree whether the emblem has to be used in order to violate this Rule. Some took the view that it does, while others maintained that any unauthorized use of an apparently authoritative indication of UN status, such as presumably its domain name, suffices.[54]

Using enemy flags, insignia, or military emblems "while engaging in attacks or in order to shield, favour, protect or impede military operations" is prohibited.[55] A majority of the Tallinn Experts considered that "it is only when the attacker's use is apparent to the enemy that the act benefits the attacker or places its opponent at a disadvantage."[56] The *Tallinn Manual* recognizes, however, that "it is permissible to feign enemy authorship of a cyber communication," an evident reflection of state practice on lawful ruses.[57]

Article 39(1) of API prohibits the "use in an armed conflict of the flags or military emblems, insignia or uniforms of neutral or other States not Parties to the conflict."[58] Any such use is unlawful, so the word "improper" is omitted; the Tallinn Experts again failed to agree whether "employment of other indicators of neutral status is prohibited. Accordingly there was no agreement among them as to whether the Rule would prohibit a cyber operation that uses the domain name of a neutral state's government but which does not use its flag, military emblems, and so on.[59]

The final kind of conventional deception operation that we should consider is espionage. The *AMW Manual* tells us that "[e]spionage consists of activities by spies" and goes on to describe a spy as "any person who, acting clandestinely or on false pretences, obtains or endeavours to obtain information of military value in territory controlled by the enemy, with the intention of communicating it to the opposing Party."[60] Cyber espionage and cyber information gathering operations that do not fulfill the criteria for espionage directed at an adverse party to the conflict do not breach the law of armed conflict.[61]

[53] For the application of this rule in cyber operations, see *Tallinn Manual*, Rule 63, citing NWP 1–14, para 12.4, the *UK Manual* (n 48), para 5.10.c, and the *AMW Manual* (n 32), Rule 112(e).

[54] *Tallinn Manual* (n 1), Commentary accompanying Rule 63, para 4. If the UN is party to an armed conflict, misuse of its emblem by an adverse party becomes misuse of an enemy emblem as distinct from misuse of the UN emblem; *AMW Manual* (n 32), commentary accompanying Rule 112(e), para 3.

[55] API, Article 39(2), *AMW Manual* (n 32), Rule 112(c), and *Tallinn Manual* (n 1), Rule 64.

[56] *Tallinn Manual* (n 1), commentary accompanying Rule 64, para 4.

[57] *Tallinn Manual* (n 1), commentary accompanying Rule 64, para 5; see also the extract from the *UK Manual* referred to earlier in the text.

[58] The *Tallinn Manual* acknowledges the customary rules relating to naval warfare and includes a customary Rule in similar terms; *Tallinn Manual* (n 1), Rule 65; see also *AMW Manual* (n 32), Rule 112(d).

[59] *Tallinn Manual* (n 1), commentary accompanying Rule 65, para 4.

[60] *AMW Manual*, Rule 118. Article 29 of the Hague Regulations, 1907 provides: "An individual can only be considered a spy if, acting clandestinely or, on false pretences, he obtains, or seeks to obtain information in the zone of operations of a belligerent, with the intention of communicating it to the hostile party. Thus, soldiers not wearing a disguise who have penetrated into the zone of operations of the hostile army, for the purpose of obtaining information, are not considered spies.... "

[61] *Tallinn Manual* (n 1), Rule 66(a), AMW Manual, Rule 119. Espionage does not breach the law of armed conflict but a combatant who commits espionage loses the right to be a prisoner of war and may be treated as a spy if captured before he reaches the army on which he depends. *Tallinn Manual*, Rule 66(b).

The word "clandestine" requires that the acts are undertaken secretly or secretively and "under false pretences" refers to acts conducted in such a way as to create the impression that the individual has the right to access the information concerned.[62] Accordingly, if a person obtains information about an adversary while the person is located outside enemy-controlled territory his activity does not amount to espionage. This means that most remotely undertaken cyber information gathering operations will not constitute espionage. However, a cyber operation such as that necessary to obtain information from a terminal of a closed computer system located in enemy controlled territory using a memory stick, for example, will constitute espionage if the act of information gathering takes place within the enemy's zone of operations and if the other elements of espionage are present.[63]

An act which is intended to gain information but which fulfils the requirements of perfidy, discussed earlier, and which results in death or injury (or, in the case of an API state party, capture), will not cease to be perfidy because of the information-gathering purpose of the operation.[64] The Commentary to the *Tallinn Manual* puts it this way:

> There is a distinction between feigning protected status and masking the originator of the attack. A cyber attack in which the originator is concealed does not equate to feigning protected status. It is therefore not perfidious to conduct cyber operations that do not disclose the originator of the operation. The situation is analogous to a sniper attack in which the location of the attacker or identity of the sniper may never be known. However, an operation that is masked in a manner that invites an adversary to conclude that the originator is a civilian or other protected person is prohibited if the result of the operation is death or injury of the enemy"[65] or, in the case of an API state party, if the result of the operation is capture of the enemy.

Heather Dinniss cites Baxter when discussing espionage and sabotage, contending that "it is primarily the act of deception for the purposes of destruction or information gathering which negates combatant status" and that "[i]n the digital age, the danger posed by spies and saboteurs to their opponents is not diminished by the lack of physical presence in the adversary's territory, which, in fact, makes it harder for the victim to detect and distinguish such attackers."[66] The present author considers that activities undertaken from outside enemy controlled territory are unlikely, as a matter of lex lata, to be regarded as constituting either espionage or sabotage.

[62] *Tallinn Manual* (n 1), commentary accompanying Rule 66, para 2 citing API Commentary, para 1779. See also *AMW Manual* (n 32), commentary accompanying Rule 120, para 2, where it is noted that an individual is not engaged in espionage if he is in the uniform of his armed forces while gathering the information and that members of military aircrew who wear civilian clothes inside a properly marked military aircraft are not spies.

[63] A civilian who undertakes remote cyber information gathering and close access cyber espionage is likely to be participating directly in the hostilities. Such participation would render the individual liable to attack while so engaged. Espionage, including cyber espionage, is also likely to breach the domestic law of the territory where the activity occurs. If apprehended by the enemy, persons engaged in either such activity are therefore liable to be tried for the relevant offences; *AMW Manual* (n 32), Rule 121.

[64] *Tallinn Manual* (n 1), Commentary accompanying Rule 60, para 11.

[65] *Tallinn Manual* (n 1), Commentary accompanying Rule 60, para 13.

[66] H H Dinniss, "Participants in Conflict—Cyber Warriors, Patriotic Hackers and the Laws of War," in D Saxon (ed), *International Humanitarian Law and the Changing Technology of War* (Leiden: Martinus Nijhoff, 2013) 251, 264–5.

V. Applying the Law to Particular Kinds of Cyber Deception Operation

The discussion in this chapter seems to suggest that future conflict is destined to feature widespread cyber deception operations the detection of which will be problematic, the originator of which will be difficult to identify, the consequences of which may be masked, and which may result in the corruption, in whole or in part, of the data on which targeting decision-making relies. International law, as we have seen, limits its prohibitions of deception operations to perfidy, which is defined relatively narrowly, and misuse of certain indicia. There is no specific legal rule that prohibits cyber activities that corrupt the enemy's understanding of the battlespace to such a degree that they are no longer able to comply with distinction, discrimination, and proportionality rules either when undertaking particular attacks or at all.

Self-evidently, and as is noted in UK statement (c), those who plan and make other decisions in relation to attack must necessarily rely on their interpretation of the available information. It, therefore, follows from this that if the information on which they rely is corrupted or in any other sense inaccurate, and if taking all possible precautionary steps will not enable the decision-makers to become aware of the corruption or other inaccuracy, or of its extent or significance for the decisions they are about to make, the decision-maker's efforts to comply with the law, specifically with the Article 57 precautions obligations, become potentially nugatory. A weapon is, therefore, fired in what is believed to be a lawful attack on what is understood to be a lawful target, but the result may be an attack that is either indiscriminate or that actually targets civilians or civilian objects. The outcome that the attacker has done everything practically possible to prevent is the outcome that in fact materializes owing to the effectiveness of the enemy's falsification of the information on which the attacker bases his decision, and on the effectiveness of the enemy's efforts to mask those falsification operations.[67]

It would seem appropriate to consider the implications of such operations by reference to remotely piloted attack operations using aircraft. The operations routinely conducted over Afghanistan and elsewhere using Predator and Reaper aircraft are controlled by ground-based operators. Those operators rely on the sensors on board the platform and on other sources of information as to what is taking place in the area in which the unmanned platform is operating. The computer systems that co-ordinate that information and present it on the screens observed by the operators, and the computer links that transmit the operators' instructions to the platform and to its weapon

[67] In a presentation given at the annual Cycon Conference in Tallinn in June 2013, the author put the issue as follows:

> If increasingly pervasive cyber capabilities are so used that deception operations become the rule rather than, relatively speaking, the exception, and if as a result little or no reliance can in future be placed on the information that would traditionally support targeting decision making, what are the consequences for the practical ability of combatants to comply with the distinction, discrimination, proportionality and precautions rules that lie at the core of targeting law? At least some concrete basis for reliable decision-making is central to the practical delivery of these protective principles. Widespread use of deception must not, it is suggested, become the cause of a slide into "anything goes." Cycon Conference, Tallinn, June 2013.

systems, are, it would seem, critical to the accurate and reliable conduct of such operations. This very criticality would suggest that, in order to comply with the Article 57(1) obligation on all parties to the conflict to spare civilians, the civilian population and civilian objects, states operating such systems must take practical steps to ensure that these links are operating properly. Once it becomes clear that the opposing party has the capacity to interfere with up and down links, with on-board navigation, weapons control, target identification, weapons guidance, and other relevant software, the Article 57 obligations to take constant care and to do everything feasible imply that manufacturers and operators of such systems will need to do all they can to make these systems robust against every known form of cyber interference.[68]

This seems to suggest that the relevant networks and systems should be so designed as to be capable of detecting and highlighting to the operator if cyber interference is affecting adversely the operation of these sensors or data processing systems. If the relevant cyber intrusions are masked or concealed, detection of them may simply not be possible; indeed, undetected data interference, as illustrated by the Stuxnet operation discovered in 2010, is likely to be a developing and enduring reality.

In his book entitled *Conflict Law*,[69] the author of this chapter discusses these issues by reference to certain kinds of attack. A critical issue would appear to be the degree to which the cyber operation wrests control of the platform and/or of the weapon system it carries. If the cyber hacker can properly be said to be in control of the weapon when it is fired and/or when it is being directed to a target, logic would suggest that the party undertaking the cyber operation becomes the attacker and, therefore, has responsibility for taking the precautions prescribed by Article 57 of API.

If, however, the intruding cyber hacker limits his or her actions to trying to deflect the attack from the originally planned target, and does not intentionally substitute a target of his or her own, it may be unsatisfactory to regard the cyber hacker as being responsible for the consequences of that deflecting, or "warding off," operation. In this regard, it must be recalled that a cyber attack must take place for the obligations in Article 57 of API to apply. As we noted earlier, the *Tallinn Manual* limits the notion of cyber attack to those cyber operations that are reasonably expected to cause death, injury, damage, or destruction. In the context of the current discussion, this suggests that only if the cyber operation is reasonably expected to cause such harm will the intrusion operation constitute a cyber attack. Accordingly, a cyber operation that is expected to result in the deflection of the weapon to an impact point where it is not expected to cause injury or damage will not amount to a cyber attack. If something goes wrong and, unexpectedly, the deflected weapon causes injury or damage, this would not, according to the *Tallinn Manual* formulation, cause the cyber operation to become a cyber attack.

It is, therefore, clear that those who are controlling weapons using computer systems will need the system to make them aware if control of the system has been lost

[68] The military justification for such action will be to ensure that the intended targets are in fact engaged. That military interest will coincide with the humanitarian concern that successful interference may cause loss of control of the weapon and consequent danger to civilians or civilian objects.

[69] William H Boothby, *Conflict Law: The Influence of New Weapons Technology, Human Rights and Emerging Actors* (The Hague: Asser Press, 2014).

or taken over. It would, in the author's view, always be useful for operators controlling such systems to be able to demonstrate publicly and in a timely way that such loss of control has occurred. An attack with profoundly unsatisfactory consequences can have strategic impact on the conduct of, and support for, an armed conflict, so it is likely to be in the interests of the party to the conflict whose weapon control system is hacked to show that the enemy was responsible for what took place.

Understandably, the analysis so far has focused on the precautionary obligations of the attacker. We should, however, also consider the obligations of the parties to the conflict to take precautions against the effects of attacks. The API, Article 58(c) requirement that parties to the conflict to the maximum extent feasible take necessary precautions to protect the civilian population, individual civilians, and civilian objects under their control against the dangers resulting from military operations is reflected in the *Tallinn Manual*, Rule 59. It should also be recalled that Articles 51(2) and 52(1) of API prohibit making the civilian population, individual civilians, or civilian objects the object of attack.

So if a cyber hacker takes control of the enemy's remotely piloted aircraft, its weapon or its systems and directs the weapon at civilians or civilian objects, he or she can properly be said to have made those civilians or civilian objects the object of the attack and will have breached Article 51 and/or Article 52, as the case may be, and will be liable to prosecution and punishment for the relevant war crime.

If the cyber hacker exerts only such control as to be able to prevent the weapon from engaging its intended military objective, he or she is unlikely to have breached Article 51(2) or 52(1), but, depending on precisely what action was taken, the hacker may have breached Articles 57(1) and 58(c). In the author's view, a cyber hacker who diminishes the reliability of the enemy weapon's precision guidance system appreciating that there is a risk that the hacker's interference will cause the weapon to attack civilians or civilian objects may have breached the Article 57(1) API duty to take constant care to spare those whom his or her operation exposes to increased risk. If the hacker does it in the reasonable expectation that this will be the consequence, there is an argument, applying the *Tallinn Manual*'s Rule 30, that he or she has become the attacker.[70]

VI. Some Emerging Propositions

Certain propositions emerge from the discussion in the present chapter. They are as follows:

1 If a cyber hacker's degree of control enables him or her to direct a weapon at a target or class of targets chosen by them, by exercising that control the hacker becomes the attacker and is legally liable for the consequences.

2 Absent such a degree of control, a hacker who interferes with a weapon control system reasonably expecting that doing so will cause the weapon to attack

[70] To deliberately diminish the reliability of the weapon guidance system, for example, in the expectation that by doing so the weapon will engage a civilian object instead of a military objective means that the hacker is undertaking a cyber operation that may be expected to have violent effects, namely death, injury, damage, or destruction, as provided for in *Tallinn Manual* (n 1), Rule 30.

civilians, civilian objects, persons, or objects entitled to specific protection or to undertake indiscriminate attacks arguably becomes the attacker and is legally liable for the consequences.

3 A cyber attacker who limits his or her interference with a weapon's guidance system to stopping the planned attack but who does not reasonably expect the weapon to attack some other person or object has not become an attacker.[71]

4 Cyber operations by which false information is conveyed to the enemy as to the party to the conflict's military posture, military strength, military plans, capabilities, or any other matters coming within the definition of ruses of war are not be rendered unlawful by being undertaken using a computer.

5 Cyber operations that invite an adversary's confidence in relation to protected status and result in death, injury (or for states party to API, capture) breach the prohibition of perfidy.

6 The prohibitions in Articles 38 and 39 of API apply to corresponding activities undertaken using cyber means. Expert opinion is divided on whether the use of the corresponding domain name would, absent any use of the relevant emblem, flag or insignia breach the rule.

VII. Conclusion

The propositions that we have identified in this chapter leave somewhat open the worrying thought that has emerged during the analysis. Cyber activities that do not breach Articles 37(1), 38, or 39 of API may nevertheless have such a considerable and adverse effect on the opposing party's understanding of the battlespace that they are unable to undertake a particular attack or attacks in general in the intended lawful manner. If the cyber operation has the effect that the cyber hacker takes control of the weapon, we have come to the conclusion that this renders the cyber hacker the attacker and thus responsible for taking the attacker's precautions as prescribed by Article 57 of API. If, however, the victim of the cyber interference operation retains control of the weapon and if the focus of the cyber interference is, say, the information on which its targeting decision-making is based, responsibility for the attack will remain with the party that is the victim of the cyber operation.

As we have noted, that cyber victim may well be unaware of the unreliability, distortion, or general inaccuracy of the information on which it is relying. The result may of course simply be that the attack is misdirected, that the intended military objective is not, therefore, attacked but that there are neither civilian casualties nor damage to civilian objects. No legal concerns would be aroused by such an outcome. Concerns would, however, arise if cyber operations were to render a party to the conflict unable to undertake hostilities that comply with the distinction and discrimination

[71] The act of a cyber hacker who is merely seeking to prevent the weapon from hitting its intended target and who exerts no other effective directional control seems difficult to equate with the "acts of violence" language employed in Article 49(1) of API. Accordingly, the cyber hacker limiting him- or herself to such "warding off" action does not become the attacker.

principles. The victim of such cyber operations would effectively be confronted with a choice between abandoning the fight and breaking the law. The question that this chapter raises is, therefore, whether such a state of affairs is acceptable. Put another way, if as the ICJ famously observed in its Nuclear Advisory opinion, the principle of distinction is "intransgressible," how can cyber operations that render compliance with that principle unrealistic be acceptable?

PART IV

RESPONSIBILITY AND ATTRIBUTION IN CYBER ATTACKS

10

Evidentiary Issues in International Disputes Related to State Responsibility for Cyber Operations

Marco Roscini[1]

I. Introduction

Evidentiary problems in inter-state litigation, in particular in relation to the attribution of certain unlawful conduct, are not peculiar to cyber operations.[2] Well before the cyber age, in the *Nicaragua* Judgment the International Court of Justice (ICJ) conceded that "the problem is not... the legal process of imputing the act to a particular State ... but the prior process of tracing material proof of the identity of the perpetrator."[3] As the United States declared in its views on information security submitted to the UN Secretary-General, then, the ambiguities of cyberspace "simply reflect the challenges... that already exists [sic] in many contexts."[4] It is undeniable, however, that these challenges are particularly evident in cyberspace, where identifying who is behind a cyber operation presents significant technical problems. As has been effectively observed, "the internet is one big masquerade ball. You can hide behind aliases, you can hide behind proxy servers, and you can surreptitiously enslave other computers to do your dirty work."[5]

[1] I am grateful to Simon Olleson for his useful comments on a previous version of this article and to Andraz Kastelic for his research assistance. All errors and omissions remain my sole responsibility. The chapter is based on developments as of June 2014 and all websites were last visited in that month. The chapter also appears in volume 50 of the *Texas International Law Journal* (2014-2015) and is reprinted here with minor modifications with the Journal's permission.

[2] Cyber operations are "the employment of cyber capabilities with the primary purpose of achieving objectives in or by the use of cyberspace." Michael N Schmitt (ed), *The Tallinn Manual on the International Law Applicable to Cyber Warfare* (New York: Cambridge University Press, 2013), 258. Cyber operations include cyber attacks and cyber exploitation. Cyber attacks are those cyber operations, whether in offence or in defense, intended to alter, delete, corrupt, or deny access to computer data or software for the purposes of (1) propaganda or deception; (2) partly or totally disrupting the functioning of the targeted computer, computer system or network, and related computer-operated physical infrastructure (if any); and/or (3) producing physical damage extrinsic to the computer, computer system, or network. Cyber exploitation refers to those operations that access other computers, computer systems, or networks, without the authorization of their owners or exceeding the limits of the authorization, in order to obtain information, but without affecting the functionality of the accessed system or amending/deleting the data resident therein. For a discussion of these definitions, see Marco Roscini, *Cyber Operations and the Use of Force in International Law* (Oxford: Oxford University Press, 2014) 10–18.

[3] *Military and Paramilitary Activities in and against Nicaragua (Nicaragua v United States of America)*, Merits [1986] ICJ Rep, para 57.

[4] UN Doc A/66/152, 15 July 2011, 18.

[5] Joel Brenner, *America the Vulnerable: Inside the New Threat Matrix of Digital Espionage, Crime, and Warfare* (London: Penguin Press, 2011) 32. As has been observed "[t]he lack of timely,

One needs only look at the three most famous cases of cyber attacks against states allegedly launched by other states to realize how thorny the problem of evidence in relation to cyber operations is. It has been claimed, in particular, that the Russian Federation was behind both the 2007 Distributed Denial of Service (DDoS) attacks against Estonia and the 2008 cyber attacks against Georgia.[6] These allegations were based on the following facts. In the Estonian case, the hackers claimed to be Russian, the tools to hack and deface were contained in Russian websites and chatrooms, and the attacks picked on May 9 (the day Russia celebrates Victory Day in Europe in World War II).[7] Furthermore, although the botnets included computers based in several countries, it seems that at least certain attacks originated from Russian IP (internet protocol) addresses, including those of state institutions.[8] According to the Estonian Defense Minister, the attacks were "unusually well-coordinated and required resources unavailable to common people."[9] The DDoS attacks also took place against the backdrop of the removal of a Russian war memorial from Tallinn city centre. Finally, Russia did not cooperate with Estonia in tracking down those responsible, and a request for bilateral investigation under the Mutual Legal Assistance Treaty between the two countries was rejected by the Russian Supreme Procurature.[10]

The cyber attacks against Georgia started immediately before and continued throughout the armed conflict between the Caucasian state and the Russian Federation in August 2008. It seems that the Russian hacker community was involved in the cyber attacks and that coordination took place in the Russian language and in Russian or Russian-related fora.[11] As in the Estonian case, it was claimed that the level of coordination and preparation suggested governmental support for the cyber attacks.[12] Finally, IP addresses belonging to Russian state-operated companies were used to launch the DDoS attacks.[13] Russia again denied any responsibility.[14]

The third case of alleged inter-state cyber operation, and possibly the most famous of the three, is that of Stuxnet. In 2012, an article published in *The New York Times*

high-confidence attribution and the possibility of 'spoofing' can create uncertainty and confusion for Governments, thus increasing the potential for crisis instability, misdirected responses and loss of escalation control during major cyberincidents." United Nations Office for Disarmament Affairs, "Developments in the Field of Information and Telecommunications in the Context of International Security," Disarmament Study Series 33, 2011, 6, at <http://www.un.org/disarmament/HomePage/ODAPublications/DisarmamentStudySeries/>.

[6] Denial of Service (DoS) attacks, of which "flood attacks" are an example, do not normally penetrate into a computer system but aim to inundate the target with excessive calls, messages, enquiries, or requests in order to overload it and force its shut down. Permanent DoS attacks are particularly serious attacks that damage the system and cause its replacement or reinstallation of hardware. When the DoS attack is carried out by a large number of computers organized in botnets, it is referred to as a DDoS attack.

[7] William A Owens, Kenneth W Dam, and Herbert S Lin, *Technology, Policy, Law, and Ethics Regarding U.S. Acquisition and Use of Cyberattack Capabilities* (Washington, DC: National Academies Press, 2009) 173.

[8] Owens et al (n 7) 173. [9] Owens et al (n 7) 173.

[10] Scott J Shackelford, "From Nuclear War to Net War: Analogizing Cyber Attacks in International Law" (2009) 27 Berkeley J. Int'l L 208; Alexander Klimburg, "Mobilising Cyber Power" (2011) 53 *Survival* 50.

[11] Eneken Tikk, Kadri Kaska, and Liis Vihul, *International Cyber Incidents. Legal Considerations* (CCDCOE, 2010) 75.

[12] Tikk et al (n 11) 75. [13] Tikk et al (n 11) 75. [14] Tikk et al (n 11) 75.

revealed that the United States, with Israel's support, had been engaging in a cyber campaign against Iran, codenamed "Operation Olympic Games," to disrupt the Islamic Republic's nuclear program.[15] Stuxnet, in particular, was allegedly designed to affect the gas centrifuges at the Natanz uranium enrichment facility.[16] Unlike other malware, the worm did not limit itself to self-replication, but also contained a "weaponized" payload designed to give instructions to other programs[17] and is, in fact, the first known use of malicious software designed to produce material damage by attacking the Supervisory Control and Data Acquisition (SCADA) system of a national critical infrastructure.[18] The allegations against the United States and Israel were based on journalistic interviews with current and former American, European, and Israeli officials and other experts, whose names are, however, not known.[19] In a recent interview, the former US National Security Agency (NSA) contractor Edward Snowden also claimed that the NSA and Israel were behind Stuxnet.[20] Symantec's researchers suggested that Stuxnet's code included references to the 1979 date of execution of a prominent Jewish Iranian businessman.[21] Other circumstantial evidence includes the fact that the worm hit primarily Iran and was specifically targeted at the Natanz nuclear facility, as the worm would activate itself only when it found the Siemens software used in that facility, and the fact that the high sophistication of the attack, the use of several zero-day hacks and the insider knowledge of the attacked system it implied required resources normally unavailable to individual hackers. Israeli and US officials have neither denied nor confirmed involvement in the operation: in response to a

[15] David E Sanger, "Obama Order Sped Up Wave of Cyberattacks Against Iran," *The New York Times*, June 1, 2012, at <http://www.nytimes.com/2012/06/01/world/middleeast/obama-ordered-wave-of-cyberattacks-against-iran.html?pagewanted=all&_r=1&>.

[16] Stuxnet presumably infiltrated the Natanz system through laptops and USB drives as, for security reasons, the system is not usually connected to the internet, and had two components: one designed to force a change in the centrifuges' rotor speed, inducing excessive vibrations or distortions that would destroy the centrifuges, and one that recorded the normal operations of the plant and then sent them back to plant operators so to make it look as everything was functioning normally. William J Broad, John Markoff, and David E Sanger, "Israeli Test on Worm Called Crucial in Iran Nuclear Delay" *The New York Times* (15 January 2011) <http://www.cfr.org/iran/nyt-israeli-test-worm-called-crucial-iran-nuclear-delay/p23850>.

[17] Jeremy Richmond, "Evolving Battlefields: Does Stuxnet Demonstrate a Need for Modifications to the Law of Armed Conflict?" (2011–12) 35 *Fordham International Law Journal* 849. Although the exact consequences of the incident are still the object of debate, the International Atomic Energy Agency (IAEA) reported that, in the period when Stuxnet was active, Iran stopped feeding uranium into a significant number of gas centrifuges at Natanz. William J Broad, "Report Suggests Problems with Iran's Nuclear Effort," *The New York Times*, November, 2010, at <http://www.nytimes.com/2010/11/24/world/middleeast/24nuke.html>. It is still unclear, however, whether this was due to Stuxnet or to technical malfunctions inherent to the equipment used. Katharina Ziolkowski, "Stuxnet–Legal Considerations," NATO Cooperative Cyber Defence Centre of Excellence (CCDCOE, 2012) 5; Ivanka Barzashka, "Are Cyber-Weapons Effective? Assessing Stuxnet's Impact on the Iranian Enrichment Programme" (2013) 158 *RUSI Journal* 52.

[18] See Thomas Rid, "Cyber War Will Not Take Place" (2012) 35 *Journal of Strategic Studies* 1, 7–20.

[19] Sanger (n 15).

[20] Jacob Appelbaum and Laura Poitras, "Edward Snowden Interview: The NSA and Its Willing Helpers," *Spiegel Online*, July 8, 2013, at <http://www.spiegel.de/international/world/interview-with-whistleblower-edward-snowden-on-global-spying-a-910006.html>.

[21] Nicolas Falliere, Liam O Murchu, and Eric Chien, *W32. Stuxnet Dossier*, Symantec, 2011, Version 1.4, February 2011, 18, at <http://www.symantec.com/content/en/us/enterprise/media/security_response/whitepapers/w32_stuxnet_dossier.pdf>.

question, President Obama's chief strategist for combating weapons of mass destruction, Gary Samore, sardonically pointed out that "I'm glad to hear they are having troubles with their centrifuge machines, and the U.S. and its allies are doing everything we can to make it more complicated."[22] According to *The Daily Telegraph*, a video that was played at a retirement party for the head of the Israel Defense Forces (IDF), Gabi Ashkenazi, included references to Stuxnet as one of his operational successes as the IDF chief of general staff.[23]

Apart from the aforementioned well-known cyber attacks, allegations of state involvement have also been made in relation to other cyber operations, including cyber exploitation activities. The US Department of Defense's 2013 Report to Congress, for instance, claims that some of the 2012 cyber intrusions into US government computers "appear to be attributable directly to the Chinese government and military," although it is not entirely clear on what grounds.[24] According to the controversial Mandiant Report, "the sheer number of [hacking group] APT1 IP addresses concentrated in these Shanghai ranges, coupled with Simplified Chinese keyboard layout settings on APT1's attack systems, betrays the true location and language of the operators."[25] The Report concludes that "APT1 is likely government-sponsored and one of the most persistent of China's cyber threat actors."[26] According to the Chinese Defence Ministry, however, the report lacked "technical proof" that linked the IP addresses used by ATP1 to a military unit of the People's Liberation Army (PLA), as the attacks employed hijacked addresses.[27] In May 2014, the US Department of Justice eventually brought charges against five members of the PLA for hacking into the computers of six organizations in western Pennsylvania and elsewhere in the United States to steal trade secrets, without providing much supporting evidence (if any at all) of the involvement of the defendants.[28]

In spite of the obvious crucial importance of evidentiary issues, works on inter-state cyber operations, both above and below the level of the use of force, have so far focused on whether such operations are consistent with primary norms of international law and on the remedies available to the victim state under the jus ad bellum and the law of state responsibility, and have almost entirely neglected a discussion of the evidence the victim state needs to produce to demonstrate, either before a judicial body or elsewhere, that an unlawful cyber operation has been conducted against it and that it is attributable to another state.[29] The first edition of the *Tallinn Manual on the*

[22] Broad et al (n 16).

[23] Christopher Williams, "Israel Video Shows Stuxnet as One of its Successes," *The Telegraph*, February 15, 2011, at <http://www.telegraph.co.uk/news/worldnews/middleeast/israel/8326387/Israel-video-shows-Stuxnet-as-one-of-its-successes.html>.

[24] United States Department of Defense, "Annual Report to Congress: Military and Security Developments Involving the People's Republic of China 2013," May 6, 2013, at 36, at <http://www.defense.gov/pubs/2013_china_report_final.pdf>.

[25] Mandiant, *APT1–Exposing One of China's Cyber Espionage Units* (2013) 39, at <http://intelreport.mandiant.com/Mandiant_APT1_Report.pdf>.

[26] Mandiant (n 25) 2.

[27] "China condemns hacking report by US firm Mandiant," *BBC*, February 20, 2013, at <http://www.bbc.co.uk/news/world-us-canada-21515259>.

[28] Read the indictment at <http://www.justice.gov/iso/opa/resources/5122014519132358461949.pdf>.

[29] Some discussion is contained in Robin Geiß and Henning Lahmann, "Freedom and Security in Cyberspace: Shifting the Focus away from Military Responses towards Non-Forcible Countermeasures

International Law Applicable to Cyber Warfare also does not discuss in depth evidentiary issues in the cyber context: the only references to evidence are contained in Rules 7 and 8.[30] The present chapter aims to fill this gap. It will start with a brief account of the international law of evidence and will then discuss who has the burden of proof in relation to claims seeking remedies (including reparation) for damage caused by cyber operations. It will then analyze the standard of proof required in the cyber context. Finally, the possible methods of proof will be examined, distinguishing between those which are admissible and those which are inadmissible. The present chapter only deals with international disputes between states, and will not discuss evidentiary issues in relation to cyber crime before domestic courts. It also does not look at evidence before international criminal tribunals, as the focus is on state responsibility for cyber operations and not on the criminal responsibility of individuals.[31]

II. The International Law of Evidence

"Evidence" is "information... with a view of establishing or disproving alleged facts."[32] It is different from "proof" in that " 'proof' is the result or effect of evidence, while 'evidence' is the "medium or means by which a fact is proved or disproved."[33] Evidence is normally required to provide proof of both the objective (be it an act or omission) and subjective elements of an internationally wrongful act, that is, its attribution to a state. A state invoking self-defense against cyber attacks, for instance, will have to produce evidence that demonstrates: (1) that the cyber attack actually occurred, that it was directed against it, and that its scale and effects reached the threshold of an "armed attack"[34] and (2) that it was attributable to a certain state.[35] For a state to invoke the right to take countermeasures, on the other hand, it may be sufficient to provide evidence that a cyber operation originated from a certain state and that that state did not exercise due diligence in terminating it, without necessarily having to prove attribution of the attack itself to the state.[36] In the *Nicaragua* case, the ICJ clearly explained

and Collective Threat-Prevention," in Katharina Ziolkowski (ed), *Peacetime Regime for State Activities in Cyberspace. International Law, International Relations and Diplomacy* (CCDCOE, 2013); and Scott J Shackelford and Richard B Andres, "State Responsibility for Cyber Attacks: Competing Standards for a Growing Problem" (2010–11) 42 Georgetown *Journal of International Law* 971–1016. In the context of law enforcement, the Council of Europe has drafted an Electronic Evidence Guide for cyber crime, available at <http://www.coe.int/t/dghl/cooperation/economiccrime/cybercrime/Documents/Electronic%20 Evidence%20Guide/default_en.asp>.

[30] *Tallinn Manual* (n 2) 34–6.

[31] The statutes and rules of international criminal tribunals provide for specific evidentiary rules. Rüdiger Wolfrum, "International Courts and Tribunals, Evidence," *Max Planck Encyclopedia of Public International Law*, vol V (Oxford: Oxford University Press, 2012) 567–9.

[32] *Max Planck Encyclopedia* (n 31) 552.

[33] Matthew J Canavan, Lawrence J Culligan, Arnold O Ginnow, Francis J Ludes, and Robert J Owens, *Corpus Juris Secundum: A Complete Restatement of the Entire American Law*, vol 31A: Evidence (St Paul: West Publishing, 1964) 820.

[34] On the distinction between "use of force" and "armed attack," see *Nicaragua*, Merits (n 3) paras 191, 195.

[35] On whether self-defense can be exercised against cyber attacks by non-state actors, see Roscini (n 2) 80–88.

[36] Geiß and Lahmann (n 28) 637.

the distinction between the objective and subjective elements from an evidentiary perspective:

> [o]ne of the Court's chief difficulties in the present case has been the determination of the facts relevant to the dispute... Sometimes there is no question, in the sense that it does not appear to be disputed, that an act was done, but there are conflicting reports, or a lack of evidence, as to who did it... The occurrence of the act itself may however have been shrouded in secrecy. In the latter case, the Court has had to endeavour first to establish what actually happened, before entering on the next stage of considering whether the act (if proven) was imputable to the State to which it has been attributed.[37]

The Court's observations were made against the backdrop of the secrecy that surrounded the US and Nicaraguan covert operations in central America,[38] which is also a quintessential characteristic of cyber operations. In this context too, then, it is likely that evidence will be required both to establish to material elements of the wrongful act and to establish its attribution. It is still unclear, for instance, not only who is responsible for Stuxnet, but also whether the worm caused any damage and, if so, to what extent, which is essential in order to establish whether the cyber operation amounted to a use of force and, more importantly, whether it was an armed attack entitling the victim state to self-defense.[39] As to establishing the subjective element of the internationally wrongful act, what is peculiar to cyber operations is that there are in fact three levels of evidence that are needed to attribute a cyber operation to a state: first, the computer(s) or server(s) from which the operations originate must be located; secondly, it is the individual that is behind the operation that needs to be identified; and thirdly, what needs to be proved is that the individual acted on behalf of a state so that his or her conduct is attributable to it.

This leads us to an important specification: the standard of proof must be distinguished from the rules of attribution. The former is "the *quantum* of evidence necessary to substantiate the factual claims made by the parties."[40] The latter, on the other hand, determine the level of connection that must exist between an individual or group of individuals and a state for the conduct of the individuals to be attributed to the state at the international level. The rules of attribution for the purposes of state responsibility have been codified in the Part One of the Articles on the Responsibility of States for Internationally Wrongful Acts adopted by the International Law Commission (ILC), as well as having been articulated in the case-law of the ICJ.[41] Evidence according to the applicable standard must be provided to demonstrate that the attribution test has been satisfied: in *Nicaragua*, for instance, the ICJ had to assess whether there was sufficient evidence that the United States had exercised "effective control" over the *contras* so that it could be held responsible for their violations of international humanitarian law.[42]

[37] *Nicaragua*, Merits (n 3) para 57. [38] *Nicaragua*, Merits (n 3) para 57.

[39] See Roscini (n 2) 45–63, 70–7.

[40] James A Green, "Fluctuating Evidentiary Standards for Self-Defence in the International Court of Justice" (2009) 58 *International and Comparative Law Quarterly* 165.

[41] For a discussion of the attribution rules in the cyber context, see Roscini (n 2) 33–40.

[42] In the *Nicaragua* case the Court did not find that there was sufficient evidence to conclude that the contras were totally dependent on the United States so to qualify as de facto organs, However, it found that a situation of partial dependency "the exact extent of which the Court cannot establish, may

The standard of proof should also be distinguished from the burden of proof. The latter does not determine how much evidence, and of what type, is necessary to prove the alleged facts, but merely identifies the litigant that must provide that evidence. In other words, the burden of proof is "the obligation on a party to show that they have sufficient evidence on an issue to raise it in a case."[43] The burden of proof includes not only the "burden of persuasion," but also the "burden of production," that is, to produce the relevant evidence before a court.[44]

Evidence may be submitted not only to an international court or tribunal, but also to political organs (for instance to secure a favorable vote), or could be disseminated more widely for the purposes of influencing public opinion and gain support for certain actions or inactions.[45] One could recall the evidence presented by the Reagan Administration before the UN Security Council to justify its 1986 strike on Tripoli as a measure of self-defense.[46] When substantiating its 2001 armed operation against Afghanistan, the US Permanent Representative to the United Nations referred to the fact that the US government had "clear and compelling information that the Al-Qaeda organization, which is supported by the Taliban regime in Afghanistan, had a central role in the [September 11, 2001] attacks," without however going into further details.[47] The same language was used by the Secretary-General of NATO.[48] Evidence was also famously one of the controversial aspects of the 2003 US–UK intervention against Iraq.[49] More recently, in the context of the proposed intervention to react against the use of chemical weapons in Syria, President Obama stated that attacking another country without a UN mandate and without "clear evidence that can be presented" would raise questions of international law.[50] The political or judicial relevance of evidence

certainly be inferred *inter alia* from the fact that the leaders were selected by the United States. But it may also be inferred from other factors, some of which have been examined by the Court, such as the organization, training and equipping of the force, the planning of operations, the choosing of targets and the operational support provided." *Nicaragua*, Merits (n 3) para 112.

[43] Anna Riddell and Brendan Plant, *Evidence before the International Court of Justice* (British Institute of International and Comparative Law, 2009) 81.

[44] Markus Benzing, "Evidentiary Issues," in Andreas Zimmermann, Karin Oellers-Frahm, Christian Tomuschat, and Christian J Tams (eds), *The Statute of the International Court of Justice. A Commentary* (Oxford: Oxford University Press, 2nd edn, 2012) 1245. As there are no parties in advisory proceedings, there is no burden of proof in this type of proceedings. Wolfrum (n 31) 565.

[45] Whether or not states have an obligation to make evidence public is a matter of debate. It has been observed that "[i]f nations are permitted to launch unilateral attacks based on secret information gained largely by inference, processed by and known only to a few individuals and not subject to international review, then Article 2(4) of the U.N. Charter is rendered virtually meaningless." Jules Lobel, "The Use of Force to Respond to Terrorist Attacks: The Bombing of Sudan and Afghanistan" (1999) 24 *Yale Journal of International Law*, 537, 547. See also George P Fletcher and Jens D Ohlin, *Defending Humanity* (Oxford: Oxford University Press, 2008) 169. Contra, see Matthew C Waxman, "The Use of Force Against States That *Might* Have Weapons of Mass Destruction" (2009) 31 Mich *Journal of International Law* 65.

[46] Lobel (n 45) 549.

[47] Letter dated October 7, 2001 from the Permanent Representative of the United States of America to the United Nations addressed to the President of the Security Council, UN Doc S/2001/946.

[48] Statement by NATO Secretary General, Lord Robertson, October 2, 2001, at <http://www.nato.int/docu/speech/2001/s011002a.htm>.

[49] See United Kingdom Foreign and Commonwealth Office, "Iraq's Weapons of Mass Destruction: The Assessment of the British Government" (2002); UN Doc S/PV.4701, 1 January 2003, 2–17 (Powell's remarks).

[50] Julian Borger, "West Reviews Legal Options for Possible Syria Intervention Without UN Mandate," *The Guardian*, August 26, 2013, at <http://www.theguardian.com/world/2013/aug/26/

may relate to the different phases of the same international dispute. For instance, the state invoking the right of self-defense against an armed attack by another state will normally try to justify the exercise of this right first before the international community and public opinion by providing evidence of the occurrence (or imminent occurrence) of the armed attack and of its attribution to the target state. If, as in the case of the *Military and paramilitary activities in and against Nicaragua*, a state subsequently brings the case before an international court, which has jurisdiction over the case, the evidence will have to be assessed by that court in order to establish international responsibility and its consequences, and in particular whether the requirements for the exercise of self-defense were met.

The problem is that there is no uniform body of rules on the production of evidence in international law.[51] There is no treaty provision that regulates evidentiary issues in non-judicial contexts and it is doubtful that international law has developed customary rules in that sense. As to the production of evidence in inter-state litigation, international courts, at least those of a non-criminal nature, normally determine their own standards in each case, which may considerably differ according to the nature of the court or the case under examination. As it is not possible to identify uniform evidentiary rules applicable in all cases and before all international courts, this chapter will focus on proceedings before the ICJ: this is because the ICJ is the main UN judicial organ, which deals, if the involved states have consented to its jurisdiction, with claims of state responsibility arising from the violation of any primary norm of international law. Our overall purpose is to establish whether rules on evidence may be identified that would apply to claims in inter-state judicial proceedings seeking remedies for damage caused by cyber operations. It should be borne in mind, however, that the conclusions reached with regard to the ICJ only apply to it and could not automatically be extended to other international courts.

Rules on the production of evidence before the ICJ are contained in the ICJ Statute, the Rules of Court (adopted in 1978), and Practice Directions for use by states appearing before the Court, first adopted in 2001 and subsequently amended. In the following pages, the relevant rules on evidentiary issues contained in those documents, as well as those elaborated by the Court in its jurisprudence, will be applied to allegations related to cyber operations.

III. Burden of Proof and Cyber Operations

The burden of proof identifies the litigant that has the onus of meeting the standard of proof by providing the necessary evidence.[52] Once the burden has been discharged

united-nations-mandate-airstrikes-syria>. Indeed, the Report of the UN Secretary-General's Investigation found "clear and convincing evidence" of the use of chemical weapons in the armed conflict. "United Nations Mission to Investigate Allegations of the Use of Chemical Weapons in the Syrian Arab Republic–Report on the Alleged Use of Chemical Weapons in the Ghouta Area of Damascus on 21 August 2013," UN Doc A/67/997–S/2013/553, September 13, 2013, 5, <http://www.un.org/disarmament/content/slideshow/Secretary_General_Report_of_CW_Investigation.pdf>.

[51] Mary Ellen O'Connell, "Evidence of Terror" (2002) *Journal of of Conflict and Security Law* 7, 21.

[52] Green (n 40)165.

according to the appropriate standard, the burden shifts to the other litigant, who has to prove the contrary. It is normally the party that relies upon a certain fact that is required to prove it (the principle, deriving from Roman law, *onus probandi incumbit actori*).[53] This general principle of law, invoked consistently by the ICJ and other international courts and tribunals,[54] "applies to the assertions of fact both by the Applicant and the Respondent."[55] The party bearing the burden of proof, therefore, is not necessarily the applicant, that is, the state that has brought the application before the tribunal, but the party "seeking to establish a fact,"[56] regardless of its procedural position.[57] For instance, the party (applicant or respondent) that relies on an exception, including self-defense, has the burden of proving the facts that are the basis for the exception.[58] It should also be recalled that the distinction between applicant and respondent may not always be clear in inter-state litigation, especially when the case is brought before an international court by special agreement between the parties.[59]

The *onus probandi incumbit actori* principle is subject to three main limitations. First, facts that are not disputed or that are agreed upon by the parties do not need to be proved.[60] Secondly, the Court has relieved a party from the burden of providing evidence of facts that are "notorious" or "of public knowledge."[61] In *Nicaragua*, for instance, the Court found that "since there was no secrecy about the holding of the manoeuvres, the Court considers that it may treat the matter as one of public knowledge, and as such, sufficiently established."[62] As has been noted, "the notion of common or public knowledge has, over the years, expanded, given the wide availability of information on current events in the press and on the internet."[63] Reports on cyber incidents have also been published by companies like McAfee, Symantec, Mandiant, and Project Grey Goose and by think-tanks like NATO's Cooperative Cyber Defence Centre of Excellence (CCDCOE). These reports essentially contain technical analysis of cyber incidents and, with the possible exception of those of the CCDCOE, do not normally investigate attribution for legal purposes of those incidents in any depth (if at all). The fact that cyber incidents have received extensive

[53] *Military and Paramilitary Activities in and against Nicaragua (Nicaragua v United States of America)*, Jurisdiction [1984] ICJ Rep., para 101.

[54] Ruth Teitelbaum, "Recent Fact-Finding Developments at the International Court of Justice," (2007) 6 *The Law and Practice of International Courts and Tribunals* 121.

[55] *Pulp Mills on the River Uruguay (Argentina v. Uruguay)* [2010] ICJ Rep, para 162.

[56] *Nicaragua*, Jurisdiction (n 53) para 101.

[57] According to Rosenne, "the tendency of the Court is to separate the different issues arising in a case, treating each one separately, applying the rule *actori incumbit probatio*, requiring the party that advances a particular contention to establish it in fact and in law. The result is that each State putting forward a claim is under the general duty to establish its case, without there being any implication that such State is 'plaintiff' or 'applicant' in the sense in which internal litigation uses those terms." Shabtai Rosenne, *The Law and Practice of the International Court, 1920–2005*, vol III (Leiden/Boston: Nijhoff, 2006) 1200–1.

[58] *Oil Platforms (Islamic Republic of Iran v. United States of America)*, Merits, [2003] ICJ Rep, para 57. See Riddell and Plant (n 43) 87.

[59] Andrés Aguilar Mawdsley, "Evidence Before the International Court of Justice," in Ronald St John Macdonald (ed), *Essays in Honour of Wang Tieya* (Dordrecht/Boston/London: Nijhoff, 1994) 538

[60] Wolfrum (n 31) 563.

[61] See, for example, *Nicaragua*, Merits (n 3) para 92. Judicial notice has been frequently invoked by international criminal tribunals. Teitelbaum (n 54) 144–5.

[62] *Nicaragua*, Merits (n 3) para 92.

[63] Riddell and Plant (n 43) 142–3.

press coverage, as in the case of Stuxnet, may also contribute to the public knowledge character of certain facts. In *Nicaragua*, however, the ICJ warned that "[w]idespread reports of a fact may prove on closer examination to derive from a single source, and such reports, however numerous, will in such case have no greater value as evidence than the original source."[64] The ICJ also held that the "massive body of information" available to the Court, including newspapers, radio, and television reports, may be useful only when it is "wholly consistent and concordant as to the main facts and circumstances of the case."[65]

Thirdly, the *onus probandi incumbit actori* principle only applies to facts, not to the law, which does not need to be proved (*jura novit curia*). It should be borne in mind, however, that, in inter-state litigation, municipal law is a fact that must be proved by the parties invoking it.[66] Furthermore, the ICJ has often distinguished between treaty law and customary international law and has held that the existence and scope of customary rules, especially when of a regional character, must be proved by the parties, as one of their two elements, state practice, is factual.[67] A party invoking national legislation or the existence of a general or cyber-specific custom in its favor, therefore, will bear the burden of producing relevant evidence before the Court.

Certain authors have suggested that the problems of identification and attribution in the cyber context could be solved by shifting the burden of proof "from the investigator and accuser to the nation in which the attack software was launched."[68] On that basis, it would be up to the state from whose cyber infrastructure the cyber operation originated to prove that it was not responsible for it and/or that it exercised due diligence not to allow the misuse of its infrastructure by others, and not to the claimant to prove the contrary. Similarly, it has been argued that "[*t*]*he fact that a harmful cyber incident is conducted via the information infrastructure subject to a nation's control is* prima facie *evidence that the nation knows of the use and is responsible for the cyber incident*."[69] This, however, is not correct. First, mere knowledge does not automatically entail direct attribution, but only a potential violation of the due diligence duty not to allow hostile acts from one's territory. What is more, the views arguing for a reversal of the burden of proof are at odds with the *jurisprudence constante* of the ICJ. In the *Corfu Channel* case, the Court famously found that the exclusive control exercised by a state over its territory "neither involves *prima facie* responsibility nor shifts the burden of proof" in relation to unlawful acts perpetrated there.[70] The Court, however, conceded that difficulties in discharging the burden of proof in such cases may allow "a more liberal recourse to inferences of fact and circumstantial evidence."[71] This

[64] *Nicaragua*, Merits (n 3) para 63.

[65] *United States Diplomatic and Consular Staff in Tehran (United States of America v Iran)* [1980] ICJ Rep, para 13.

[66] Wolfrum (n 31) 557.

[67] *Asylum Case (Colombia/Peru)*, Merits [1950], ICJ Rep, at 276–7; *Rights of Nationals of the United States of America in Morocco (France v United States of America)*, Merits [1952] ICJ Rep, 200.

[68] Richard A Clarke and Robert K Knake, *Cyber War: The Next Threat to National Security and What to Do About It* (New York: HarperCollins 2012) 249.

[69] Daniel J Ryan, Maeve Dion, Eneken Tikk, and Julie J Ryan, "International Cyberlaw: A Normative Approach" (2011) 42 Georgetown *Journal of International Law* 1185 (emphasis in the original).

[70] *Corfu Channel Case (UK v Albania)* [1949] ICJ Rep, at 18.

[71] *Corfu Channel* (n 70).

point will be further explored in section VI.[72] In *Armed Activities (DRC v Uganda)*, the ICJ also did not shift the burden of proving that Zaire had been in a position to stop the armed groups' actions originating from its border regions, as claimed by Uganda in its counter-claim, from Uganda to the DRC, and, therefore, found that it could not "conclude that the absence of action by Zaire's Government against the rebel groups in the border area is tantamount to 'tolerating' or 'acquiescing' in their activities."[73]

If one applies these findings in the cyber context, the fact that a state has exclusive "territorial" control of the cyber infrastructure from which the cyber operation originates does not per se shift the burden of proof and it is, therefore, still up to the claimant to demonstrate that the territorial state is responsible for the cyber operation or that it failed to comply with its due diligence duty of vigilance, and not to the territorial state to demonstrate the contrary.[74]

Even beyond the principle of territorial control, the fact that relevant evidence is in the hands of the other party does not per se shift the burden of proof. In the *Avena* case, the ICJ held that it could not

> accept that, because such information may have been in part in the hands of Mexico, it was for Mexico to produce such information. It was for the United States to seek such information, with sufficient specificity, and to demonstrate both that this was done and that the Mexican authorities declined or failed to respond to such specific requests.... The Court accordingly concludes that the United States has not met its burden of proof in its attempt to show that persons of Mexican nationality were also United States nationals.[75]

The fact that cyber operations were conducted in the context of an armed conflict, as was the case of those against Georgia in 2008, also does not affect the normal application of the burden of proof. In *Nicaragua*, the ICJ recalled the *Corfu Channel* and *Tehran Hostages* judgments and found that "[a] situation of armed conflict is not the only one in which evidence of fact may be difficult to come by, and the Court has

[72] See section VI.

[73] *Armed Activities on the Territory of the Congo (Democratic Republic of the Congo v Uganda)* [2005] ICJ Rep, para 301. Judge Kooijmans dissented and argued that "[i]t is for the State under a duty of vigilance to show what efforts it has made to fulfill that duty and what difficulties it has met" and concluded that the DRC had not provided evidence to show that it had adopted "credible measures" to prevent trans-border attacks. Separate Opinion of Judge Kooijmans, *Armed Activities on the Territory of the Congo*, para 82.

[74] It should not be forgotten that cyberspace consists of a physical and syntactic (or logical) layer: the former includes the physical infrastructure through which the data travel wired or wireless, including servers, routers, satellites, cables, wires, and the computers, while the latter includes the protocols that allow data to be routed and understood, as well as the software used and the data. David J Betz and Tim Stevens, "Analogical Reasoning and Cyber Security" (2013) 44 *Security Dialogue* 151. Cyber operations can then be seen as "the reduction of information to electronic format and the actual movement of that information between physical elements of cyber infrastructure." Nils Melzer, *Cyberwarfare and International Law* (UNIDIR, 2011) 5, at <http://www.isn.ethz.ch/Digital-Library/Publications/Detail/?lng=en&id=134218>. In its 2013 Report, the Group of Governmental Experts established by the UN General Assembly confirmed that "State sovereignty and international norms and principles that flow from sovereignty apply to State conduct of ICT-related activities, and to their jurisdiction over ICT infrastructure within their territory." UN Doc A/68/98, June 24, 2013, at 8.

[75] *Avena and Other Mexican Nationals (Mexico v United States of America)* [2004] ICJ Rep 2004, para 57.

in the past recognized and made allowance for this."[76] Even in such circumstances, therefore, "it is the litigant seeking to establish a fact who bears the burden of proving it."[77] In the *El Salvador/Honduras* case, the Court stated that it "fully appreciates the difficulties experienced by El Salvador in collecting its evidence, caused by the interference with governmental action resulting from acts of violence. It cannot, however, apply a presumption that evidence which is unavailable would, if produced, have supported a particular party's case; still less a presumption of the existence of evidence which has not been produced."[78]

The application of the *onus probandi incumbit actori* principle is also not affected by the possible asymmetry in the position of the litigants in discharging the burden of proof due to the fact that one has acted covertly (as is virtually always the case of cyber operations).[79] As Judge Owada points out in his Separate Opinion attached to the *Oil Platforms* judgment, however, the Court should "take a more proactive stance on the issue of evidence and that of fact-finding" in such cases so to ensure that the rules of evidence are applied in a "fair and equitable manner" to both parties.[80]

Finally, it has been argued that a reversal of the burden of proof may derive from an application of the precautionary principle originating from international environmental law in cyberspace.[81] The precautionary principle entails "the duty to undertake all appropriate regulatory and other measures at an early stage, and well before the (concrete) risk of harm occurs."[82] On this view, states would have an obligation to implement measures to prevent the possible misuse of their cyber infrastructure, in particular by establishing a national cyber security framework.[83] Regardless of whether or not the precautionary principle, with its uncertain normativity, extends to cyberspace,[84] it still would not lead to a reversal of the burden of proof from the claimant to the state from where the cyber operation originates. In the *Pulp Mills* case, the ICJ concluded that "while a precautionary approach may be relevant in the interpretation and application of the provisions of the Statute [of the River Uruguay], it does not follow that it operates as a reversal of the burden of proof."[85] The Court, however, did not specify whether the precautionary principle may result in at least a lowering of the standard of proof.[86]

In light of what we have just discussed, it can be concluded that it is unlikely that the ICJ would accept that there is a reversal of the burden of proof in the cyber context. As has been correctly argued, "suggesting a reversal of the burden of proof could easily lead to wrong and even absurd results given the possibility of routing cyber operations through numerous countries, and to the denouncing of wholly uninvolved and innocent States."[87] In the case of the 2007 DDoS campaign against Estonia, for instance,

[76] *Nicaragua*, Jurisdiction (n 53) para 101. [77] *Nicaragua*, Jurisdiction (n 53) para 101.
[78] *Land, Island and Maritime Frontier Dispute (El Salvador/Honduras: Nicaragua intervening)* [1992] ICJ Rep, para 63.
[79] *Oil Platforms* (n 58), Separate Opinion of Judge Owada, para 46.
[80] *Oil Platforms* (n 58), para 47.
[81] See, critically, Thilo Marahun, "Customary Rules of International Environmental Law–Can they Provide Guidance for Developing a Peacetime Regime for Cyberspace?," in Ziolkowski (n 29) 475.
[82] Ziolkowski (n 17) 169. [83] Ziolkowski (n 17) 169.
[84] Marauhn doubts it. Marauhn (n 81) 475–6. [85] *Pulp Mills* (n 55) para 164.
[86] Benzing (n 44) 1247. [87] Geiß and Lahmann (n 28) 628.

the botnets included computers located not only in Russia, but also in the United States, Europe, Canada, Brazil, Vietnam, and other countries.[88] Difficulties in discharging the burden of proof, which are particularly significant in the context under examination, may, however, result in an alleviation of the standard of proof required to demonstrate a particular fact. It is to this aspect that the analysis now turns.

IV. Standard of Proof and Cyber Operations

It is well-known that, while in civil law systems there are no specific standards of proof that judges have to apply as they are authorized to evaluate the evidence produced according to their personal convictions on a case-by-case basis, common law jurisdictions employ a rigid classification of standards, including (from the most stringent to the least) beyond reasonable doubt (i.e., indisputable evidence, used in criminal trials), clear and convincing (or compelling) evidence (i.e., more than probable but short of indisputable), and preponderance of evidence or balance of probabilities (more likely than not, reasonably probable, normally used in civil proceedings).[89] A fourth standard is that of prima facie evidence, which merely requires indicative proof of the correctness of the contention made.[90]

The Statute of the ICJ and the Rules of Court neither require specific standards of proof, nor indicate what methods of proof the Court will consider as being probative in order to meet a certain standard. The ICJ has to date avoided clearly indicating the standards of proof expected from the litigants during the proceedings.[91] It has normally referred to the applicable standard of proof in the judgments, but at that point it is of course too late for the parties to take it into account in pleading their cases.[92]

There is then no agreement on what standard of proof the ICJ should expect from the parties in the cases before it. If, because of their nature, international criminal courts use the beyond reasonable doubt standard in their proceedings,[93] the most appropriate analogy for inter-state litigation is not with criminal trials, but with certain types of civil litigation.[94] In his Dissenting Opinion in the *Corfu Channel* case, Judge Krylov suggested that "[o]ne cannot condemn a State on the basis of probabilities. To establish international responsibility, one must have clear and indisputable facts."[95] Wolfrum argued that, while the jurisdiction of an international court over a case should be established beyond reasonable doubt, the ICJ has generally applied a standard comparable to that of preponderance of evidence used in domestic civil proceedings when

[88] Owens, Dam, Lin, *supra* (n 7) 173.

[89] Mary Ellen O'Connell, "Rules of Evidence for the Use of Force in International Law's New Era" (2006) 100 *American Society of International Law Proceedings* 45; Marko Milanović, "State Responsibility for Genocide" (2006) 17 *European Journal of International Law* 594; Green (n 40) 167.

[90] Green (n 40) 166; Geiß and Lahmann (n 28) 624.

[91] This approach has been criticized by judges from common law countries. See, for example, *Oil Platforms* (n 58), Separate Opinion of Judge Buergenthal, para 41–6 and Separate Opinion of Judge Higgins, para 30–39.

[92] Teitelbaum (n 54) 124. It has been suggested that "the Court might consider whether, either prior to the submission of written pleadings, after the first round of written pleadings, or prior to the oral hearings, it should ask the parties to meet a specific burden of proof for certain claims." Teitelbaum (n 54) 128.

[93] Wolfrum (n 31) 569.

[94] Waxman (n 45) 59.

[95] *Corfu Channel* (n 70), Dissenting Opinion of Judge Krylov, at 72.

deciding disputes involving state responsibility.[96] Others have maintained that such a standard only applies to cases not concerning attribution of international wrongful acts, such as border delimitations, while when international responsibility is at stake, the standard is stricter and requires clear and convincing evidence.[97]

It is, therefore, difficult, and perhaps undesirable,[98] to identify a uniform standard of proof generally applicable in inter-state litigation, or even a predominant one: the Court "tends to look at issues as they arise."[99] This case-by-case approach, however, does not exclude that a standard of proof may be identified having regard to the primary rules in dispute, that is, "the substantive rules of international law through which the Court will reach its decision."[100] Indeed, when the allegation is the same, it seems logical that the evidentiary standard should also be the same.[101] There are indications, for instance, that claims related to jus ad bellum violations, in particular in relation to the invocation of an exception to the prohibition of the use of force in international relations, have been treated as requiring "clear and convincing evidence."[102] In the *Nicaragua* judgment, the Court referred to "convincing evidence" of the facts on which a claim is based and to the lack of "clear evidence" of the degree of control exercised by the United States over the contras.[103] In the *Oil Platforms* case, the ICJ rejected evidence with regard to Iran's responsibility for mine laying that was "highly suggestive, but not conclusive" and held that "evidence indicative of Iranian responsibility for the attack on the *Sea Isle City*" was insufficient.[104] In *DRC v Uganda*, the ICJ referred again to facts "convincingly established by the evidence," "convincing evidence," and "evidence weighty and convincing."[105] Beyond the ICJ, the Eritrea-Ethiopia Claims Commission also found that there was "clear" evidence that events in the vicinity of Badme were minor incidents and did not reach the magnitude of an armed attack.[106] All the above suggests that at least clear and convincing evidence is expected for claims related to the use of force. As self-defense is an exception to the prohibition of the use of force, in particular, the standard of proof should be high enough to limit its invocation to exceptional circumstances and, thus, avoid abuses.[107]

[96] Wolfrum (n 31) 566. [97] Riddell and Plant (n 43) 133. [98] Green (n 40) 167.

[99] Sir Arthur Watts, "Burden of Proof, and Evidence before the ICJ," in Friedl Weiss (ed), *Improving WTO Dispute Settlement Procedures. Issues and Lessons from the Practice of Other International Courts and Tribunals* (Folkestone: Cameron May, 2000) 289, 294.

[100] Rosenne (n 57) 1043. In *Ahmadou Sadio Diallo (Republic of Guinea v Democratic Republic of the Congo)* [2010] ICJ Rep, para 54, the ICJ mades a similar point with regard to the burden of proof.

[101] Green (n 40) 170–1. The author suggests that one consistent standard should apply to all cases of self-defense (whatever magnitude the consequences of the violation of the prohibition of the use of force might have) and both to the objective and subjective elements of the internationally wrongful act (Green (n 40) 169).

[102] O'Connell (n 51) 22ff.; Teitelbaum (n 54) 125ff.

[103] Green (n 40) 172. See *Nicaragua*, Merits (n 3) paras 29, 109.

[104] *Oil Platforms* (n 58) para 71, 61. See Green (n 40) 172–3; Teitelbaum (n 54) 125–6.

[105] *DRC v Uganda* (n 73) paras 72, 91, 136. Confusingly, however, in other parts of the judgment the Court seemed to employ a prima facie or preponderance of evidence standard, in particular when it had to determine whether the conduct of armed groups against the DRC was attributable to Uganda. Green (n 40) 175–6.

[106] *Jus Ad Bellum—Ethiopia's Claims 1–8* [2005] Eritrea-Ethiopia Claims Commission, Partial Award, Reports of International Arbitral Awards, XXVI, para 12. See O'Connell (n 89) 45.

[107] Mary Ellen O'Connell, "Lawful Self-Defense to Terrorism" (2002) 63 *University of Pittsburgh Law Review* 898.

If clear and convincing evidence is required at least in relation to claims related to the use of armed force, the question arises whether there is a special, and lower, standard in the cyber context, in particular for claims of self-defense against cyber operations. Indeed, "evidentiary thresholds that might have worked well in a world of conventional threats—where capabilities could be judged with high accuracy and the costs of false negatives to peace and security were not necessarily devastating—risk exposing states to unacceptable dangers."[108] There is of course no case-law in relation to claims arising out of inter-state cyber operations, so possible indications in this sense have to be found elsewhere. The Project Grey Goose Report on the 2008 cyber operations against Georgia, for instance, relies on the concordance of various items of circumstantial evidence to suggest that the Russian government was responsible for the operations.[109] In its reply to the UN Secretary-General on issues related to information security, the United States claimed that "high-confidence attribution of identity to perpetrators cannot be achieved in a timely manner, if ever, and success often depends on a high degree of transnational cooperation."[110] In a Senate questionnaire in preparation for a hearing on his nomination to head of the US Cyber Command, General Alexander argued that "some level of mitigating action" can be taken against cyber attacks "even when we are not certain who is responsible."[111] Similar words were employed by his successor, Vice Admiral Michael S Rogers: "International law does not require that a nation know who is responsible for conducting an armed attack before using capabilities to defend themselves from that attack."[112] However, Vice Admiral Rogers also cautioned that "from both an operational and policy perspective, it is difficult to develop an effective response without a degree of confidence in attribution."[113] Overall, these views seem to suggest an evidentiary standard, based on circumstantial evidence, significantly lower than clear and convincing evidence and even than a preponderance of the evidence, on the basis that identification and attribution are more problematic in a digital environment than in the analogue world.[114]

It is difficult, however, to see why the standard of proof should be lower simply because it is more difficult to reach it. The standard of proof exists not to disadvantage the claimant, but to protect the respondent against false attribution, which, thanks to tricks like IP spoofing, onion routing and the use of botnets, is a particularly serious

[108] Waxman (n 45) 62. The author argues that "the required degree of certainty about capability ought to vary with certainty about intent" (Waxman (n 45) 61): transposed in the cyber context, the higher the likelihood that an adversary will be able and willing to use cyber weapons, the lower the evidence required to prove it.

[109] *Russia/Georgia Cyber War—Findings and Analysis*, Project Grey Goose: Phase I Report, October 17, 2008, at <http://www.scribd.com/doc/6967393/Project-Grey-Goose-Phase-I-Report>.

[110] UN Doc A/66/152, 15 July 2011, at 16–17.

[111] Responses to advance questions, Nomination of Lt Gen Keith Alexander for Commander, US Cyber Command, US Sen Armed Serv Committee, April 15, 2010, 12, at <http://armed-services.senate.gov/statemnt/2010/04%20April/Alexander%2004-15-10.pdf>.

[112] Advance questions for Vice Admiral Michael S. Rogers, USN, Nominee for Commander, United States Cyber Command, at <http://www.armed-services.senate.gov/imo/media/doc/Rogers_03-11-14.pdf>.

[113] Advance questions (n 112).

[114] See, for example, David E Graham, "Cyber Threats and the Law of War" (2010) 4 *Journal of National Seccurity Law & Policy* 93 (the author seems, however, to confuse attribution criteria and standards of proof).

risk in the cyber context. The aforementioned views are also far from being unanimously held, even within the US government: the Air Force Doctrine for Cyberspace Operations, for instance, states that attribution of cyber operations should be established with "sufficient confidence and verifiability."[115] A report prepared by Italy's Parliamentary Committee on the Security of the Republic goes further and requires it to be demonstrated "unequivocally" ("in modo inequivocabile") that an armed attack by cyber means originated from a state and was undertaken on the instruction of governmental bodies.[116] The document also suggests that attribution to a state requires "irrefutable digital «evidence»" ("«prove» informatiche inconfutabili"), which—the Report concedes—is a standard which is very difficult to meet.[117] Germany also highlighted the need for "reliable attribution" of malicious cyber activities in order to avoid "'false flag' attacks," misunderstandings and miscalculations.[118] In relation to the DDoS attacks against Estonia, a UK House of Lords document lamented that "the analysis of today is really very elusive, not *conclusive* and it would still be very difficult to act on it."[119] Finally, the AIV/CAVV Report, which has been endorsed by the Dutch government, requires "reliable intelligence... before a military response can be made to a cyber attack" and "sufficient certainty" about the identity of the author of the attack.[120] In relation to the Report, the Dutch government also argued that self-defense can be exercised against cyber attacks "only if the origin of the attack and the identity of those responsible are sufficiently certain."[121]

All in all, clear and convincing evidence seems the appropriate standard not only for claims of self-defense against traditional armed attacks, but also for those against cyber operations: a prima facie or preponderance of evidence standard might lead to specious claims and false or erroneous attribution, while a beyond reasonable doubt

[115] United States Air Force, Cyberspace Operations: Air Force Doctrine Document 3–12 (2010), 10, at <http://static.e-publishing.af.mil/production/1/af_cv/publication/afdd3-12/afdd3-12.pdf>.

[116] Comitato Parlamentare per la Sicurezza della Repubblica (COPASIR), Relazione sulle possibili implicazioni e minacce per la sicurezza nazionale derivanti dall'utilizzo dello spazio cibernetico, Doc XXXIV, no 4, July 7, 2010, 26, at <http://www.parlamento.it/documenti/repository/commissioni/bicamerali/COMITATO%20SICUREZZA/Doc_XXXIV_n_4.pdf>.

[117] COPASIR (n 116).

[118] Permanent Mission of the Federal Republic of Germany to the United Nations, "Note Verbale No. 516/2012," November 5, 2015, 1, at <http://www.un.org/disarmament/topics/informationsecurity/docs/Germany_Verbal_Note_516_UNODA.pdf>. Laurie Blank also observes that "the victim State must tread carefully and seek as much clarity regarding the source of the attack as possible to avoid launching a self-defense response in the wrong direction." Laurie Blank, "International Law and Cyber Threats from Non-State Actors" (2013) 89 *Naval War College International Law Studies* 417.

[119] House of Lords, European Union Committee, "Protecting Europe against large-scale cyber-attacks" (HL Paper 68, 5th Report of Session 2009–10), 42, at <http://www.publications.parliament.uk/pa/ld200910/ldselect/ldeucom/68/68.pdf> (emphasis added).

[120] Advisory Council on International Affairs and the Advisory Committee on Issues of Public International Law, "Cyber Warfare," (No 77, AIV/No 22, CAVV December 2011),22, at <http://www.aiv-advies.nl/ContentSuite/upload/aiv/doc/webversie__AIV77CAVV_22_ENG.pdf>.

[121] Government of the Netherlands, Government response to the AIV/CAVV report on cyber warfare, 5, at <http://www.rijksoverheid.nl/bestanden/documenten-en-publicaties/rapporten/2012/04/26/cavv-advies-nr-22-bijlage-regeringsreactie-en/cavv-advies-22-bijlage-regeringsreactie-en.pdf>. The CCDCOE Report on Georgia also concludes that "there is no *conclusive* proof of who is behind the DDOS attacks, even though finger pointing at Russia is prevalent by the media." Eneken Tikk, Kadri Kaska, Kristel Rünnimeri, Mari Kert, Anna-Maria Talihärm, and Liis Vihul, *Cyber Attacks Against Georgia: Legal Lessons Identified* (CCDCOE, 2008) 12 (emphasis added).

standard would be unrealistic: in the *Norwegian Loans* case, Judge Lauterpacht emphasized that "the degree of burden of proof... adduced ought not to be so stringent as to render the proof unduly exacting."[122] As explained by Michael Schmitt, a clear and convincing standard "obliges a state to act reasonably, that is, in a fashion consistent with the normal state practice in same or similar circumstances. Reasonable states neither respond precipitously on the basis of sketchy indications of who has attacked them nor sit back passively until they have gathered unassailable evidence."[123]

Those who criticize a clear and convincing evidence standard for the exercise of self-defense against cyber operations rely on the fact that, due to the speed at which such operations may occur and produce their consequences, the requirement of a high level of evidence may in fact render it impossible for the victim state safely to exercise its right of self-defense. Such concerns, however, are exaggerated. Indeed, if the cyber attack was a standalone event that instantaneously produced its damaging effects, a reaction in self-defense would not be necessary. If, on the other hand, the cyber attack were continuing or formed of a series of smaller scale cyber attacks,[124] the likelihood that clear and convincing evidence could be collected would considerably increase.[125]

Having said that, there are also indications that the most serious allegations, such as those involving international crimes, require a higher standard to discharge the burden of proof.[126] As Judge Higgins wrote in her Separate Opinion attached to the *Oil Platforms* judgment, "the graver the charge the more confidence there must be in the evidence relied on."[127] In *Corfu Channel*, the Court appeared to suggest that the standard of proof is higher for charges of "exceptional gravity against a State."[128] In the *Bosnian Genocide* case, the ICJ confirmed that "claims against a State involving charges of exceptional gravity must be proved by evidence that is *fully conclusive*.... The same standard applies to the proof of attribution for such acts" (and accordingly applies both to the objective and subjective elements of an international crime).[129] The Court also found that assistance provided by Yugoslavia to the Bosnian Serbs had not

[122] *Certain Norwegian Loans (France v Norway)* [1957] ICJ Rep, Separate Opinion of Judge Sir Hersch Lauterpacht, at 39.

[123] Michael N Schmitt, "Cyber Operations and the *Jus ad Bellum* Revisited" (2011–12) 56 *Villanova Law Review* 595.

[124] On the application of the doctrine of accumulation of events to cyber operations, see Roscini (n 2) 108–10.

[125] A similar reasoning is utilised, in relation to the identification of the state responsible for the cyber attack, by Yoram Dinstein, "Cyber War and International Law: Concluding Remarks at the 2012 Naval War College International Law Conference" (2013) 89 *Naval War College International Law Studies* 282.

[126] Contra, see *Prisoners of War—Eritrea's Claim 17* [2003] Eritrea-Ethiopia Claims Commission, Partial Award, Reports of International Arbitral Awards, XXVI, paras 45–7.

[127] *Oil Platforms* (n 58), Separate Opinion of Judge Higgins, para 33.

[128] *Corfu Channel* (n 70) 17. This interpretation of the Court's judgment, however, is not uncontroversial: see Andrea Gattini, "Evidentiary Issues in the ICJ's Genocide Judgment," (2007) 5 *Journal of International Criminal Justice* 889, 896.

[129] *Application of the Convention on the Prevention and Punishment of the Crime of Genocide (Bosnia and Herzegovina v Serbia and Montenegro)*, Merits, [2007] ICJ Rep, para 209 (emphasis added). See also *Corfu Channel* (n 70) 17. It is not entirely clear whether the Court linked the notion of gravity to the importance of the norm allegedly breached or the magnitude of the violation. It would seem more correct to refer the gravity to the former, as, if the evidentiary standard depended on the latter, "some States could have a perverse incentive to sponsor more devastating attacks so as to raise the necessary burden of proof and potentially defeat accountability." Shackelford and Andres (n 29) 990.

been "established beyond any doubt."[130] Gravity is of course inherent in any jus cogens violation. Claims of reparation for cyber operations qualifying as war crimes, crimes against humanity or acts of genocide, therefore, will require "fully conclusive" evidence, not just evidence that is "clear and convincing." As has been aptly suggested, however, "[a] higher standard of proof may only be justified if the Court is willing to balance this strict approach with a more active use of its fact-finding powers to make sure that claims for breaches of *jus cogens* norms are not doomed to fail merely on evidential grounds."[131]

In the *Bosnian Genocide* judgment, the Court also appeared to make a distinction between a violation of the prohibition of committing acts of genocide, for which evidence must be "fully conclusive," and a violation of the obligation to prevent acts of genocide, where the Court required "proof at a high level of certainty appropriate to the seriousness of the allegation,"[132] even though not necessarily fully conclusive.[133] Such an approach appears justified by the different nature of the obligation breached: indeed, presumptions and inferences necessarily play a more significant role when the wrongful act to be proved consists of an omission, as in the case of the breach of an obligation to prevent.[134] By the same token, it may be suggested that the standard of proof required to prove that a state has conducted cyber operations amounting to international crimes is higher than that required to prove that it did not exercise the necessary due diligence to stop its cyber infrastructure from being used by others to commit international crimes.

V. Methods of Proof and Cyber Operations

What type of evidence may be relied on in order to meet the required standard of proof and establish that a cyber operation has occurred, has produced damage, and is attributable to a certain state or non-state actor? The production of evidence before the ICJ is regulated by Articles 48 to 52 of its Statute and by the Rules of Court. There is, however, no list of the methods of proof available to parties before the Court nor any indication of their different probative weight. Article 48 of the ICJ Statute provides only that "[t]he Court shall...make all arrangements connected with the taking of evidence," while Article 58 of the Rules of Court confirms that "the method of handling the evidence and of examining any witnesses and experts...shall be settled by the Court after the views of the parties have been ascertained in accordance with Article 31 of these Rules."

As a leading commentator has observed, "[t]he International Court of Justice has construed the absence of restrictive rules in its Statute to mean that a party may generally produce any evidence as a matter of right, so long as it is produced within the time limits fixed by the Court."[135] Although it is primarily the parties' responsibility

[130] *Bosnian Genocide* (n 129) para 422.
[131] Benzing (n 44) 1266.
[132] *Bosnian Genocide* (n 129) para 210.
[133] Benzing (n 44) 1266.
[134] Gattini (n 128) 899. In *Nicaragua*, the Court had already found that the fact that Nicaragua had to prove a negative (the non-supply of arms to rebels in neighboring countries) had to be "borne in mind" when assessing the evidence. *Nicaragua* (n 3) para 147.
[135] Durward V Sandifer, *Evidence Before International Tribunals* (Virginia: University Press of Virginia, 1975) 184.

to produce the evidence necessary to prove the facts alleged, the Court may also order the production of documents, call experts and witnesses, conduct site visits, and request relevant information from international organizations.[136] In *Nicaragua*, for instance, the Court found that it was "not bound to confine its consideration to the material formally submitted to it by the parties."[137] In that judgment, the ICJ also emphasized the principle of free assessment of evidence, stating that "within the limits of its Statute and Rules, [the Court] has freedom in estimating the value of the various elements of evidence."[138]

In the next pages, methods of proof that may be relevant in relation to cyber operations will be examined.

(a) Documentary evidence

Although there is no formal hierarchy between different sources, the ICJ has taken a civil law court approach and has normally given primacy to written documents over oral evidence.[139] Documentary evidence includes "all information submitted by the parties in support of the contentions contained in the pleadings other than expert and witness testimony."[140] According to Rosenne, documentary evidence can be classified in four categories: "published treaties included in one of the recognized international or national collections of treaty texts; official records of international organizations and national parliaments; published and unpublished diplomatic correspondence and communiqués and other miscellaneous materials, including books, maps, plans, charts, accounts, archival material, photographs, films, legal opinions and opinions of experts, etc.; and affidavits and declarations."[141]

Although the Court has the power to call upon the parties to produce any evidence it deems necessary or to seek such evidence itself, it has normally refrained from doing so and has relied on that spontaneously produced by the litigants. All documents not "readily available" must be produced by the interested party. A "publication readily available" is a document "available in the public domain... in any format (printed or electronic), form (physical or on-line, such as posted on the internet) or on any data medium (on paper, on digital or any other media) [that] should be accessible in either of the official languages of the Court" and which it is possible to consult "within a reasonably short period of time."[142] The accessibility should be assessed in relation to the

[136] Articles 49 and 50 of the ICJ Statute; Articles 62, 66, 67, and 69 of the Rules of Court.

[137] *Nicaragua*, Merits (n 3) para 30. See Article 49 of the ICJ Statute and Article 62 of the Rules of Court.

[138] *Nicaragua*, Merits (n 3) para 60. See also *DRC v Uganda* (n 73) para 59.

[139] Riddell and Plant (n 43) 232; Aguilar Mawdsley (n 59) 543.

[140] Wolfrum (n 31) 558.

[141] Rosenne (n 57) 1246 (footnotes omitted). In the *Bosnian Genocide* judgment, the Court noted that the parties had produced "reports, resolutions and findings by various United Nations organs, including the Secretary-General, the General Assembly, the Security Council and its Commission of Experts, and the Commission on Human Rights, the Sub-Commission on the Prevention of Discrimination and Protection of Minorities and the Special Rapporteur on Human Rights in the former Yugoslavia; documents from other inter-governmental organizations such as the Conference for Security and Co-operation in Europe; documents, evidence and decisions from the ICTY; publications from governments; documents from non-governmental organizations; media reports, articles and books." *Bosnian Genocide* (n 129) para 211. See also *DRC v Uganda* (n 73) para 60.

[142] Practice Direction IX bis (2)(i) and (ii).

Court and the other litigant.[143] The fact that a publication is "readily available" does not necessarily render the concerned facts public knowledge, but rather relieves the party from the burden of having to produce it (Article 56(4) of the Rules of Court).[144] The facts, however, still need to be proved.

Official state documents, such as national legislation, cyber doctrines, manuals, strategies, directives, and rules of engagement, may become relevant in establishing state responsibility for cyber operations. In *Nicaragua*, for instance, the responsibility of the United States for encouraging violations of international humanitarian law was established on the basis of the publication of a manual on psychological operations. According to the Court, "[t]he publication and dissemination of a manual in fact containing the advice quoted above must... be regarded as an encouragement, which was likely to be effective, to commit acts contrary to general principles of international humanitarian law reflected in treaties."[145] Not all state documents, however, have the same probative value: in *DRC v Uganda*, the Court dismissed the relevance of certain internal military intelligence documents because they were unsigned, unauthenticated, or lacked explanation of how the information was obtained.[146]

Military cyber documents are frequently classified in whole or in part for national security reasons. According to the doctrine of privilege in domestic legal systems, litigants may refuse to submit certain evidence to a court on confidentiality grounds. No such doctrine exists before the ICJ.[147] One could actually argue that there is an obligation on the litigants to cooperate in good faith with the Court in the proceedings before it, and, therefore, to produce all requested documents.[148] There is, however, no sanction for failure to do so: Article 49 of the ICJ Statute limits itself to providing that "[t]he Court may, even before the hearing begins, call upon the agents to produce any document or to supply any explanations. *Formal note shall be taken of any refusal.*"[149] While the ICTY has found that "to grant States a blanket right to withhold, for security purposes, documents necessary for trial might jeopardize the very function of the International Tribunal, and 'defeat its essential object and purpose',"[150] the ICJ has been reluctant to draw inferences from the refusal of a party to produce confidential documents. The problem has arisen twice before the Court: in the *Corfu Channel* and in the *Bosnian Genocide* cases. In the former, the ICJ called the United Kingdom, pursuant to Article 49 of the Statute, to produce an admiralty order. The United Kingdom refused to produce the document on grounds of naval secrecy,[151] and witnesses also

[143] Practice Direction IX bis (2)(i)(ii). [144] Benzing (n 44) 1241.

[145] *Nicaragua*, Merits (n 3) para 256.

[146] *DRC v Uganda* (n 73) paras 125, 127, 128, 133, 134, 137.

[147] One of the problems with applying the doctrine of privilege in inter-state litigation is that international courts are unlikely to be able to verify whether state security interests are genuinely jeopardized by the document disclosure. Riddell and Plant (n 43) 208.

[148] It has been observed that "when a State becomes a party to the Statute of the ICJ, it necessarily accepts the obligation to produce before the Court all evidence available to it in any case it contests." Riddell and Plant (n 43) 49.

[149] Emphasis added.

[150] *Blaškić (Lašva Valley)* [1997] Case IT-95-14, Judgment on the Request of the Republic of Croatia for Review of the Decision of Trial Chamber II of July 18, 1997, 29 October 1997, para 65.

[151] Anthony Carty, "The Corfu Channel Case and the Missing Admiralty Order," *The Law and Practice of International Courts and Tribunals* 3 (2004) 1.

refused to answer questions in relation to the document. The ICJ decided not to "draw from this refusal to produce the orders any conclusions differing from those to which the actual events gave rise."[152] In the *Bosnian Genocide* case, even though Bosnia and Herzegovina had called upon the Court to request Serbia and Montenegro to produce certain documents classified as military secrets, the Court decided not to proceed with the request, although it reserved the right subsequently to request the documents *motu proprio*.[153] In its judgment, the ICJ limited itself to noting "the Applicant's suggestion that the Court may be free to draw its own conclusions' from the fact that Serbia and Montenegro had not produced the document voluntarily.[154] However, it does not seem that the Court ultimately drew any inferences from Serbia's non-disclosure of the classified documents.[155] It should be noted that, in both the aforementioned cases, alternative evidence was available to the Court. It has been suggested that "it remains a matter of conjecture how the ICJ might respond in cases where a confidential communication is the only possible evidence to determine the veracity of a factual assertion, and no alternative materials are available."[156] A possible solution is that any classified information be produced in closed sittings of the court.[157]

Documents of international organizations may also be presented as evidence. Overall, the Court has given particular credit to UN reports, Security Council resolutions and other official UN documents.[158] In *Bosnian Genocide*, the ICJ stated that the probative value of reports from official or independent bodies "depends, among other things, on (1) the source of the item of evidence (for instance, partisan, or neutral), (2) the process by which it has been generated (for instance an anonymous press report or the product of a careful court or court-like process), and (3) the quality of the character of the item (such as statements against interest, and agreed or uncontested facts)."[159] Several documents of international organizations address cyber issues. In particular, information security has been on the UN agenda since 1998 when the Russian Federation introduced a draft resolution in the First Committee of the UN General Assembly. Since then, the General Assembly has adopted a series of annual resolutions on the topic. The resolutions have called for the views of the UN member states on information security and established three Groups of Governmental Experts that have examined threats in cyberspace and discussed cooperative measures to address them. While the first Group, established in 2004, did not produce a substantive report, the second, created in 2009, issued a report in 2010,[160] and the third Group, which met between 2012 and 2013, also adopted a final report containing a set of recommendations.[161] In addition, the views of UN member states

[152] *Corfu Channel* (n 70) 32. [153] *Bosnian Genocide* (n 129) para 44.
[154] *Bosnian Genocide* (n 129) para 206. See Teitelbaum (n 54) 131.
[155] Riddell and Plant (n 43) 214. [156] Riddell and Plant (n 43) 217.
[157] Riddell and Plant (n 43) 218; Benzing (n 44) 1243. [158] Teitelbaum (n 54) 146.
[159] *Bosnian Genocide* (n 129) para 227. In the case of the "Fall of Srebrenica" Report of the Secretary-General pursuant to General Assembly resolution 53/35 (UN Doc A/54/549), the Court concluded that "the care taken in preparing the report, its comprehensive sources and the independence of those responsible for its preparation all lend considerable authority to it." *Bosnian Genocide* (n 129) para 230.
[160] UN Doc A/65/201, July 30, 2010. [161] UN Doc A/68/98, June 24, 2013.

on information security are contained in the annual reports of the UN Secretary-General on developments in the field of information and telecommunications in the context of international security.[162]

The Court has also relied on fact-finding from commissions and other courts.[163] In *DRC v Uganda*, the Court considered the Report of the Porter Commission, observing that neither party had challenged its credibility.[164] Furthermore, the Court accepted that "evidence [included in the Report] obtained by examination of persons directly involved, and who were subsequently cross-examined by judges skilled in examination and experienced in assessing large amounts of factual information, some of it of a technical nature, merits special attention."[165] For these reasons, facts alleged by the parties that found confirmation in the Report were considered clearly and convincingly proved.[166] There are, however, no examples of reports by judicial commissions in relation to cyber operations. One can at best recall the 2009 Report of the Independent Fact-Finding Mission on the Conflict in Georgia established by the Council of the European Union, which briefly addressed the cyber operations against Georgia. The Report, however, is not of great probative weight, as it did not reach any conclusion on their attribution or legality and simply noted that "[i]f these attacks were directed by a government or governments, it is likely that this form of warfare was used for the first time in an inter-state armed conflict."[167] Even if not of use to establish attribution, however, the Report could be relied on to establish that the cyber operations against Georgia did in fact occur.

Documents produced by NGOs and think-tanks may also play an evidentiary role, albeit limited. In relation to cyber operations, the CCDCOE has prepared reports containing technical and legal discussion of the Estonia, Georgia, and Iran cases, as well as of other cyber incidents.[168] Project Grey Goose produced an open source investigation into cyber conflicts, including the 2008 cyber attacks on Georgia. In this case, the Report concluded "with high confidence that the Russian government will likely continue its practice of distancing itself from the Russian nationalistic hacker community thus gaining deniability while passively supporting and enjoying the strategic benefits of their action."[169] Information security companies like Symantec, McAfee, and Mandiant also regularly compile detailed technical reports on cyber threats and specific incidents.[170] In general, however, reports from NGOs and other non-governmental bodies have been considered by the ICJ as having less probative value than publications of states and international organizations, and have been used in a corroborative role

[162] United Nations Office for Disarmament Affairs, Developments in the Field of Information and Telecommunications in the Context of International Security, at <http://www.un.org/disarmament/topics/informationsecurity>.

[163] Teitelbaum (n 54) 152.

[164] *DRC v Uganda* (n 73) para 60.

[165] *DRC v Uganda* (n 73) para 61.

[166] Teitelbaum (n 54) 153.

[167] Report of the Independent Fact-Finding Mission on the Conflict in Georgia, September 2009, vol II, at 217–19, at <http://www.ceiig.ch/Report.html>.

[168] The CCDCOE is a think tank based in Tallinn that was created after the 2008 DDoS attacks against the Baltic state. It is not integrated into NATO's structure or funded by it. Its reports can be accessed at <https://www.ccdcoe.org/publications.html>.

[169] *Russia/Georgia Cyber War—Findings and Analysis* (n 109) 3.

[170] See, for example, Mandiant Report (n 25). Mandiant is a US-based information security company.

only.[171] In *DRC v Uganda*, for instance, the ICJ considered a report by International Crisis Group not to constitute "reliable evidence."[172] Similarly, in *Oil Platforms* the Court did not find publications such as *Lloyd's Maritime Information Service*, the *General Council of British Shipping* and *Jane's Intelligence Review* to be authoritative public sources, as it had no "indication of what was the original source, or sources, or evidence on which the public sources relied."[173] This "unequal treatment" of documents of international organizations and NGOs has been criticized: "the correct approach is for the Court to apply its general evaluative criteria to documents produced by NGOs just as it does to those generated by UN actors."[174]

As far as press reports and media evidence are concerned, one may recall, in the cyber context, the already mentioned article in *The New York Times* attributing Stuxnet to the United States and Israel.[175] The ICJ, however, has been very reluctant to accept press reports as evidence and has treated them "with great caution."[176] Press reports that rely only on one source, upon an interested source, or give no account of their sources have, therefore, been treated as having no probative value.[177] In the *Bosnian Genocide* case, the Court dismissed an article in *Le Monde* qualifying it as "only a secondary source."[178] In *Nicaragua*, the Court held that, even when they meet a "high standard of objectivity," it would regard the reports in press articles and extracts from books presented by the parties "not as evidence capable of proving facts, but as material which can nevertheless contribute, in some circumstances, to corroborating the existence of a fact, i.e., as illustrative material additional to other sources of evidence."[179] This was so provided that they were "wholly consistent and concordant as to the main facts and circumstances of the case."[180] It has been suggested that this expression means that "the press reports in question would have to confirm the facts as alleged by both of the parties, or confirm the facts that have not been denied or contested by the parties."[181]

Apart from this, press reports may contribute, together with other sources, to demonstrate public knowledge of facts of which the Court may take judicial notice, thus relieving a party from having to discharge the burden of proof with regard to those facts.[182] The fact that cyber incidents like Stuxnet have received extensive media coverage and that the article in *The New York Times* has been followed by many others, including in *The Washington Post*,[183] would not, however, as such increase their probative weight or mean that the covered facts are of public knowledge: as already mentioned, in *Nicaragua* the ICJ noted that "[w]idespread reports of a fact may prove on closer

[171] Riddell and Plant (n 43) 249. [172] *DRC v Uganda* (n 73) para 129.
[173] *Oil Platforms* (n 58) para 60. [174] Riddell and Plant (n 43) 250.
[175] See nn 15, 16, and 17. [176] *Nicaragua, Merits* (n 3) para 62.
[177] *DRC v Uganda* (n 73) para 68. [178] *Bosnian Genocide* (n 129) para 357.
[179] *Nicaragua*, Merits (n 3) para 62.
[180] *Nicaragua*, Merits (n 3) (quoting *United States Diplomatic and Consular Staff in Tehran* (n 65) 13). See also *DRC v Uganda* (n 73) para 68.
[181] Teitelbaum (n 54) 140. [182] *Nicaragua*, Merits (n 3) para 63.
[183] Ellen Nakashima, Greg Miller and Julie Tate, "U.S., Israel developed Flame computer virus to slow Iranian nuclear efforts, officials say," *The Washington Post*, June 19, 2012, at <http://www.washingtonpost.com/world/national-security/us-israel-developed-computer-virus-to-slow-iranian-nuclear-efforts-officials-say/2012/06/19/gJQA6xBPoV_story.html>.

examination to derive from a single source, and such reports, however numerous, will in such case have no greater value as evidence than the original source."[184]

(b) Official statements

Statements made by official authorities outside the context of the judicial proceedings may play an important evidentiary role. In the *Tehran Hostages* case, for instance, the ICJ recalled that it had "a massive body of information from various sources concerning the facts and circumstances of the present case, including numerous official statements of both Iranian and United States authorities."[185] Statements "emanating from high-ranking official political figures, sometimes indeed of the highest rank, are of particular probative value when they acknowledge facts or conduct unfavourable to the State represented by the person who made them."[186] However, all depends on how those statements were made public: "evidently, [the Court] cannot treat them as having the same value irrespective of whether the text is to be found in an official national or international publication, or in a book or newspaper."[187] In other words, statements that can be directly attributed to a state are of more probative value.

The US Department of Defense's *Assessment of International Legal Issues in Information Operations* confirms that "[s]tate sponsorship might be persuasively established by such factors as...public statements by officials."[188] There do not seem to be, however, any official statements by Russian or Chinese authorities directly or even indirectly acknowledging responsibility for the cyber operations against Estonia, Georgia, and the United States: on the contrary, involvement was denied. With regard to Stuxnet, when asked questions about the incident, US and Israeli authorities neither admitted nor denied attribution. Whether this allows inferences to be drawn is discussed in section VI.[189]

(c) Witness testimony

Witnesses may be called to provide direct oral evidence by the Court and by the litigants: the latter case is conditioned upon the absence of objections by the other litigant or the recognition by the Court that the evidence is likely to be relevant.[190] The Court may also put questions to the witnesses and experts called by the parties.[191] The ICJ has not made extensive use of oral evidence.[192] In *Corfu Channel*, for instance, naval officers were called to testify by the United Kingdom about the damage suffered by the Royal Navy ships and the nature and origin of the mines.[193] Albania also called

[184] *Nicaragua*, Merits (n 3) para 63.
[185] *United States Diplomatic and Consular Staff in Tehran* (n 65) para 13.
[186] *Nicaragua*, Merits (n 3) para 64. [187] *Nicaragua*, Merits (n 3) para 65.
[188] Department of Defense, *An Assessment of International Legal Issues in Information Operations*, May 1999, 21, at <http://www.au.af.mil/au/awc/awcgate/dod-io-legal/dod-io-legal.pdf>.
[189] See section VI.
[190] Articles 62(2) and 63 of the Rules of Court. It should be recalled that international courts and tribunals do not normally have the authority or the capability to issue *subpoena* to coercively bring a witness before them. Wolfrum (n 31) 560.
[191] Rule 65 of the Rules of Court. [192] See the cases in Aguilar Mawdsley (n 59) 543.
[193] *Corfu Channel* (n 70) 8.

witnesses to testify the absence of mines in the Channel.[194] Five witnesses were called by Nicaragua to testify in the *Nicaragua* case.[195] In the same case, the Court noted that "testimony of matters not within the direct knowledge of the witness, but known to him only from hearsay" is not "of much weight."[196]

It is worth recalling that the Court has also accepted witness evidence given in written form and attached to the written pleadings, but it has treated it "with caution"[197] and has generally considered it of a probative value inferior to that of direct oral witness testimony.[198] Factors to be considered in assessing the probative weight of affidavits include time, purpose, and context of production, whether they were made by disinterested witnesses and whether they attest to the existence of facts or only refer an opinion with regard to certain events.[199]

(d) Enquiry and experts

According to Article 50 of the ICJ Statute, "[t]he Court may, at any time, entrust any individual body, bureau, commission, or other organization that it may select, with the task of carrying out an enquiry or giving an expert opinion."[200] Enquiries have never been commissioned by the Court, which has rather relied on fact-finding reports from other sources.[201] Experts may be necessary in cases of a highly technical nature or that involve expertise not possessed by the judges:[202] it is likely, therefore, that the Court will appoint experts in cases involving cyber technologies. The Court, however, would not be bound by their report.

Experts may also be called by the parties.[203] As to the form of their participation in the oral proceedings, in *Pulp Mills* the ICJ reminded the parties that "those persons who provide evidence before the Court based on their scientific or technical knowledge and on their personal experience should testify before the Court as experts, witnesses or in some cases in both capacities, rather than counsel, so that they may be submitted to questioning by the other party as well as by the Court."[204] In the *Whaling in the Antarctic* case, therefore, the experts called by both Australia and Japan gave evidence as expert witnesses and were cross-examined,[205] and the Court relied heavily on their statements to conclude that the special permits granted by Japan for the killing, taking, and treating of whales had not been granted "for purposes of scientific research."[206]

[194] *Corfu Channel* (n 70) 8. [195] *Nicaragua*, Merits (n 3) para 13.

[196] *Nicaragua*, Merits (n 3) para 68.

[197] *Territorial and Maritime Dispute Between Nicaragua and Honduras in the Caribbean Sea (Nicaragua v Honduras)*, [2007] ICJ Rep, para 244.

[198] Riddell and Plant (n 43) 280–1. [199] *Nicaragua v Honduras* (n 197) para 244.

[200] See also Article 67 of the Rules of Court.

[201] Benzing (n 44) 1259. For criticism of this practice, see Daniel Joyce, "Fact-Finding and Evidence at the International Court of Justice: Systemic Crisis, Change or More of the Same?" (2007) XVIII *Finnish Yearbook of International Law* 283–306.

[202] In the *Corfu Channel* Case, the Court appointed a Committee of Experts because of the insurmountable differences of opinion between the parties on certain facts. *Corfu Channel* (n 70) 9.

[203] Rules of Court, Article 63. [204] *Pulp Mills* (n 55) para 167.

[205] *Whaling in the Antarctic (Australia v Japan: New Zealand Intervening)*, judgment, March 31, 2014, paras 20–1, at <http://www.icj-cij.org/docket/files/148/18136.pdf>.

[206] *Whaling in the Antarctic* (n 205) para 227.

(e) Digital evidence

Digital forensics "deals with identifying, storing, analyzing, and reporting computer finds, in order to present valid digital evidence that can be submitted in civil or criminal proceedings."[207] It includes the seizure, forensic imaging, and analysis of digital media, and the production of a report on the evidence so collected.[208] It seems that most countries "do not make a legal distinction between electronic evidence and physical evidence. While approaches vary, many countries consider this good practice, as it ensures fair admissibility alongside all other types of evidence."[209] Of course, not only do data have to be collected, but they also need to be interpreted, and the parties may disagree on their interpretation.

Digital evidence, however, is unlikely to play, on its own, a decisive role to establish state responsibility for cyber operations, for several reasons. First, digital evidence is "volatile, has a short life span, and is frequently located in foreign countries."[210] Secondly, the collection of digital evidence can be very time consuming and requires the cooperation of the relevant internet service providers, which may be difficult to obtain when the attack originates from other states. Thirdly, although digital evidence may lead to the identification of the computer or computer system from where the cyber operation originates, it does not necessarily identify the individual(s) responsible for the cyber operation (as the computer may have been hijacked, or the IP spoofed). In any case, it will say nothing about whether the conduct of those individuals can be attributed to a state under the law of state responsibility.[211]

VI. Presumptions and Inferences in the Cyber Context

As Judge ad hoc Franck emphasized in *Sovereignty over Pulau Ligitan and Pulau Sipadan*, "[p]resumptions are necessary and well-established aspects of both common and civil law and cannot but be a part of the fabric of public international law."[212] Previously, in his Dissenting Opinion in *Corfu Channel*, Judge Azevedo had

[207] Italy's Presidency of the Council of Ministers, "National Strategy Framework for Cyberspace Security," December 2013, at 42, at <http://www.sicurezzanazionale.gov.it/sisr.nsf/wp-content/uploads/2014/02/italian-national-strategic-framework-for-cyberspace-security.pdf>.

[208] See the traceback technology described in Jay P Kesan and Carol M Hayes, "Mitigative Counterstriking: Self-Defense and Deterrence in Cyberspace" (2011–12) 25 *Harvard Journal of Law & Technology* 482ff. The US Department of Defense is apparently seeking to improve attribution capabilities through behaviour-based algorithms. US Department of Defense, Cyberspace Policy Report—A Report to Congress Pursuant to the National Defense Authorization Act for Fiscal Year 2011, Section 934 (November 2011) 4, at <http://www.defense.gov/home/features/2011/0411_cyberstrategy/docs/NDAA%20Section%20934%20Report_For%20webpage.pdf>.

[209] United Nations Office on Drugs and Crime, Comprehensive Study on Cybercrime (February 2013), XXIV, at <http://www.unodc.org/documents/organized-crime/UNODC_CCPCJ_EG.4_2013/CYBERCRIME_STUDY_210213.pdf>.

[210] Fred Schreier, "On Cyberwarfare," The Geneva Centre for the Democratic Control of Armed Forces, DCAF Horizon 2015 Working Paper No 7, 2012, 65.

[211] In this sense, cf Rules 7 and 8 of the *Tallinn Manual* (n 2) 34, 36.

[212] *Sovereignty over Pulau Ligitan and Pulau Sipadan (Indonesia/Malaysia)* [2001] ICJ Rep, Dissenting Opinion of Judge ad hoc Franck, para 11.

argued that "[i]t would be going too far for an international court to insist on direct and visual evidence and to refuse to admit, after reflection, a reasonable amount of human presumptions with a view to reaching that state of moral, human certainty with which, despite the risk of occasional errors, a court of justice must be content."[213]

Although the difference is often blurred in inter-state litigation, presumptions may be prescribed by law ("legal presumptions," or presumptions of law), or be reasoning tools used by the judges (presumptions of fact, or inferences).[214] In other words, "[p]resumptions of law derive their force from *law*, while presumptions of fact derive their force from *logic*."[215] In international law, presumptions of law can derive from treaties, international customs, and general principles of law.[216] According to Judge Owada in his Dissenting Opinion in the *Whaling in the Antarctic* case, for instance, good faith on the part of a contracting state in performing its obligations under a treaty "has necessarily to be presumed,"[217] although the presumption is subject to rebuttal.

Inferences, or presumptions of fact, are closely linked to circumstantial evidence.[218] In the *Corfu Channel* case, Judge Padawi Pasha defined "circumstantial evidence" as "facts which, while not supplying immediate proof of the charge, yet make the charge probable with the assistance of reasoning."[219] Inferences "convincingly" establishing state sponsorship for cyber operations are suggested in the US Department of Defense's *Assessment of International Legal Issues in Information Operations*, including "the state of relationships between the two countries, the prior involvement of the suspect state in computer network attacks, the nature of the systems attacked, the nature and sophistication of the methods and equipment used, the effects of past attacks, and the damage which seems likely from future attacks."[220] In its reply to the UN Secretary-General on issues related to information security, the United States also claimed that "the identity and motivation of the perpetrator(s) can only be inferred from the target, effects and other circumstantial evidence surrounding an

[213] *Corfu Channel* (n 70), Dissenting Opinion of Judge Azevedo, paras 90–1.

[214] Chittharanjan F Amerasinghe, "Presumptions and Inferences in Evidence in International Litigation," (2004) 3 *The Law & Practice of International Courts and Tribunals* 395. Irrebuttable presumptions (juris et de jure) must be distinguished from rebuttable ones (juris tantum): the former are immune to evidence proving facts that contradict them, while the latter shift the burden of demonstrating the opposite to the other litigant.

[215] Thomas M Franck and Peter Prows, "The Role of Presumptions in International Tribunals," (2005) 4 *The Law & Practice of International Courts and Tribunals* 197, 203.

[216] Mojtaba Kazazi, *Burden of Proof and Related Issues: A Study on Evidence Before International Tribunals* (Boston: Kluwer, 1996) 245.

[217] *Whaling in the Antarctic* (n 205), Dissenting Opinion of Judge Owada, para 21.

[218] Riddell and Plant (n 42) 113. According to Judge Bustamante's Separate Opinion in the *Barcelona Traction* case, it is "possible to arrive at a conclusion on the basis merely of inferences or deductions forming part of a logical process." *Barcelona Traction, Light and Power Company, Limited (Belgium v Spain)*, Preliminary Objections, [1964] ICJ Rep, Judge Bustamante's Separate Opinion, at 84.

[219] *Corfu Channel* (n 70), Dissenting Opinion of Judge Badawi Pasha, at 59.

[220] *Assessment of International Legal Issues* (n 188) 21. For a critique of the use of the sophistication criterion to establish attribution, see Clement Guitton and Elaine Korzak, "The Sophistication Criterion for Attribution. Identifying the Perpetrators of Cyber-Attacks" (2013) 158 RUSI Journal 2–68.

incident."[221] The commentary to Rule 11 of the Tallinn Manual refers to inferences from "the prevailing political environment, whether the operations portends the future use of military force, the identity of the attacker, any record of cyber operations by the attacker, and the nature of the target (such as critical infrastructure)," in order to determine whether a cyber operation qualifies as a use of force under Article 2(4) of the UN Charter.[222]

The ICJ, however, "has demonstrated an increasing resistance to the drawing of inferences from secondary evidence."[223] Only inferences to protect state sovereignty are normally drawn by the Court, while others are treated with great caution.[224] Examples of situations where the ICJ has explored the possibility of drawing inferences include exclusive control of territory and non-production of documents.[225] As to the first, it has been argued that the state from which the cyber operation originates has presumptive knowledge of such operation. US officials have claimed, for instance, that, from the control that the Iranian government exercises over the internet, it is "hard to imagine" that cyber attacks originating from Iran against US oil, gas, and electricity companies could be conducted without governmental knowledge, even in the absence of direct proof of state involvement.[226] The same considerations may be extended to cyber operations originating from China and other states where access to the internet is under strict governmental control. The US Department of Defense's *Assessment of International Legal Issues in Information Operations* also claims that "[s]tate sponsorship might be persuasively established by such factors as…the location of the offending computer within a state-controlled facility."[227] In the literature, Garnett and Clarke have claimed that "in a situation where there have been repeated instances of hostile computer activity emanating from a State's territory directed against another State, it seems reasonable to presume that the host State had knowledge of such attacks and so should incur responsibility."[228] At least certain of the cyber attacks against Estonia and Georgia originated from Russian IP addresses, including those of state institutions.[229] The Mandiant Report also traced the cyber intrusions into US computers back to Chinese IP addresses.[230] As has been

[221] UN Doc A/66/152, July 15, 2011, at 16–17. See the almost identical words in Government of India, Ministry of Communications and Information Technology, "Discussion draft on National Cyber Security Policy" (draft v1.0, 2011), at 4, at <http://deity.gov.in/sites/upload_files/dit/files/ncsp_060411.pdf>.

[222] *Tallinn Manual* (n 2) 51–2.

[223] Teitelbaum (n 54) 157.

[224] Riddell and Plant (n 43) 413.

[225] Waxman has highlighted the need to use "propensity inferences," which are based on the past behaviour of a regime and its inclination to undertake certain actions. Waxman (n 45) 66. He concludes that "there is no escaping some reliance on propensity inferences because of the limits of forensic evidence" (Waxman (n 45) 66). As the author himself points out, however, previous conduct can be misleading when the regime in question bluffs about its capabilities to intimidate or deter, as in the case of Saddam Hussein's Iraq (Waxman (n 45) 68).

[226] Nicole Perlroth and David E Sanger, "New Computer Attacks Traced to Iran, Officials Say," *The New York Times*, May 24, 2013, at <http://www.nytimes.com/2013/05/25/world/middleeast/new-computer-attacks-come-from-iran-officials-say.html?_r=0>.

[227] *Assessment of International Legal Issues* (n 188) 21.

[228] Richard Garnett and Paul Clarke, "Cyberterrorism: A New Challenge for International Law," in Andrea Bianchi (ed) *Enforcing International Law Norms against Terrorism* (Oxford and Portland: Hart, 2004) 479.

[229] Owens et al (n 7) 173; Tikk et al (n 11) 75.

[230] Mandiant Report (n 25) 4.

seen, however, in the *Corfu Channel* case the ICJ held that "it cannot be concluded from the mere fact of the control exercised [by a state] over its territory...that that State necessarily knew, or ought to have known, of any unlawful act perpetrated therein."[231] Only if there are other indications of state involvement may territorial control contribute to establish knowledge.[232] In *Oil Platforms*, the ICJ also refused to accept the US argument that the territorial control exercised by Iran over the area from which the missile against the Sea Isle City had been fired was sufficient to demonstrate Iran's responsibility.[233] These conclusions are transposed in the cyber context by Rules 7 and 8 of the *Tallinn Manual*, according to which neither the fact that a cyber operation originates from a state's governmental cyber infrastructure nor that it has been routed through the cyber infrastructure located in a state are sufficient evidence for attributing the operation to those states, although it may be "an indication that the State in question is associated with the operation."[234] The *Manual* does not clarify what probative value this "indication" would have.

If control of cyber infrastructure is not on its own sufficient to prove knowledge of the cyber operations originating therefrom, and even less direct attribution, it may, however, have "a bearing upon the methods of proof available to establish the knowledge of that State as to such events."[235] In particular,

> [b]y reason of this exclusive control [within its frontiers], the other State, the victim of a breach of international law, is often unable to furnish direct proof of facts giving rise to responsibility. Such a State should be allowed a *more liberal recourse to inferences of fact and circumstantial evidence*. This indirect evidence is admitted in all systems of law, and its use is recognized by international decisions.[236]

According to the Court, then, inferences become particularly valuable, and assume a probative value higher than normal, when a litigant is unable to provide direct proof of facts because the evidence is under the exclusive territorial control of the other litigant. Such indirect evidence "must be regarded as of special weight when it is based on a series of facts linked together and leading logically to a single conclusion."[237] The ICJ, therefore, coupled the exclusive territorial control by Albania with its silence about the mine laying and other circumstantial evidence, and concluded that Albania had knowledge of the mines.[238] Transposed to the cyber context, the presence or origination of

[231] *Corfu Channel* (n 70) 18. See contra the Individual Opinion of Judge Alvarez, according to whom "every State is considered as having known, or as having a *duty* to have known, of prejudicial acts committed in parts of its territory where local authorities are installed."*Corfu Channel* (n 70) 18, Individual Opinion of Judge Alvarez, 44.

[232] *Corfu Channel* (n 70) 22.

[233] *Oil Platforms* (n 58) para 61.

[234] *Tallinn Manual* (n 2) 34–6.

[235] *Corfu Channel* (n 70) 18.

[236] *Corfu Channel* (n 70) 18 (emphasis added).

[237] *Corfu Channel* (n 70) 18. This may for instance be the case when a large number of cyber operations originate from the governmental cyber infrastructure of the same country.

[238] *Corfu Channel* (n 70) 22. In the *United States Diplomatic and Consular Staff in Tehran Case*, the Court also found that "[i]n the governing circumstances..., in which the United States is unable to gain access to its diplomatic and consular representatives in Iran, or to its Embassy and consular premises in Iran and to the files which they contain, the Court will appreciate that certain factual details, particularly those relating to the current condition of United States personnel in Tehran, are unavailable to the United States Government at this time." *United States Diplomatic and Consular Staff in Tehran* (n 65) 125.

the hazard in the cyber infrastructure controlled by a state does not per se demonstrate knowledge by that state, but may contribute to such a finding if it is accompanied by other circumstantial evidence pointing in that direction. In *Corfu Channel*, the Court, however, specified that, when proof is based on inferences, these must "leave *no room* for reasonable doubt."[239] In the *Bosnian Genocide* case, the Court confirmed that in demonstrating genocidal intent "for a pattern of conduct to be accepted as evidence of its existence, it would have to be such that it could only point to the existence of such intent."[240] In any case, "no inference can be drawn which is inconsistent with facts incontrovertibly established by the evidence."[241]

Of course, the Court will first have to determine whether the party has "exclusive territorial control" of the cyber infrastructure from which the cyber operations originated (and, therefore, potentially of the evidence of who was responsible for them) before allowing the more "liberal" recourse to inferences. This may cause particular difficulties in case of armed conflict: in the *DRC v Uganda* case, for instance, whether Uganda had had control over Congolese territory was one of the issues in dispute.[242] In the cyber context, determining whether a litigant has "territorial control" of the cyber infrastructure and whether such control is "exclusive" may be equally difficult to establish and is linked to the ongoing debate on the states' creeping jurisdiction over the internet and cyberspace in general.[243] In this context, it should be recalled that Rule 1 of the *Tallinn Manual* accepts that "[a] State may exercise control over cyber infrastructure and activities within its sovereign territory."[244]

It should also be noted that the ICJ has not always allowed the "more liberal recourse to inferences of fact and circumstantial evidence" in cases of exclusive territorial control. In the *Bosnian Genocide* case, Bosnia and Herzegovina argued that, because of Serbia and Montenegro's geographical situation, the standard of proof should be lower and that the respondent "had a special duty of diligence in preventing genocide and the proof of its lack of diligence can be inferred from fact and circumstantial evidence."[245] The Court rejected this reasoning and established Serbia and Montenegro's responsibility for failure to prevent genocide not on the basis of inferences but on documentary evidence and ICTY testimony.[246]

Does refusal to disclose evidence allow negative inferences? Article 38 of the Rules of Procedure of the Inter-American Commission on Human Rights provides that the facts alleged in the petition "shall be presumed to be true if the State has not provided responsive information during the period set by the Commission under the provisions of Article 37 of these Rules of Procedure, as long as other evidence does not lead to a different conclusion."[247] This is due to the different nature of human

[239] *Corfu Channel* (n 70) 18 (emphasis in the original).
[240] *Bosnian Genocide* (n 129) para 373.
[241] *Case Concerning the Temple of Preah Vihear (Cambodia v Thailand)*, Merits, [1962] ICJ Rep, Dissenting Opinion of Sir Percy Spender, at 109.
[242] Teitelbaum (n 54) 136. [243] See Roscini (n 2) 23–4. [244] *Tallinn Manual* (n 2) 15.
[245] *Bosnian Genocide* (no 129), Reply of Bosnia and Herzegovina, April 23, 1998, at 839, at <http://www.icj-cij.org/docket/files/91/10505.pdf>.
[246] Teitelbaum (n 54) 138–9. For critical comments on the ICJ's reliance on ICTY evidence, see Joyce (n 201) 298–305.
[247] Text at <http://www.oas.org/en/iachr/mandate/Basics/rulesiachr.asp>.

rights tribunals, where one of the parties is an individual and the other is a government, while disputes before the ICJ are between sovereign states. According to Article 49 of its Statute, the ICJ may only take "formal note" of the refusal to disclose evidence: this provision authorizes the Court to draw inferences but does not create a presumption of law.[248] In any case, as has already been seen, in the *Corfu Channel* and the *Bosnian Genocide* cases the Court declined to draw any inferences from refusal to produce evidence, in the former case because there was a series of facts contrary to the inference sought. Of course, if the litigant decides not to produce certain evidence, it will bear the risk that the facts it claims will not be considered sufficiently proved.

VII. Inadmissible Evidence

There are no express rules on the admissibility of evidence in the ICJ Statute. Therefore, "[t]he general practice of the Court has been to admit contested documents and testimony, subject to the reservation that the Court will itself be the judge of the weight to be accorded to it."[249] Evidence may, however, be declared inadmissible because it has been produced too late or not in the prescribed form.[250] Another example of inadmissible evidence is provided by the decision of the Permanent Court of International Justice (PCIJ) in the *Factory at Chórzow* case, where the ICJ's predecessor held that it "cannot take account of declarations, admissions or proposals which the parties may have made in the course of direct negotiations when the negotiations in questions have not led to an agreement between the parties."[251] The underlying reason for the inadmissibility of such material is to facilitate the diplomatic settlement of international disputes through negotiations, so that the negotiating parties do not have to fear that what they say in the negotiating context may be used against them in subsequent judicial proceedings.[252]

Is evidence obtained through a violation of international law also inadmissible? Traditional espionage and cyber exploitation, used in support of traceback technical tools, may be a helpful instrument to establish proof of state responsibility for cyber operations.[253] India has noted that "[c]yber security intelligence forms an integral component of security of cyber space in order to be able to anticipate attacks, adopt suitable counter measures and attribute the attacks for possible counter action."[254]

[248] Riddell and Plant (n 43) 205.

[249] Keith Highet, "Evidence, the Court, and the Nicaragua Case" (1987) 81 *American Journal of International Law* 1, 13.

[250] See Article 52 of the ICJ Statute. Late evidence may be admissible if the other litigant consents to it or if the Court does not reject it. Christian J Tams, "Article 52," in Zimmermann, Oellers-Frahm, Tomuschat and Tams, *The Statute* (n 44) 1312, 1316.

[251] *Factory at Chorzów* (Jurisdiction) [1927] PCIJ (Ser. A, No. 09), at 19. The ICJ referred to this limit in *Frontier Dispute (Burkina Faso/Republic of Mali)* [1986] ICJ Rep, para 147, and *Maritime Delimitation and Territorial Questions between Qatar and Bahrain (Qatar v Bahrain)*, Jurisdiction, [1994] ICJ Rep, para 40.

[252] Benzing (n 44) 1242.

[253] Nicholas Tsagourias, "Cyber Attacks, Self-Defence and the Problem of Attribution" (2012) 17 *Journal of Conflict and Security Law* 234; Owens et al (n 7) 140–1.

[254] *Government of India, Ministry of Communications and Information Technology* (n 213) 4.

It is doubtful whether the above activities constitute internationally wrongful acts, although one commentator has argued, for instance, that cyber espionage may be a violation of the sovereignty of the targeted state whenever it entails an unauthorized intrusion into cyber infrastructure located in another state (be it governmental or private).[255] Data monitoring and interceptions may also be a violation of international human rights law.[256]

Assuming, *arguendo*, that espionage and cyber exploitation are, at least in certain instances, internationally wrongful acts, what is the probative value of the evidence so collected? There is no express rule in the Statute of the ICJ providing that evidence obtained through a violation of international law is inadmissible. It is also not a general principle of law, as it seems to be a rule essentially confined to the US criminal system.[257] As Thirlway argues, the rule in domestic legal systems is motivated by the need to protect the defendant against the wider powers of the prosecutor and its possible abuses: in inter-state litigation, there is no criminal trial and no dominant party, as the litigants are states in a position of sovereign equality.[258] In the *Corfu Channel* case, the ICJ did not dismiss evidence illegally obtained by the United Kingdom in Operation Retail; on the contrary, it relied on it in order to determine the place of the accident and the nature of the mines.[259] In fact, Albania never challenged the admissibility of the evidence acquired by the British Navy,[260] and the Court did not address the question. What it found was not that the evidence had been illegally obtained, but that the purpose of gathering evidence did not exclude the illegality of certain conduct.[261] In general, "the approach of the Court is to discourage self-help in the getting of evidence involving internationally illicit acts, not by seeking to impose

[255] Wolff Heintschel von Heinegg, "Territorial Sovereignty and Neutrality in Cyberspace" (2013) 89 *Naval War College International Law Studies* 129 ("It could be argued…that damage is irrelevant and the mere fact that a State has intruded into the cyber infrastructure of another State should be considered an exercise of jurisdiction on foreign territory, which always constitutes a violation of the principle of territorial sovereignty"). More cautiously, an early study of the US Department of Defense concluded that "[a]n unauthorized electronic intrusion into another nation's computer systems may very well end up being regarded as a violation of the victim's sovereignty. It may even be regarded as equivalent to a physical trespass into a nation's territory, but such issues have yet to be addressed in the international community….If an unauthorized computer intrusion can be reliably characterized as intentional and it can be attributed to the agents of another nation, the victim nation will at least have the right to protest, probably with some confidence of obtaining a sympathetic hearing in the world community." *An Assessment of International Legal Issues* (n 188) 19–20. On the other hand, Doswald-Beck argues that, when the individual conducts intelligence gathering from outside the adversary's territory through cyber exploitation, "the situation should be no different from someone gathering data from a spy satellite." Louise Doswald-Beck, "Some Thoughts on Computer Network Attack and the International Law of Armed Conflict" (2002) 76 *Naval War College International Law Studies* 172.

[256] Jann K Kleffner and Heather A Harrison Dinniss, "Keeping the Cyber Peace: International Legal Aspects of Cyber Activities in Peace Operations" (2013) 89 *Naval War College International Law Studies* 512.

[257] Hugh Thirlway, "Dilemma or Chimera?—Admissibility of Illegally Obtained Evidence in International Adjudication" (1984) 78 American Journal of International Law 622, 627–8; Nasim Hasan Shah, "Discovery by Intervention: The Right of a State to Seize Evidence Located Within the Territory of the Respondent State" (1959) 53 American Journal of International Law 595, 607–9. See contra Wolfrum (n 31) 563.

[258] Thirlway (n 257) 628–9.

[259] Shah (n 257) 606–7.

[260] Thirlway (257) 632.

[261] *Corfu Channel* (n 70) 34–5.

any bar on the employment of evidence so collected, but by making it clear that such illicit activity is not necessary, since secondary evidence will be received and treated as convincing in appropriate circumstances."[262] In a cyber context, this means that the fact that direct evidence is located in the computers or networks of another state does not entitle the interested litigant to access them without authorization to submit it in the proceedings, but allows the Court to give more probative weight to circumstantial evidence.

VIII. Conclusions

There are a number of main conclusions that can be drawn from the application to cyber operations of the ICJ's rules and case-law on evidence:

- The burden of proof does not shift in the cyber context and continues to rest on the party that alleges a certain fact.
- Whilst it is uncertain that a uniform standard of proof applicable to *all* cases involving international responsibility for cyber operations can be identified, it appears that claims of self-defense against cyber operations, like those against kinetic attacks, must be proved with clear and convincing evidence. On the other hand, fully conclusive evidence is needed to prove that a litigant conducted cyber operations amounting to international crimes, and a slightly less demanding standard seems to apply when what needs to be proved is that the state did not exercise due diligence to stop its cyber infrastructure from being used by others to commit international crimes.
- The Court may take "formal note" of the refusal of a party to present classified cyber documents, but it has so far refrained from drawing negative inferences from the non-production of documents. In any case, any such negative inferences could not contradict factual conclusions based on consistent evidence produced by the parties.
- The Court gives more probative weight to official documents of states and international organizations such as the United Nations. NGO reports and press articles on cyber incidents are only secondary sources of evidence that may be useful to corroborate other sources or to establish the public knowledge of certain facts, providing they are sufficiently rigorous and only when they are "wholly consistent and concordant as to the main facts and circumstances of the case".[263]
- The drawing of inferences is approached by the ICJ with great caution. When there are objective difficulties for a litigant to discharge the burden of proof because the direct evidence lies within the exclusive territorial control of the other litigant,

[262] *Corfu Channel* (n 70) 641. It has been argued, however, that evidence obtained through a jus cogens violation, for instance torture, should be deemed inadmissible. Wolfrum (n 31) 563.

[263] *United States Diplomatic and Consular Staff in Tehran* (n 65) 13.

including its cyber infrastructure, a more liberal recourse to inferences of fact is admissible providing that they leave no room for reasonable doubt.

- Even if a litigant obtains evidence illegally, for example, through an unauthorized intrusion into the computer systems of another state, the evidence so obtained may be taken into account by the Court, although the purpose of collecting evidence does not exclude the illegality of the conduct.

11

Low-Intensity Cyber Operations and the Principle of Non-Intervention

Sean Watts

I. Introduction

The principle of non-intervention is an important, though poorly understood and implemented aspect of international law. A legal outgrowth of sovereignty and territorial control, non-intervention prohibits states from coercively imposing their will on the internal and external matters of other states. The pressing need for the principle is clear. It would be difficult to imagine a peaceful system of sovereigns that did not include such a norm. However, the level of clarity and compliance connected with the principle has never been equal to its seeming importance in the international system. Considerable ambiguity and regular breach have long accompanied the principle of non-intervention, leading some even to question its status as law.

It is difficult to explain precisely why this critical international norm is in such a state. Perhaps its current condition reflects the limits of substantive consensus in a system comprising legally equal though politically, economically, and militarily unequal and diverse sovereigns? Perhaps intervention as a means of exerting international influence is too attractive to definitively proscribe in the fiercely competitive realm of international politics? Or perhaps competing norms and rules have bested intervention in the competition for states' finite international legal attention span?

Whatever the forces behind the underdeveloped state of the principle of non-intervention and whatever states' compliance record, non-intervention undoubtedly occupies an identifiable place on the spectrum of internationally wrongful acts. Somewhere between the prohibition on use of force or armed attack, one of the, if not the most severe, wrongful acts between states, and the international law prohibition of simple violations of sovereignty, a comparatively less grave form of internationally wrongful acts, one likely finds the prohibition of intervention. Non-intervention seems to lie at a middle ground of international wrongs—not insignificant, but also not supreme among wrongful acts. As states consider and weigh the merits and costs of various modes of interaction in the international system, charting options on this legal spectrum with some specificity becomes a prudent, if not always simple exercise.

Cyber operations, increasingly common both in the communal and routine senses, have emerged as a vital mode of interaction between states. When, where, and how states conduct themselves in cyberspace now significantly shapes the condition of their international relations. Early investigations of cyber relations between states focused on the possibility and prospect of massive cyber attacks producing debilitating and destructive consequences.[1] Though certainly feasible, such cyber cataclysms have proved unlikely events. The interconnected, inter-reliant, and networked aspects of cyberspace that make cyber attacks possible simultaneously counsel states against their haphazard use. An attacking state may stand to lose as much or more than it gains from crippling a competitor's electronic infrastructure. Low-intensity cyber operations, actions taken short of destructive or violent attacks, present a far more likely picture of future state cyber interactions. In addition to being highly feasible and often inexpensive, low-intensity cyber operations offer attractive prospects for anonymity, appear to frustrate attack correlation by targets, and may also reduce the likelihood of provoking severe retaliation. In short, low-intensity cyber operations offer states appealing opportunities to degrade adversaries while avoiding the likely strategic and legal costs of massively destructive cyber attacks.

How these low-intensity cyber operations fare under international legal analysis is a topic of increasing importance. Whether the low-level impact, relatively small scale effects, and de minimis intrusions of low-intensity cyber operations implicate established international legal norms such as the principle of non-intervention deserves increased attention. This chapter makes the case for closer examination of the principle of non-intervention and its operation in low-intensity cyber operations. Likely increases in the frequency and range of low-intensity cyber operations between states will highlight doctrinal gaps in the principle of non-intervention. While it is unclear whether states will choose to resolve these gaps, it is hoped that this chapter contributes a useful awareness whether for purposes of decision-making or reform.

II. Non-Intervention in the International Legal System

(a) Customary status

Non-intervention constitutes a long-standing[2] and fundamental norm of customary international law confirmed on multiple occasions by the International Court of Justice (ICJ).[3] The principle derives primarily from the international law notions of sovereignty and territory.[4] Non-intervention not only permits states to operate free from outside interference by other states, but also reinforces the notion of sovereign equality among states.[5] States regularly express or resort to the principle of non-intervention in legal statements, briefs, and memorials to international tribunals.[6] International

[1] ET Jensen, "Computer Attacks on Critical National Infrastructure: A Use of Force Invoking the Right of Self-Defense" (2002) 38 *Stanford Journal of International Law* 207.

[2] The earliest, multilateral positive expression of the non-intervention principle is found in the 1933 Montevideo Convention, Article 8 which states, "[n]o state has the right to intervene in the internal or external affairs of another." Montevideo Convention on the Rights and Duties of States, December 26, 1933, 49 Stat. 3097, T.S. 881, Article 8.

[3] *Armed Activities in DRC* Judgment, paras 161–3; *Nicaragua* Judgment, para 202; Corfu Channel case, 35.

[4] *Nicaragua* Judgment, para 251 (noting, "The effects of the principle of respect for territorial sovereignty inevitably overlap with those of the principles of the prohibition of the use of force and of non-intervention.").

[5] *Nicaragua* Judgment, para 202.

[6] See, for example, United States State Department, Briefing Remarks, "Secretary Clinton's Meeting with Colombian Foreign Minister Jaime Bermúdez," (August 18, 2009) at <http://www.state.gov/r/

publicists, in particular, have long identified the principle of non-intervention as an essential aspect of the principles of sovereignty and equality among states.[7]

The *Nicaragua* Judgment of the ICJ represents an important moment in the development of the customary international law principle of non-intervention. Publicists had previously expressed confusion concerning the doctrinal details of the principle of non-intervention throughout the early and mid-twentieth century.[8] In particular, the principle had been difficult to square with frequent interventions by powerful states. The *Nicaragua* Judgment added a significant degree of clarity to the principle.

The Court's first clarification concerned the principle's status as customary international law.[9] The *Nicaragua* Court drew significant support from the Declaration on the Principles of International Law concerning Friendly Relations and Co-operation among States.[10] The Declaration identifies and affirms a duty on the part of states "not to intervene in matters within the domestic jurisdiction of any State, in accordance with the Charter [of the UN]." The *Nicaragua* Court also cited other UN General Assembly resolutions, including the Declaration on the Inadmissibility of Intervention in the Domestic Affairs of States and the Protection of their Independence and Sovereignty.[11]

The *Nicaragua* Court dismissed both the non-binding nature of General Assembly resolutions, as well as state declarations made at the adoption of the Declaration on Inadmissibility of Intervention, as limitations on the support offered by these provisions to the customary or binding nature of the principle. The Court noted that while upon adoption the United States had indicated the Declaration on Inadmissibility of Intervention constituted "only a statement of political intention and not a formulation

pa/prs/ps/2009/aug/128079.htm> (explaining Defense Cooperation Agreement compliance with "the principle of non-intervention"); Statement of Defense of the United States, Iran–United States Claims Tribunal, Claim No. A/30 n 33, 52, 56 (1996) (reciting reference to the principle of non-interference in bilateral agreement and defending against alleged violation of general principle of non-interference); Miami Plan of Action: First Summit of the Americas, para 1 (December 11, 1994) at <http://www.state.gov/p/wha/rls/59683.htm> (reciting Organization of American States members' commitment to the "principle of non-intervention") accessed October 28, 2014; Remarks of the President of the United States, Gerald R Ford, *Remarks in Helsinki* (August 1, 1975) at <http://www.millercenter.org/president/speeches/detail/3393> (affirming inclusion of principle of non-intervention in Helsinki Final Act).

7 James Crawford, *Brownlie's Principles of Public International Law* (Oxford: Oxord University Press, 8th edn, 2012) 447; MN Shaw, *International Law* (Cambridge: Cambridge University Press, 7th edn, 2014) 719; L Oppenheim, 1 *International Law: Peace* (London: Longmans, Greeen and Company, 2d edn, 1912) 188 (identifying the "International Personality of States" as the source of the principle of non-intervention); RJ Vincent, *Nonintervention and International Order* (Princeton: Princeton University Press, 1974) 310; JL Brierly, *The Law of Nations* (Oxford: Oxford University Press, 5th edn, 1955) 308; GG Wilson, *Handbook on International Law* (St. Paul: West Publishing Company, 3d edn, 1939) 58–9 (noting, however, that in practice nonintervention often amounts to a matter of policy); ED Dickinson, *The Equality of States in International Law* (Cambridge, MA: Harvard University Press, 1920) 260.

8 Fenwick observed, for instance, "Of all the terms in general use in international law none is more challenging than that of 'intervention.'" CG Fenwick, "Intervention: Individual and Collective" (1945) 39 *American Journal of International Law* 645. Von Glahn estimates, "Few terms in international law have led to more acrimonious arguments than has the prohibition of intervention." G von Glahn, *Law among Nations* (London: Longman, 7th edn, 1996) 578. A 1974, book-length study of the principle observed, "the conception of the principle of nonintervention... is perhaps too general to be helpful." RJ Vincent, *Nonintervention and International Order* (Princeton: Princeton University Press, 1974) 311.

9 *Nicaragua* Judgment, para 202.

10 *Nicaragua* Judgment, para 192 (citing General Assembly Res 2625 (XXV)).

11 *Nicaragua* Judgment, para 203 (citing General Assembly Res 2131 (XX)).

of law,"[12] the US statement did not constitute a limit on the Declaration's legal force. The Court insisted that the Declaration restated principles committed to without reservation by the United States and other states in previous resolutions.[13] The prevalence of non-intervention provisions in other international instruments, such as the Montevideo Convention[14] and the Helsinki Final Act,[15] also influenced the Court's conclusion with respect to the customary status of the principle.[16]

As noted already, frequent violations by states have called into question the status of the principle of non-intervention. Some commentators have raised serious objections to the existence of the rule.[17] Breaches, however, have generally not undermined the integrity of the rule as a fundamental principle of international law. The ICJ specifically addressed the legal effect of states' non-compliance with the principle of non-intervention in the *Nicaragua* case. The Court observed, "It is not to be expected that in the practice of States the application of the rules in question should have been perfect, in the sense that States should have refrained, with complete consistency, from the use of force or from intervention in each other's internal affairs."[18] The Court then noted, "Expressions of an *opinio juris* regarding the existence of the principle of non-intervention in customary international law are numerous and not difficult to find."[19] Despite frequency of breaches in state practice, the *Nicaragua* Court was willing to examine nearly each of the acts attributable to the parties to the case for compliance with the principle of non-intervention.

Support for the principle is not limited to the work of the ICJ. The International Law Commission (ILC)'s Declaration on the Rights and Duties of States, provides similar support for the customary status of non-intervention. Article 3 of the Declaration states, "Every State has the duty to refrain from intervention in the internal or external affairs of any other State." The Declaration attracted insufficient state commentary in the General Assembly in sessions immediately following submission of the Draft Declaration to warrant consideration for adoption.[20] Delay appears to have been attributable to a deficit of attention, however, rather than to substantive objections to the Declaration or to Article 3 specifically. As noted by the ICJ, later meetings of the General Assembly adopted provisions that included the ILC's Article 3 expression of the principle of non-intervention.

[12] Official Records of the General Assembly, Twentieth Session, First Committee, A/C. 1/SR. 1423, 436).

[13] *Nicaragua* Judgment, para 203.

[14] Convention on Rights and Duties of States, December 26, 1933, 49 Stat. 3097, 165 LNTS 19.

[15] Conference on Security and Co-operation in Europe Final Act, August 1, 1975, 14 ILM 1292, at <http://www.osce.org/mc/39501> accessed October 28, 2014.

[16] *Nicaragua* Judgment, para 204.

[17] A D'Amato, "There Is No Norm of Intervention or Non-intervention in International Law" (2001) 7 *Internatioanl Legal Theory* 33 (arguing for a theory of non-intervention limited to efforts to compromise states' territorial integrity).

[18] *Nicaragua* Judgment, para 186.

[19] *Nicaragua* Judgment, para 202. Curiously, the Court omitted reference to any such instances or expressions of state opinio juris.

[20] GA Res 596 (VI) of December 7, 1951. The General Assembly postponed consideration of the Draft Declaration "until a sufficient number of States have transmitted their comments and suggestions, and in any case to undertake consideration as soon as a majority of the Member States have transmitted such replies." GA Res 596 (VI).

It is clear that, as an accepted customary norm, the principle of non-intervention extends to states' actions in cyberspace. The argument that cyberspace constitutes a law-free zone is no longer taken seriously. Therefore, that cyberspace and cyber means present states with greater opportunities for intervention in other states' domestic and foreign affairs does not excuse violations of the principle in cyberspace. Nor, for the moment, does the fact that states might frequently intervene through cyber means undermine the binding nature of the principle of non-intervention. To the extent cyber interventions take place without justifications, they clearly remain violations of customary international law.

As a final point with respect to its customary status, it should be noted that the principle of non-intervention operates exclusively with respect to states.[21] Non-intervention does not operate on private actors as such. Although international law increasingly extends its prohibitions and protections to the actions of private persons and entities, there is no evidence of states' intention for the principle of non-intervention to follow suit. For instance, while the Rome Statute of the International Criminal Court and subsequent meetings of states parties have devoted significant attention to converting the prohibition of the use of force into the criminal act of aggression, the Rome diplomatic conference did not even consider intervention as a possible substantive offence chargeable against individuals.[22]

(b) Relationship to expressions in treaties

Despite its central legal role in the maintenance of international peace, the principle of non-intervention applicable to states received no explicit mention in the UN Charter. Commentators have claimed support from UN Charter Articles 2(1) and (4) nonetheless though neither provision expressly states the principle.[23] Article 2(1) states: "The Organization is based on the principle of the sovereign equality of all its Members." Article 2(4) states: "All Members shall refrain in their international relations from the threat or use of force against the territorial integrity or political independence of any state, or in any manner inconsistent with the Purposes of the United Nations."

A highly formal interpretation of these Charter provisions might deduce that Members *may* take action against the territorial integrity or political independence short of the threat or use of force. That is, one might read Article 2(4) to suggest that states' territorial integrity and political independence are *exclusively* protected against the threat or use of force. The effect, of course, would be significantly to undermine the existence of non-intervention as a legal norm. States, jurists, and the majority of commentators have not adopted this view, however. The accepted view instead regards

[21] United States Department of Justice, Office of Legal Counsel, Memorandum Opinion for the Attorney General, Intervention by States and Private Groups in the Internal Affairs of Another State, April 12, 1961 (observing, "international law with respect to intervention in the internal affairs of another state, by force or other means, is designed to set standards for the conduct of states.").

[22] M Jamnejad and M Wood, "The Principle of Non-Intervention" (2009) 22 *Leiden Journal of International Law* 348, n 50.

[23] G Nolte, "Article 2(7)," in Bruno Simma et al (eds), *The Charter of the United Nations: A Commentary* (Oxford: Oxford University Press, 3rd edn 2012) 280, 284.

Articles 2(1) and 2(4) as merely two components of a general international legal system protecting states' territorial integrity and political independence from interference through both treaty provisions such as the UN Charter and customary international law principles such as non-intervention. Where the Charter prohibits extreme forms of intervention such as the threat or use of force, customary international law separately prohibits both threat and use of force as well as less coercive forms of intervention.

Despite codification in a number of international instruments, the ICJ has insisted the principle of non-intervention retains a separate and independent identity as a norm of customary international law. In the *Nicaragua* Judgment, the Court rejected arguments by the United States that the UN Charter and other treaties entirely subsumed customary law concerning intervention and the use of force. The Court observed, "The fact that principles, recognized as such, have been codified or embodied in multilateral conventions does not mean that they cease to exist and to apply as principles of customary law...."[24] The Court added that owing to their separate existence in international law, the content of such customary norms might not in all cases be identical to those captured in treaties.[25] Thus, the principle of non-intervention may vary between treaty-based expressions and that found in customary international law. Therefore, developments in interpretation and application to treaty-based understandings of non-intervention may not necessarily apply to customary understandings. The soundest conception appreciates the UN Charter, and specifically its provisions on the threat and use of force, as complementing and often informing rather than displacing the customary principle of non-intervention.

Although the principle expresses a concept nearly identical to UN Charter Article 2(7), the latter is not a formal source for the Rule. Article 2(7) states: "Nothing in the present Charter shall authorize the United Nations to intervene in matters which are essentially within the domestic jurisdiction of any state...." From its text, it is clear Article 2(7) addresses the United Nations as an organization rather than individual member states as such. Still, the concerns that gave rise to Article 2(7) are similar to those supporting the principle of non-intervention. Concepts and understandings applicable to Article 2(7), therefore, may be informative as to the operation of the principle of non-intervention.

(c) Ius cogens status

Some jurists attribute ius cogens status to the principle of non-intervention. In a separate concurring opinion in the ICJ *Nicaragua* case, Judge Sette-Camara observed that non-intervention, alongside non-use of force, could "be recognized as peremptory rules of customary international law which impose obligations on all States."[26] Writings of select publicists support this assessment, noting that states could not modify the non-interference principle by mere treaty or new patterns of state practice.[27]

[24] *Nicaragua* Judgment, para 174. [25] *Nicaragua* Judgment, para 175.

[26] Sep. Concurring Op., Judge Sette-Camara, *Nicaragua* Judgment, 199.

[27] Jianming Shen, "The Non-intervention Principle and Humanitarian Interventions," (2001) 7 *International Legal Theory* 1, 7; Dino Kritsiotis, "Reappraising Policy Objections to Humanitarian Intervention," (1998) 19 *Michigan Journal of International Law* 1005, 1042–3; A Cassese, *International*

All the same, the strong evidentiary burden required to unequivocally attribute ius cogens status to the principle does not appear to have been met.

(d) Terminology

Finally, with respect to the legal standing of the principle, one finds regrettably inconsistent use of terms addressing the principle of non-intervention.[28] In particular, publicists occasionally use the term "interference" somewhat interchangeably with intervention.[29] Instruments adopted by states and the UN as well as judgments of the ICJ, however, employ the term intervention more consistently. Accordingly, this chapter uses the term intervention to refer to wrongful and coercive acts by a state that intrude into internal or external matters of another state. This chapter reserves the term "interference" for acts that intrude into affairs reserved to the sovereign prerogative of another state irrespective of the level of coercion or compulsion associated required of a prohibited intervention.[30]

III. The Meaning of "Intervention"

States are in constant contact and communication with one another—no more obviously or extensively so than through their actions in cyberspace. As Wright noted long ago:

> [states] find their interests affected by the acts of others and attempt to influence those acts. They do so by internal development of culture, economy, and power; by achievements in technology, science, literature, and the arts; by international communication utilizing radio, press, popular periodicals and technical journals; by the travel and trade of their citizens; and by official utterances, legislative action, and diplomatic correspondence. International law is faced with the issue: When does proper influence become illegal intervention?[31]

While agreement on the existence of a principle of non-intervention has been fairly settled for some time, commentators and jurists have struggled to identify the precise contours of the principle and to apply those delineations to ever-evolving and increasingly inter-tangled international relations. This section will showcase current

Law in a Divided World (Oxford: Clarendon Press, 1986) (identifying non-intervention as "part and parcel of *jus cogens*"). *But see* Anthony D'Amato, "There is No Norm of Intervention or Non-intervention in International Law," (2001) 7 *International Legal Theory* 33.

[28] Philip Kunig, "Prohibition of Intervention," *Max Planck Encyclopedia of Public International Law* (Oxford: Oxford University Press, April 2008) (noting us by scholars of the term "*interference* by a state in the internal or foreign affairs of another state") para 1 (emphasis added).

[29] See JL Brierly, *The Law of Nations* (Oxford: Clarendon Press, 5th edn, 1955) 308; Katja S Ziegler, "Domaine Réservé," *Max Planck Encyclopedia of Public International Law* (n 28) para 1.

[30] Oppenheim, 1 *International Law: Peace* (n 7) 189. Oppenheim explains intervention as a "dictatorial interference," suggesting the former exists as a sub-species of sort of interference. Oppenheim further identifies so-called "intercession" as interference of a friendly nature, such as advice or offers of support concerning another state's domestic or external affairs. Oppenheim (n 7).

[31] Quincy Wright, "Espionage and the Doctrine of Non-Intervention in Internal Affairs," in Q Wright, J Stone, RA Falk, and RJ Stanger (eds) *Essays on Espionage* (Ohio: Ohio University Press,1962) 4–5.

statements on the scope and elements of the non-intervention principle and briefly apply these precepts to various cyber contexts.

It is clear that not all violations of territory or sovereignty immediately concern the principle of non-intervention. Typically, intervention involves a state's effort to coercively influence outcomes in or conduct with respect to a matter reserved to a target state. That is, intervention involves an act or campaign by a state to force another state's hand with respect to a choice or decision in a matter reserved to the target state's discretion.[32] As Kunig observes, intervention "aims to impose certain conduct of consequence on a sovereign state."[33]

In a cyber context, a mere intrusion into another state's networks to gather information would certainly amount to a violation of sovereignty. However, without evidence that the effort to gain information formed part of a campaign to coercively influence an outcome or course of conduct in the target state, the intrusion would not be properly characterized as an intervention. On the other hand, if the network intrusion and extraction of information were conducted in order to assist a resistance or opposition movement's effort to influence political events in the target state, the intrusion would properly be considered for characterization as a violation of the principle of non-intervention. In this sense, it is likely that the best understanding of non-intervention appreciates a nuanced and particularized notion of coercion.

(a) Coercion

Twentieth-century commentators observed that non-intervention prohibits only acts that are "dictatorial" by nature or effect.[34] Dickinson noted helpfully that coercion is present if an intervention "cannot be terminated at the pleasure of the state that is subject to the intervention."[35] Later, the ICJ confirmed coercion as an element of prohibited intervention in the *Nicaragua* case. The Court observed, "Intervention is wrongful when it uses methods of coercion.... The element of coercion... defines, and indeed forms the very essence of prohibited intervention. . .."[36]

Publicists seeking to elaborate the notion of coercion in international relations, law, and politics often envision a spectrum of action.[37] The simplest conceptions of coercion depict a range of action from minimally invasive acts on one

[32] See, generally, John Norton Moore (ed), *Law & Civil War in the Modern World* (Baltimore: Johns Hopkins Press, 1974). Moore's volume compiles a collection of essays addressed to intervention in internal armed conflicts. The majority of the essays focus on intervention as a means for states to influence outcomes in other states' internal struggles for power.

[33] Philip Kunig, "Prohibition of Intervention," in *Max Planck Encyclopedia of Public International Law* (n 28)) para 1. Oppenheim observed similarly that intervention is "dictatorial interference by a state in the affairs of another state *for the purpose of maintaining or altering the actual condition of things.*" Oppenheim, (n 7) 188 (emphasis added).

[34] Wright, "Espionage and the Doctrine of Non-Intervention in Internal Affairs," (n 31) 5; Wright, "Legality of Intervention under the UN Charter" (1957) 51 *Proceedings of the American Society of International Law* 79, 79.

[35] Edwin DeWitt Dickinson, *The Equality of States in International Law* (Boston: Harvard University Press, 1920) 260.

[36] *Nicaragua* Judgment, para 205.

[37] See, for example, Rosalyn Higgins, "Intervention and International Law," in Hedley Bull (ed), *Intervention in World Politics* (Oxford: Clarendon Press, 1984) 30.

end to exceptionally aggressive acts on the other. Still, as former ICJ Judge Higgins observes, such linear notions of coercion may be oversimplifications for purposes of determining legality.[38] Judge Higgins further explains, not all maximally invasive acts are unlawful and not all minimally invasive acts are lawful.[39] Accordingly, resort to factors beyond the degree of invasiveness or intrusion involved in an act may be required in order to determine whether an act is sufficiently coercive to constitute intervention. Higgins suggests consideration of whether an act amounts to a violation of jurisdiction to understand whether the threshold of intervention is met. Perhaps more helpfully, McDougal and Feliciano suggest coercion determinations account for "consequentiality."[40] Specifically, they propose consideration of three dimensions of consequentiality including, "the importance and number of values affected, the extent to which such values are affected, and the number of participants whose values are so affected."[41]

Applied to cyber means and acts, Myres and Feliciano's dimensions of coercion might consider the nature of state interests affected by a cyber operation, the scale of effects the operation produces in the target state, and the reach in terms of number of actors involuntarily affected by the cyber operation in question. Although their work predates the use of cyber operations by states in international relations, Myres's and Feliciano's conceptions of methods of state coercion are also helpful, advocating that determinations of coercion go beyond mere consideration of degrees of intensity. They note that each dimension of state power, "the diplomatic, the ideological, the economic, and the military instruments" can be used to achieve coercion. Cyber operations now may be employed in each of Myres and Feliciano's four instruments of state power. Thus, one might envision a diplomatic exercise of coercion through cyber means, an ideological exercise of coercion through cyber means, and so on. The important point for purposes of cyber operations is to appreciate that coercion sufficient to constitute intervention may occur not only in military cyber operations but also through states' ideological, diplomatic, or economic cyber enterprises.

Direct forms of coercion are certainly sufficient to support a finding of unlawful intervention. Acts amounting to direct threats or uses of force, particularly armed force, by a state against another constitute the clearest examples of direct coercion sufficient to establish prohibited intervention. In the *Nicaragua* case, the ICJ determined the US placed naval mines in waters belonging to Nicaragua.[42] The Court also determined the United States had attacked a number of Nicaraguan facilities.[43] The Court held emplacement of naval mines and attacks amounted to uses of force in violation of the principle of non-use of force.[44] Although the Court's later analysis noted that uses of force might simultaneously constitute violations of the principle of

[38] Higgins, "Intervention and International Law" (n 37). [39] Higgins (n 37).

[40] Myres S McDougal and Florentino P Feliciano, "International Coercion and World Public Order: The General Principles of the Law of War," in Myres S McDougal et al (eds), *Studies in World Public Order* (New Haven: New Haven Press, 1987) 263.

[41] McDougal et al, *Studies in World Public Order* (n 40) 263–68.

[42] *Nicaragua* Judgment, para 80 [43] *Nicaragua* Judgment, paras 81, 86.

[44] *Nicaragua* Judgment, para 227.

non-intervention, the Court curiously declined to characterize the mining and attacks as such explicitly in the remainder of the judgment.[45]

The *Nicaragua* Court's approach might suggest that where an act simultaneously constitutes a use of force and an intervention, it is preferable to resort to the more grave use-of-force characterization. The Court repeated this pattern of characterization later in the *Nicaragua* Judgment, noting that US attacks and naval mining operations constituted uses of force as well as violations of Nicaraguan sovereignty.[46] However, the Court again declined to characterize either the attacks or mining as interventions, although by logic and the Court's earlier observations they would seem to amount to such.

In this respect, it may be appropriate to understand the principle of non-use of force as a form of lex specialis with respect to intervention, particularly with respect to acts simultaneously constituting direct uses of force. The Court's analysis in the *Armed Activities in DRC* case confirms the use-of-force lex specialis approach somewhat. Although in that case the Court determined that Ugandan troop presence and support to rebels constituted both violations of the non-use of force and the non-intervention principles, the Court ultimately characterized Uganda's interventions as "a grave violation of the prohibition on the use of force. . .." owing to their "magnitude and duration."[47] Still, neither the *Nicaragua* findings nor the *Armed Activities in DRC* findings preclude a determination that direct threats of uses of force qualify as violations of the customary international law principle of non-intervention and, in a cyber context, violations of this Rule.

Extended to a cyber context, the *Nicaragua* Court's apparent lex specialis approach would counsel characterizing cyber operations that amount to uses of force as violations of United Nations Charter Article 2(4) rather than as mere violations of the principle of non-intervention. Thus, a cyber operation by a state against another that resulted in physical destruction of target systems or property or even death or injury resulting in an undesired domestic political outcome would be better characterized as a use of force than an intervention, although the technical elements of the latter would be satisfied.

Indirect forms of coercion may also be sufficient to support a finding of unlawful intervention. In its findings of fact, the ICJ *Nicaragua* Court determined the United States had supplied assistance to rebels including, "training, arming, equipping... military and paramilitary actions in and against Nicaragua."[48] Notwithstanding that such acts did not constitute direct uses of force or intervention by the United States against Nicaragua, the Court held that such acts amounted to uses of force inconsistent with the customary principle of non-use of force. Like the Court's decision with respect to direct use of force discussed earlier, the Court's later analysis supports the conclusion that indirect uses of force may simultaneously qualify as violations of the principle of non-intervention and should be characterized simultaneously as such.

[45] In a later case, the Court confirmed its determination that that acts involving the use of force constitute breaches of the principle of non-intervention, as well as the principle of non-use of force. *Armed Activities in DRC* Judgment, para 164.

[46] *Nicaragua* Judgment, para 251.

[47] *Armed Activities in DRC* Judgment, para 165.

[48] *Nicaragua* Judgment, para 228.

The Court held, "the [non-intervention] principle forbids all States or groups of States to intervene directly *or indirectly* in internal or external affairs of other States."[49]

However, unlike the Court's aforementioned analysis with respect to direct use of force, the Court affirmatively characterized the United States' indirect uses of force as *both* violations of the principle of non-use of force and as prohibited interventions. The Court held that support given by the United States to the military and paramilitary activities of the contras in Nicaragua including financial support, training, supply of weapons, intelligence and logistic support, constitutes a clear breach of the principle of non-intervention'.[50] The Court's characterizations perhaps suggest a limit to the lex specialis nature of the non-use of force principle in cases involving *indirect* use of force.

To qualify as coercion for purposes of the principle of non-intervention, the acts of a state need not involve physical coercion or force. The *Nicaragua* Court observed that acts amounting to force, such as military action or acts supporting military action, merely constitute "particularly obvious" forms of intervention. It is clear that acts prohibited by the principle of non-use of force may simultaneously amount to violations of the principle of non-intervention, however, nothing in the Court's judgment limits intervention to such acts. In fact, the Court explicitly limited its efforts with respect to defining non-intervention to the facts presented by the case.[51] As the Court noted, its limited discussion of non-intervention was simply a function of the nature of the complaints made by the parties to the *Nicaragua* case.

To appreciate the full range of acts constituting intervention, in addition to threats or uses of force, one must also envision a range of acts short of the threat or use of force yet involving a degree of coercion sufficient to constitute intervention. For instance, in its findings of fact the *Nicaragua* Court determined the United States had funded military and paramilitary activities by rebels against Nicaragua.[52] While the Court did not regard mere funding as a threat or use of force in violation of the principle of non-use of force, the Court did regard the supply of funds to rebels conducting military and paramilitary activities in and against Nicaragua as "undoubtedly an act of intervention in the internal affairs of Nicaragua."[53]

A critical distinction between a violation of the principle of non-use of force and a violation of the principle of non-intervention, therefore, is the level of coercion involved. Where violations of the former involve more severe or damaging scale and effects, violations of the latter may be found in cases lacking resort to armed force or lacking such scale and effects. Actions merely restricting a state's choice with respect to a course of action or compelling a course of action may be sufficient to amount to violations of the principle of non-intervention.

A state's technical support for cyber operations hostile to another state is perhaps the best illustration of prohibited indirect intervention in a cyber context. Where supporting measures have historically been understood to lie outside the use of force, such measures may constitute prohibited intervention if they involve or result in coercion.

[49] *Nicaragua* Judgment, para 205.
[50] *Nicaragua* Judgment, para 242.
[51] *Nicaragua* Judgment, para 205.
[52] *Nicaragua* Judgment, para 99.
[53] *Nicaragua* Judgment, para 228.

Recall that the *Nicaragua* Court did not find that logistical support amounted to a use of force but held that such acts did constitute prohibited intervention. Thus, a state's significant logistical support for hostile cyber operations of an insurgent group against another state would likely constitute a prohibited intervention. Although the supporting state would not have committed a coercive act directly against the target state, the principle of non-intervention appears to be broad enough to capture indirect involvement in coercion.

(b) Economic acts

Courts and commentators have considered whether economic acts by states may constitute prohibited intervention as well. In the *Nicaragua* case, the ICJ addressed allegations that the United States had ceased economic aid to Nicaragua in order to inflict economic damage and to weaken the Nicaraguan political system. Specifically, Nicaragua established that the United States had reduced a sugar import quota by 90% and later instituted a trade embargo.

The ICJ summarily determined neither act amounted to a breach of the customary law principle of non-intervention.[54] The Court did not provide reasoning for these determinations, apparently regarding each as self-evidently outside the scope of prohibited intervention. Recalling the element of coercion, the *Nicaragua* Court's determination that the sugar quota reduction did not constitute prohibited intervention is easily understood. The sugar quota likely constituted preferential economic treatment on behalf of the United States toward Nicaragua. Withdrawal or reduction of the quota would not seem to involve the coercion anticipated by intervention. The quota reduction did not in fact coerce Nicaragua in any significant respect but rather altered unilateral, preferential treatment.

The Court's determination that the trade embargo did not constitute an intervention also seems consistent with established notions of intervention. While an economic blockade or other forcible effort to prevent a state's participation in global markets would likely constitute a prohibited intervention, a unilateral embargo on imports or exports does not seem to involve the requisite coercion of a prohibited intervention. As a general matter of international law, states remain free to choose their trading partners.[55] Only through acceptance of multilateral or bi-lateral treaties or other positive agreements are states subject to limits on their conduct of commerce and trade. The *Nicaragua* Court noted as much when it declined, on a jurisdictional basis, to evaluate United States' economic and trade obligations toward Nicaragua.[56] These conclusions with respect to economic acts also appear to be correct upon comparison with a number of instruments purporting to formulate a non-intervention

[54] *Nicaragua* Judgment, para 245.

[55] See Defense of the United States, Iran-United States Claims Tribunal, Claim A/30, 57 (1996) (stating, "Every state has the right to grant or deny foreign assistance, to permit or deny exports, to grant or deny loans or credits, and to grant or deny participation in national procurement or financial management, on such terms as it finds appropriate.") (citing *Iran v United States*, AWD No. 382-B1-FT, 62, 19 Iran-US CTR 273 292 (August 31, 1988)); Higgins, "Intervention and International Law" (n 37) 32.

[56] *Nicaragua* Judgment, para 245.

norm applicable to economic acts. The usual practice for non-intervention agreements seeking to prohibit economic interference appears to recite economic measures of coercion explicitly rather than impliedly.[57]

While open access and liberal connectivity are undoubtedly prominent aspects of the culture of the internet and cyberspace, the principle of non-intervention does not convert these values into legal obligations. As with economic policies, states remain free as a general matter to decide with whom they will and will not maintain electronic communications. At present, only positive obligations, specifically undertaken to guarantee open communications or electronic access with another state, appear to restrain a state's options with respect to cyber embargo. It seems clear for now that a state may undertake a cyber embargo, a refusal to receive or forward cyber communications, against another state without fear of violating the principle of non-intervention. While such an act might indeed appear coercive in nature, would likely require coercive measures to enforce, and may even actually force the embargoed state to take steps it would not otherwise adopt, the embargo would not amount to intervention as presently understood because the coercion in question would be taken primarily with respect to a matter on which the acting state retains prerogative.

(c) Propaganda

Propaganda has been defined as a communication "to shape attitudes and behavior... for political purposes."[58] To constitute an intervention, propaganda must involve not merely persuasion but rather coercion. To discern coercive propaganda, publicists have asked "[i]s the audience's" choice of alternatives severely restricted as a result of the use of the instrument?"[59] It might be said with respect to cyber propaganda that legitimacy depends less on means than "upon its outcome, actual or expected."[60] Acts that amount merely to persuasion remain legitimate.[61]

Thus, a state-sponsored internet campaign attempting to persuade a state to adopt a particular treaty regime would not constitute a violation of the principle of non-intervention. Similarly, a state's offer of incentives or favorable treatment to induce a course of action by another state would not constitute intervention. Only where propaganda could be said to constitute significant support for coercion on a level commensurate with the logistical and financial support recognized by the

[57] See, for example, Charter of the Organization of American States, 2 UST 2394 (April 30, 1948) Article 16 (stating "No State may use or encourage the use of coercive measures of an economic or political character in order to force the sovereign will of another State and obtain from it advantages of any kind"); Declaration on Principles of International Law Concerning Friendly Relations and Co-operation Among States in Accordance with the Charter of the United Nations, GA Res 2625, 25th Sess, Supp No 18, at 121, UN Doc No A/8018 (1970) (stating, "No State may use or encourage the use of economic, political or any other type of measures to coerce another State to obtain from it the subordination of the exercise of its sovereign rights to secure from it advantages of any kind.").

[58] Bhagevatula S. Murty, *Propaganda and World Public Order* (New Haven and London: Yale University Press, 1968) 1.

[59] Murty, *Propaganda and World Public Order*, 29.

[60] Murty, 129.

[61] Wright, "Espionage and the Doctrine of Non-Intervention in Internal Affairs" (n 31) 5.

Nicaragua Court, could cyber propaganda amount to a violation of the principle of non-intervention.

(d) UN Charter, Art. 2(7) intervention distinguished

While, as noted already, UN Charter, Article 2(7) does not operate on states as such, states' understandings of and publicists' scholarship on Article 2(7) may serve as helpful guides to understanding the customary international law principle of non-intervention and this Rule. With respect to Article 2(7), one finds conflicting indications concerning the meaning or threshold of intervention. Nolte notes "strong indications" that the Charter's drafters intended the term "intervene" would be understood broadly and to prohibit a wide range of conduct against states' sovereignty.[62] Merely including an issue concerning domestic conditions of a particular state as an item for committee discussion or formation of a committee itself has not been regarded as a prohibited intervention. Yet Nolte notes that issuing instructions to that state or even recommendations by a UN organ may be sufficiently intrusive to constitute a violation of Article 2(7).[63] Ultimately, Nolte advocates an understanding of intervention informed by reference to the matter or issue concerned rather than by resort to analyzing the particular action taken.[64]

It seems that intervention for purposes of Article 2(7) applicable to the UN prohibits a far broader range of acts than does customary international law applicable to states. Overall, Article 2(7) intervention does not appear to regard coercion as a prerequisite to the extent that the customary principle of non-intervention does. Where a minimally or even non-coercive recommendation by a UN body may constitute a prohibited intervention for purposes of Article 2(7), it is unlikely a similar act by a state in its sovereign capacity would constitute a violation of the principle of non-intervention. In this respect, intervention for purposes of Article 2(7) might more closely approximate the concept of interference in customary international law. Thus, while use of cyber means by a UN organ to criticize or recommend a course of action by a state might constitute an intervention for purposes of 2(7), the same would not constitute an intervention if carried out by a state as such.

(e) Declaration on the Inadmissibility of Intervention

Finally, with respect to the scope and meaning of prohibited intervention, the UN General Assembly's Declaration on the Inadmissibility of Intervention and Interference in the Internal Affairs of states purports to define the principle on non-intervention with greater precision than previous instruments.[65] The Declaration prohibits a particularly broad range of acts and appears even to

[62] Georg Nolte, "Article 2(7)," in Bruno Simma et al (eds), *The Charter of the United Nations: A Commentary* (New York: Oxford University Press, 3d edn, 2012) 280, 285 (citing UNCIO VI, 486).

[63] Nolte, "Article 2(7)," 289. [64] Nolte, 290.

[65] GA Res A/RES/36/103, December 9, 1981.

preclude the possibility of traditional justifications operating to excuse interventions. Commentators have seriously challenged the Declaration's accuracy as a restatement of the customary principle of non-intervention.[66] The ICJ has also not relied on the Declaration in any significant respect. At present then, the Declaration appears to reflect at best a minority view with respect to the legal notion of prohibited intervention.

IV. "Internal or External Affairs"

In addition to requiring a conception of which acts constitute prohibited, coercive interventions, the principle of non-intervention requires an understanding of the objects protected, that is, the aspects of states' sovereignty the principle insulates from intervention. As the ICJ explained in the *Nicaragua* case, the principle of non-intervention "forbids all States or groups of States to intervene directly or indirectly *in internal or external affairs of other States.* A prohibited intervention must accordingly be one bearing on matters in which each State is permitted, by the principle of sovereignty, to decide freely."[67] The range of protection offered by the Rule is generally coextensive with the range of matters reserved to states by the international law principle of sovereignty. The ICJ has cited, in particular, "choice of a political, economic, social, and cultural system, and the formulation of foreign policy."[68] Thus, the Rule prohibits cyber acts by states intended to eliminate or reduce another state's prerogatives on these matters.

Reference to views of the framers of the UN Charter have been made to illustrate the scope of internal matters protected by the principle:

> matters taking place within the territory of a state, personally as matters concerning individuals within the jurisdiction of a state, functionally as matters which could be dealt with conveniently and efficiently by states individually, or politically as matters which could be dealt with by states individually without affecting the interests of others.[69]

(a) Domaine réservé

The concept of domaine réservé, developed to describe activities and matters belonging to the internal or domestic affairs of states, clarifies the scope of activities protected by the principle of non-intervention. Like non-intervention, domaine réservé is said to proceed from sovereignty as well as the notion of sovereign equality.[70] Indeed, some publicists consider the reach of state sovereignty and domaine réservé to be coextensive.[71]

[66] Kunig, "Prohibition of Intervention" (n 28) para 20. Kunig notes, "[t]he broad definition of the non-intervention principle given by this resolution was passed against the will of many states and does not reflect general international opinion on the topic." Kunig, para 20.

[67] *Nicaragua* Judgment, para 205 (emphasis added).

[68] *Nicaragua* Judgment, para 205.

[69] Vincent, *Nonintervention and International Order* (n 8) 7.

[70] Katja S Ziegler, "Domaine Réservé" (n 28) para 1.

[71] Ziegler, "Domaine Réservé,", para 1.

The Permanent Court of International Justice (PCIJ) first addressed domaine réservé in its *Nationality Decrees* advisory opinion.[72] In that case, the PCIJ urged an understanding of domaine réservé based largely on the content and reach of international law. The Court observed that the scope of domain réservé consisted of matters "not, in principle, regulated by international law."[73] Thus, matters concerning rights and duties clearly committed by states to international law would not form part of domaine réservé under the PCIJ's advice.[74] Conversely, matters on which international law does not speak or that international law leaves solely to the prerogative of states constitute domaine réservé and are, therefore, to be regarded as protected from state intervention by the principle.

The precise scope of matters forming the domaine réservé of states has always been in flux and is likely to remain so in the future. Both treaty law and developments in customary international law can affect the content of the domaine réservé.[75] States may surrender matters previously regarded as within their exclusive jurisdiction to the international legal system by treaty. Similarly development of customary international law norms or principles may divest a state of exclusive jurisdiction over what was previously regarded as an internal matter protected by sovereignty or the principle of non-intervention. A number of issues formerly regarded within the domaine réservé is now thought to have been removed by subsequent international law. For instance, "jurisdiction over and treatment of foreign nationals on a states territory" has been displaced significantly from domaine réservé by human rights law.[76] "Discretion regarding the admission of foreign nationals to a State's territory" has been displaced by international asylum and refugee provisions. And international human rights law appears to have displaced significantly treatment of a state's own nationals as well as the issue of nationality itself from domaine réservé.[77]

Some publicists consider domaine réservé to be of reduced importance.[78] Interdependence and increased connectivity between states have diminished the operation of domaine réservé as an impediment to international action. These factors appear especially relevant in the interconnected world of cyberspace. Analogies to the interconnection of the environment may provide useful analogies. The work of some UN bodies also suggests that fewer matters are exclusively reserved to domestic prerogative, although state reaction to this view is uncertain.[79]

[72] Nationality Decrees Issued in Tunis and Morocco, Advisory Opinion, 1923 PCIJ Rep. Ser. B No. 4, at 22- (February 7). The Court judged the terms "solely within the domestic jurisdiction," "d'ordre intérieur," and "à la compétence exclusive" to have the same meaning for purposes of addressing domaine réservé.

[73] Nationality Decrees Opinion, at 24.

[74] Nationality Decrees Opinion, at 27. Because the question put before the Court in the request for an advisory opinion concerned state jurisdiction with regard to nationality of persons born in protected territory and because the answer to that question derived from international law, the Court advised that issue was not within the domaine réservé of France. Nationality Decrees Opinion, at 27.

[75] Ziegler, "Domaine Réservé" (n 28) para 2. [76] Ziegler (n 28) para 5.

[77] Ziegler (n 28) para 5. [78] Ziegler (n 28) para 5.

[79] UNHCR Report, *United Nations Strategies to Combat Racism and Racial Discrimination* (February 26, 1999) UN Doc. E/CN.4/1999/WG.1/BP.7. (Observing, the domestic jurisdiction exception "has over the years lost its political and legal weight to the point that it has no more than an unsuitable residual value.")

The matter most clearly within the domaine réservé appears to be choice of states' political systems and their means of political organization. Although notions of self-determination and human rights have made international inroads to state prerogative in governance, selection of a political system remains at the heart of sovereignty and, therefore, a core aspect of domaine réservé. Thus, cyber means generally may not be used to alter or suborn modification of a state's political organs. For instance, a state may not use cyber means to support domestic or foreign agents' efforts to alter another state's governmental or social structure. Nor may a state provide to such agents cyber means for doing so. Other matters likely included in a concept of domaine réservé might be those of education or official language.

Nonetheless, employing the ICJ's notions of domaine réservé to states' activities in cyberspace, one expects a dynamic environment. As the scope of issues related to cyberspace committed by states to international law increases, one would expect the cyber domaine réservé to decrease. For instance, proposals to standardize network maintenance and security through international agreements might displace these topics as internal matters. The consequence for the principle of non-intervention might be a reduced expectation of exclusivity or even protection from foreign meddling in these areas.

As a final point, it is worth emphasizing that displacement of a matter or issue from the domaine réservé does not constitute an overall eradication or waiver of the principle of non-interference, nor an open season on influencing conditions in another state's territory. For instance, were a state to commit to international regulation of network or operating system security, such a commitment would not waive or reduce protection of that state's network infrastructure from foreign state intrusions. Infrastructure, including networks, hardware, and information resident thereon would remain protected by international rules protecting property of the state in question and its nationals. To be sure, a state might consent to enforcement of the security standards through an agreement establishing such standards.[80] However, as a default rule, in the absence of such provisions, rules derived from sovereignty and territorial integrity would protect even offending networks and infrastructure. In other words, exclusion of an issue from domaine réservé does not in any way constitute a general justification for cyber intervention.

(b) Subversive intervention

Commentators have identified so-called "subversive intervention" as a form of intervention clearly satisfying the requirement that prohibited intervention concerns a matter of internal affairs.[81] The Declaration on Principles of International Law

[80] Note also that enforcement efforts involving a state asserting rights on behalf of an individual are limited by the duty to give the offending state an opportunity to remedy or a requirement to exhaust domestic remedies, especially with respect to initiating international litigation. See Interhandel, Judgment, Preliminary Objections (*Switzerland v US*) 1959 ICJ 6, 27 (March 21).

[81] Gerhard von Glahn, *Law Among Nations* (Boston: Longman, 7th edn, 1996); Philip Kunig, "Prohibition of Intervention," *Max Plank Encyclopedia of Public International Law* (n 28) para 25; Q Wright, "Subversive Intervention" (1960) 54 *American Journal of International Law* 521.

Concerning Friendly Relations and Co-operation among states in accordance with the Charter of the United Nations elaborates stating,

> Every state has the duty to refrain from organizing, instigating, assisting or participating in acts of civil strife or terrorist acts in another state or acquiescing in organized activities within its territory directed towards the commission of such acts, when the acts referred to in the present paragraph involve a threat or use of force.
>
> ...
>
> no state shall organize, assist, foment, finance, incite or tolerate subversive, terrorist or armed activities directed towards the violent overthrow of the regime of another state, or interfere in civil strife in another state.. . .[82]

A word about the International Law Commission Draft Declaration on the Rights and Duties of states is appropriate to the topic of what matters constitute internal affairs for purposes of the principle of non-intervention. As noted previously, the Declaration holds less-than-authoritative status as international law.[83] The Declaration, nonetheless, appears to provide further support for a conception of non-intervention that prohibits fomenting civil strife in another state.[84] Article 4 of the Declaration states, "Every State has the duty to refrain from fomenting civil strife in the territory of another State and to prevent the organization within its territory of activities calculated to foment such civil strife." Article 4 includes two aspects. First, the Article articulates a duty to refrain from instigating and supporting civil strife in the territory of other states. The Article's second duty then is in the nature of a prophylactic duty to prevent the organization on a state's own territory of activities instigating, conducting, or supporting civil strife in another state. Only the first duty appears directly relevant to the principle of non-intervention. It may be useful to conceive of the first duty as an elaboration of the concept of "internal affairs," protected from intervention by the preceding Article's general statement of the principle of non-intervention. The latter duty of Article 4 may be more appropriately characterized as an elaboration of the duty of vigilance or diligence. Notwithstanding the Declaration's elaborations and general respect for the work of the ILC, the Declaration cannot be regarded as providing definitive support for the meaning of "internal affairs" or for the concept of "civil strife." Upon its completion in 1949, the Declaration provoked insufficient state response to warrant consideration by the UN General Assembly and no state has yet submitted the Declaration for consideration.[85]

(c) UN Charter, Article 2(7) on internal matters

Scholarship on UN Charter, Article 2(7) clarifies not only the meaning and scope of acts constituting intervention but also the range of state interests protected by the principle of non-intervention. While much of the spirit of the general principle of

[82] GA Res 2625 (XXV), October 24, 1970.

[83] See discussion in this chapter, supported by n 20.

[84] Draft Declaration on Rights and Duties of States,

[85] Sergio M. Carbone and Lorenzo Schiano di Pepe, "Fundamental Rights and Duties of States," in *Max Planck Encyclopedia of Public International Law* (n 28) (January 2009) para 14.

non-intervention is captured in Article 2(7), a word of caution is appropriate. Consistent with the understanding that Article 2(7) applies only to UN entities as such, commentators acknowledge that the range of matters held to be free from involvement of international organizations by virtue of being internal matters may differ from the range of matters insulated from state intervention.[86] Still, the divergence between understandings of internal matters for purposes of Article 2(7) and for purposes of the customary principle of non-intervention may not be so great as divergence concerning understandings of what acts constitute intervention under Article 2(7) and its customary counterpart treated in this chapter.[87] In fact, drafting history suggests largely parallel notions of protected matters between both Article 2(7) and the customary international principle of non-intervention.[88]

UN practice identifies a shrinking range of matters deemed "essentially within the domestic jurisdiction of a State."[89] Addressing what matters "are essentially within the domestic jurisdiction of any State" and, therefore, beyond the reach of the UN under Article 2(7), commentators have also acknowledged that preservation of domestic sovereignty may be of waning importance.[90] Some understandings of the Article 2(7) reservation of matters "essentially within the jurisdiction of any State..." would even appear to render the phrase a nullity.[91] Scholarly comment and UN practice notwithstanding, states routinely rely on sovereignty, and Article 2(7) specifically, as shields against UN interference in internal matters.[92] A number of UN General Assembly resolutions have reiterated state support for the notion of certain matters lying exclusively within state domestic jurisdiction and, therefore, outside the reach of UN organs and by extension other states as well.[93] States have invoked Article 2(7) especially frequently and vociferously with respect to matters pertaining to national governmental systems and elections.[94]

(d) External affairs

The principle of non-intervention protects the integrity of states' external relations as well as internal or domestic matters reserved to state prerogative.[95] External matters

[86] Ziegler, "Domaine Réservé," in *Max Planck Encyclopedia of Public International Law* (n 28) para 24.

[87] See discussion in this chapter, supported by n 62–3.

[88] Georg Nolte, "Article 2(7),"(n 62) 280, 289.

[89] Nolte, "Article 2(7) (n 62) 294.

[90] Robert Jennings and Arthur Watts (eds), 1 *Oppenheim's International Law* (Oxford: Oxford University Press, 9th edn, 1992) 428; Christoph Shreuer, "The Waning of the Sovereign State: Towards a New Paradigm of International Law" (1993) 4 *European Journal of International Law* 447. Some commentators have apparently gone so far as to ask whether Article 2(7) is obsolete. Ian Brownlie, *Principles of Public International Law* (Oxford: Oxford University Press, 7th edn, 2008) 296; Rudolph Bernhardt, "Domestic Jurisdiction of States and International Human Rights Organs" (1986) 7 *Human Rights Law Journal* 205, 205.

[91] Comparing Article 2(7) to Article 15, paragraph 8 of the Covenant of the League of Nations, Kelsen observed, "there are no matters which are 'essentially' within the domestic jurisdiction ofs a state, just as there are no matters which by their very nature are 'solely' within the domestic jurisdiction of a state." Hans Kelsen, *Principles of International Law* (New York: Holt, Rinehart and Winston, 2d edn, 1966) 299.

[92] Nolte, "Article 2(7)" (n 62)) 280, 283. Nolte identifies recent recitations of Article 2(7) by China, The Russian Federation, Brazil, Libya, Sudan, Belarus, and Sri Lanka. Nolte (n 62) 294, n 101.

[93] Nolte, "Article 2(7)" (n 62)) 280, 294, n 97–8.

[94] Nolte, 280, 306.

[95] Oppenheim, 1 *International Law* (n 90) 305.

protected by the principle likely include the choice of extending diplomatic and consular relations, membership in international organizations, and the formation or abrogation of treaties.

Ordinary practices of diplomacy appear not to constitute intervention.[96] Threats, however, of acts that constitute intervention or that constitute intervention themselves are prohibited.[97] Protests alleging violations of international law, however, are legitimate and would not constitute intervention.[98]

(e) Motive or intent

Finally, it is appropriate to say a word about intent. Whether an intervention concerns the internal or external affairs of a state appears to be an objective, de facto inquiry. Motive or intent of the intervening state is of little relevance. It is the fact of coercion with respect to an internal or external affair that establishes a prohibited intervention. In the *Nicaragua* case, the ICJ noted evidence that the United States intended to overthrow the government of Nicaragua and to inflict economic damage. However, the Court declined to make findings with respect to US intentions or motives. The Court observed, "in international law, if one State, with a view to the coercion of another State, supports and assists armed bands in that State whose purpose is to overthrow the government of that State, that amounts to an intervention by the one State in the internal affairs of the other, whether or not the political objective of the State giving such support and assistance is equally far-reaching."[99] The Court's ruling suggests that states providing support to third parties may violate non-intervention unwittingly. That is, where a state provides assistance to third parties seeking to influence or coerce another state with respect to an internal matter, the supporting state will be held liable for intervention regardless of its subjective motive for supporting the third party.

In the *Armed Activities in DRC* case, the ICJ held that both provisions of the Declaration on Friendly Relations were declaratory of customary international law.[100] Applying both provisions, the Court characterized Ugandan training and military support for rebels operating in the territory of the Democratic Republic of Congo as violations of the principle of non-intervention.[101] The *Armed Activities* Court reiterated its statement from the *Nicaragua* case that non-intervention prohibits a state "to intervene, directly or indirectly, with or without armed force, in support of an internal opposition in another State."[102] The Court declined to make a determination of Uganda's motive for supporting the rebels. While Uganda sought to justify its training and military support to the rebels as necessary measures to guard its own security

[96] Gaetano Arangio-Ruiz, Human Rights and Non-Intervention in the Helsinki Final Act, (1977 IV) 157 Recueil Des Cours 195, 261, 264 (stating "the ordinary practices of diplomacy will not be hit by the prohibition [of intervention]").

[97] Kunig, "Prohibition of Intervention," *Max Planck Encyclopedia of Public International Law* (n 81) para 27.

[98] Q Wright, "Legality of Intervention under the UN Charter" (1957) 51 *Proceedings of the American Society of International Law* 79, 79–80.

[99] *Nicaragua* Judgment, para 241.

[100] *Armed Activities in DRC* Judgment, para 162.

[101] *Armed Activities in DRC* Judgment, para 163.

[102] *Armed Activities in DRC* Judgment, para 164 (quoting *Nicaragua* Judgment, para 206).

rather than to overthrow the DRC President, the Court held the fact of support for the rebels itself was sufficient to establish a violation of the non-intervention principle.[103] Thus, motive of the supporting state appears to be of little consequence in determining support for subversive activities as violations of the principle of non-intervention. The effect or consequence of a state's activities supporting subversive activity, even as a collateral matter, in another state appears from the *Armed Activities* case to be sufficient.[104]

In a cyber context then, suppose a state that provides technical and financial support to a private hacktivist group in another state. Suppose further that the supporting state does so out of ethnic or religious sympathy for the group and without specifically intending the group coerce the host state with respect to any of the latter's internal or external matters. By the logic of the *Nicaragua* Judgment, if the hacktivist group converts the state's support to coerce the host state, the former state will have violated the principle of non-intervention. It is not the intention or motive but rather the fact of the supporting state's assistance affecting the host state's sovereign prerogative that establishes the violation.

V. Justifications

As the ICJ made clear in the *Nicaragua* case, there is no general right of intervention. Rather, the Court has limited lawful interventions to specifically justified circumstances. Potential justifications for intervention include consent, self-defense, participation in a UN-mandated Chapter VI pacific settlement action or Chapter VII enforcement action, and countermeasures.

In the *Corfu Channel* case, the ICJ rejected application of a general "theory of intervention," which the British claimed justified violation of the territory of another state, in that case Albania, to preserve evidence and to ensure compliance with international law.[105] Rather than triggering a general right of intervention, the *Corfu Channel* Court determined that alleged Albanian violations of international law constituted "extenuating circumstances" for the British intervention, appropriate for consideration at the damages stage of litigation.[106] Indeed, the Court ultimately determined that a mere declaration of the British intervention as unlawful was sufficient satisfaction given the Albanian breaches of international law that had provoked that intervention.[107]

The *Corfu Channel* facts illustrate a critical distinction between the general right to intervention rejected by the Court and the specific justifications later approved by the Court in the *Nicaragua* case. In *Corfu Channel*, as characterized by the British advocates, the British intervention into Albanian waters was not undertaken by consent, self-defence, countermeasures, or any other justification. Rather the British advocates appear to have claimed a general right to intervene, which the Court rejected out of hand. Had the British intervention been argued to be necessary as a means of self-defence or a counter-measure in response to Albanian attacks in the form of naval mines, perhaps the Court would have considered the British intervention justified. In the *Nicaragua*

[103] *Armed Activities in DRC* Judgment, para 163.
[104] *Armed Activities in DRC* Judgment, para 163. [105] *Corfu Channel* case, 34–5.
[106] *Corfu Channel* case, 35. [107] *Corfu Channel* case, 35.

case, the ICJ confirmed the *Corfu Channel* holding. The *Nicaragua* Court held that, as with other prohibitions in international law, the customary principle of non-intervention operates subject to specific justifications rather than a general right of abrogation.[108]

As a final general matter with respect to justifications for intervention, an intervening state must first exhaust or deem ineffective non-coercive measures such as diplomacy, negotiation, mediation, or arbitration.[109]

VI. Conclusion

It is perhaps understandable if, at present, the principle of non-intervention does not figure prominently in states' legal evaluations of diplomatic, political, economic, and military endeavors. International legal instruments, including the UN Charter, decisions of the ICJ, and international law scholarship devote far greater attention to regulation of more grave violations of sovereignty. The more pressing breaches of peace and sovereignty reflected in acts of aggression, uses of force, and armed attacks surely still warrant pride of place.

Yet the advent and proliferation of state sponsored cyber operations, especially low-intensity cyber campaigns producing effects short of destruction and injury, seem likely to augment the importance of non-intervention as a legal means of maintaining international peace and respect for sovereignty. Cyberspace offers states enticing inroads to exploit critical vulnerabilities of their peer competitors. The temptation seems already to have proved nearly impossible for some states to resist. Whether the principle of non-intervention, as presently comprised would be up to the task of regulating this alluring and disruptive means of intervention is another matter.

As presently understood, the customary international law principle of non-intervention includes significant doctrinal gaps and ambiguities. States have not achieved or expressed consensus on a notion of coercion sufficient to constitute intervention. Nor have states managed to muster a clear notion of the internal and external matters protected by the principle's prohibition.

If this chapter has appeared to offer clarifications to the principle of non-intervention, they should be understood as interpretive possibilities rather than definitive conclusions—options to be considered by states rather than to bind the final arbiters of international law. The author recalls well that legal ambiguity calls for respect as often as it calls for resolution. It is entirely possible that the current state of the principle reflects the full extent to which states are willing to commit the issue to international law. As the lengthy catalog of attempts to refine the principle through resolutions and declarations makes clear, states have not wanted for opportunities to clarify the principle of non-intervention. The critical role the principle plays in supporting a peaceful system of legally equal sovereigns and the emergence of cyber operations, however, suggest, at a minimum, a burgeoning interest in either addressing or confirming the ambiguities of non-intervention.

[108] *Nicaragua* Judgment, para 246.

[109] Q Wright, "Legality of Intervention under the UN Charter" (1957) 51 *Proceedings of the American Society of International Law* 79, 88.

Index